Making the Modern World

Making the Modern World

Materials and Dematerialization

VACLAV SMIL

Distinguished Professor Emeritus,
University of Manitoba, Canada

WILEY

Registered office

John Wiley & Sons Ltd, The Atrium, Southern Gate, Chichester, West Sussex, PO19 8SQ, United Kingdom

For details of our global editorial offices, for customer services and for information about how to apply for permission to reuse the copyright material in this book please see our website at www.wiley.com.

Library of Congress Cataloging-in-Publication Data

Smil, Vaclav.
  Making the modern world : materials and dematerialization / Vaclav Smil.
    pages cm
  Includes index.
   ISBN 978-1-119-94253-5 (pbk.)
  1. Waste minimization. 2. Materials. 3. Raw materials. I. Title.
  TD793.9.S64 2014
  306.3–dc23

                                  2013024672

A catalogue record for this book is available from the British Library.

ISBN: 9781119942535

Set in 10/12pt Times by Laserwords Private Limited, Chennai, India
Printed and bound in Malaysia by Vivar Printing Sdn Bhd

2 2014

About the Author

Vaclav Smil conducts interdisciplinary research in the fields of energy, environmental and population change, food production and nutrition, technical innovation, risk assessment and public policy. He has published more than 30 books and close to 500 papers on these topics. He is a Distinguished Professor Emeritus at the University of Manitoba, a Fellow of the Royal Society of Canada (Science Academy), the first non-American to receive the American Association for the Advancement of Science Award for Public Understanding of Science and Technology, and in 2010 he was listed by *Foreign Policy* among the top 50 global thinkers.

Previous works by author

China's Energy

Energy in the Developing World
  (edited with W. Knowland)

Energy Analysis in Agriculture
  (with P. Nachman and T. V. Long II)

Biomass Energies

The Bad Earth

Carbon Nitrogen Sulfur

Energy Food Environment

Energy in China's Modernization

General Energetics

China's Environmental Crisis

Global Ecology

Energy in World History

Cycles of Life

Energies

Feeding the World

Enriching the Earth

The Earth's Biosphere

Energy at the Crossroads

China's Past, China's Future

Creating the 20th Century

Transforming the 20th Century

Energy: A Beginner's Guide

Oil: A Beginner's Guide

Energy in Nature and Society

Global Catastrophes and Trends

Why America Is Not a New Rome

Energy Transitions

Energy Myths and Realities

Prime Movers of Globalization

Japan's Dietary Transition and Its Impacts
  (with K. Kobayashi)

Harvesting the Biosphere

Should We Eat Meat?

Contents

Preface: Why and How

The story of humanity – evolution of our species; prehistoric shift from foraging to permanent agriculture; rise and fall of antique, medieval, and early modern civilizations; economic advances of the past two centuries; mechanization of agriculture; diversification and automation of industrial protection; enormous increases in energy consumption; diffusion of new communication and information networks; and impressive gains in quality of life – would not have been possible without an expanding and increasingly intricate and complex use of materials. Human ingenuity has turned these materials first into simple clothes, tools, weapons, and shelters, later into more elaborate dwellings, religious and funerary structures, pure and alloyed metals, and in recent generations into extensive industrial and transportation infrastructures, megacities, synthetic and composite compounds, and into substrates and enablers of a new electronic world.

This material progress has not been a linear advance but has consisted of two unequal periods. First was the very slow rise that extended from pre-history to the beginnings of rapid economic modernization, that is, until the eighteenth century in most of Europe, until the nineteenth century in the USA, Canada, and Japan, and until the latter half of the twentieth century in Latin America, the Middle East, and China. An overwhelming majority of people lived in those pre-modern societies with only limited quantities of simple possessions that they made themselves or that were produced by artisanal labor as unique pieces or in small batches – while the products made in larger quantities, be they metal objects, fired bricks and tiles, or drinking glasses, were too expensive to be widely owned.

The principal reason for this limited mastery of materials was the energy constraint: for millennia our abilities to extract, process, and transport biomaterials and minerals were limited by the capacities of animate prime movers (human and animal muscles) aided by simple mechanical devices and by only slowly improving capabilities of the three ancient mechanical prime movers: sails, water wheels, and wind mills. Only the conversion of the chemical energy in fossil fuels to the inexpensive and universally deployable kinetic energy of mechanical prime movers (first by external combustion of coal to power steam engines, later by internal combustion of liquids and gases to energize gasoline and Diesel engines and, later still, gas turbines) brought a fundamental change and ushered in the second, rapidly ascending, phase of material consumption, an era further accelerated by generation of electricity and by the rise of commercial chemical syntheses producing an enormous variety of compounds ranging from fertilizers to plastics and drugs.

And so the world has become divided between the affluent minority that commands massive material flows and embodies them in long-lasting structures as well as in durable and ephemeral consumer products – and the low-income majority whose material possessions amount to a small fraction of material stocks and flows in the rich world. Now the list of products that most Americans claim they cannot live without includes cars, microwave ovens, home computers, dishwashers, clothes dryers, and home air conditioning (Taylor *et al.*, 2006) – and they have forgotten how recent many of these possessions are because just 50 years ago many of them were rare or nonexistent. In 1960 fewer than 20% of all US households had a dishwasher, a clothes dryer, or air conditioning, the first color TVs had only just appeared, and there were no microwave ovens, VCRs, computers, cellphones, or SUVs.

In contrast, those have-nots in low-income countries who are lucky enough to have their own home live in a poorly-built small earthen brick or wooden structure with as little inside as a bed, a few cooking pots, and some worn clothes. Those readers who have no concrete image of this great material divide should look at Peter Menzel's *Material World: A Global Family Portrait* in which families from 30 nations are photographed in front of their dwellings amidst all of their household possessions (Menzel, 1995). And this private material contrast has its public counterpart in the gap between the extensive and expensive infrastructures of the rich world (transportation networks, functioning cities, agricultures producing large food surpluses, largely automated manufacturing) and their inadequate and failing counterparts in poor countries.

These contrasts make it obvious that a huge material mobilization and transformation will be needed just to narrow the gap between these two worlds. At the same time, material consumption has been a major cause of environmental pollution and degradation and further multiplication of current demand may pose a worrisome threat to the integrity of the biosphere. These impacts also raise questions of analytical boundaries: their reasoned choice is inevitable because including every conceivable material flow would be impractical and because there is no universally accepted definition of what should be included in any fairly comprehensive appraisal of modern material use. This lack of standardization is further complicated by the fact that some analyses have taken the maximalist (total resource flow) approach and have included every conceivable input and waste stream, including "hidden" flows associated with the extraction of minerals and with crop production as well as oxygen required for combustion and the resulting gaseous emissions and wastes released into waters or materials dissipated on land.

In contrast, others studies have restricted their accounts to much more reliably quantifiable direct uses of organic and inorganic material inputs required by national economies. I will follow the latter approach, focusing in some detail on key (because of their magnitude or their irreplaceable quality) materials consumed by modern economies. Their huge material claims lead us to ask a number of fundamental questions. How much further should the affluent world push its material consumption? Are any further increases associated with genuine improvements in quality of life? To what extent is it possible to divorce economic growth and improvements in the average standard of living from increased material consumption? In other words, does relative dematerialization (reduced material use per unit of product or performance) lead to absolute decline in demand for materials?

In order to answer these questions in a convincing manner I must review the evolution of human material uses; describe all the principal materials, their extraction, production, and their dominant applications; and take a closer look at the evolving productivities of material extraction, processing, synthesis, finishing, and distribution and at the energy costs and environmental impact of rising material consumption. And, as always in my books, I will not offer any time-specific forecasts regarding future global and national use of materials. Instead, I will look at possible actions that could reduce our dependence on materials while maintaining a good quality of life and narrowing the gap between affluent and low-income economies.

We must realize that in the long run even the most efficient production processes, the least wasteful ways of design and manufacturing, and (for those materials that can be recycled) the highest practical rates of recycling may not be enough to result in dematerialization rates great enough to negate the rising demand for materials generated by continuing population growth, rising standards of living, and the universal human preference for amassing possessions. This makes it highly likely that in order to reconcile our wants with the preservation of the biosphere's integrity we will have to make deliberate choices that will help us to reduce absolute levels of material consumption, and thereby redefine the very notion of modern societies whose very existence is predicated on incessant and massive material flows.

1

What Gets Included

Any study aiming to elucidate the complexity of material flows of modern societies, their prerequisites and their consequences, should be as comprehensive as possible, indeed its coverage should be truly all-encompassing. But this easily stated aspiration runs immediately into the key categorical problem: what constitutes the complete set of modern material uses? There is no self-evident choice, no generally accepted list, only more or less liberally (and also more or less defensively) defined boundaries of a chosen inclusion; a reality best illustrated by reviewing the selections made by the past comprehensive studies and adopted by leading international and national databases of material flows.

The first comparative study of national resource flows (Adriaanse *et al.*, 1997), subtitled *The Material Basis of Industrial Economies*, excluded water and air but included not only all agricultural harvests (not just raw materials but all food and feed as well), all forestry products, aquatic catches, extraction of minerals and fossil fuels, but also hidden flows accounting for extraction, movement, or losses of materials that create environmental impacts but have no acknowledged economic value. These hidden flows are dominated by overburden materials that have to be removed during the exploitation of mineral deposits (above all in open-cast coal and ore mining), processing wastes (particularly massive flows associated with the separation of relatively rare metals from rocks), soil, sand, and rocks that have to be removed and shifted during large construction projects, and soil erosion originating from fields and permanent plantations. Hidden flows are not monitored and their quantification is, at best, a matter of approximate estimates; more often of just informed guesses.

This is even more the case with the annual totals for hidden flows associated with imported raw materials: obviously, these estimates will be particularly uncertain in the case of large affluent economies (USA, Japan, Germany) that import a wide range of materials from scores of countries. Not surprisingly, the study resorted to using worldwide averages for these calculations: for example, it applied the rate of 0.48 t of overburden for a ton of bauxite and 2 t of overburden per ton of iron ore – global generalizations that must result in considerable errors when used as national averages.

Making the Modern World: Materials and Dematerialization, First Edition. Vaclav Smil.
© 2014 John Wiley & Sons, Ltd. Published 2014 by John Wiley & Sons, Ltd.

Erosion rates are even more variable, their detailed national studies are rare and annual soil losses can differ by up to an order of magnitude even within relatively small regions, and yet the study used only the rates derived from the US inventory. Another highly uncertain inclusion was quantifying the mass of grass grazed by cattle (other animal feed was included in crop harvests).

Three years after this first comparative study came another project led by the World Resources Institute (WRI), *The Weight of Nations* (Matthews *et al.*, 2000). This study presented material flows for the four nations included in the original work (the USA, Japan, Germany, and the Netherlands) as well as for Austria and extended the accounting period from 1975 to 1996 (the original ended in 1993). Its subtitle, *Material Outflows from Industrial Economies*, indicated the report's concern with outputs produced by the metabolism of modern societies. As its predecessor, this study included all fossil fuels, hidden material flows (dominated by surface coal mining overburden), as well as the processing wastes from oil and coal industries.

Similarly, estimates for process losses and overburden removal were made for all nonfuel minerals and metals, and the report also quantified earth moved during all construction activities (highway, public, and private and also for dredging), soil erosion losses in agriculture, and waste from synthetic organic chemicals and from the pharmaceutical industry. But, unlike the original study, the 2000 report also included data on additional inputs (oxygen in combustion and in respiration) and outputs, including the total output of CO_2 from respiration and water vapor from all combustion, and it separated waste streams into three gateways: air, land, water. The air gateway quantified gaseous emissions (CO_2, CO, SO_x and NO_x, volatile organic carbohydrates) including oxygen from all combustion, the outputs to land included municipal solid waste, industrial wastes, and dissipative flows to land (manure, fertilizers, salt spread on roads, worn tire rubber, evaporated solvents), and water outputs, trace organic load, and total nitrogen and phosphate burdens.

Eurostat has been publishing annual summaries of domestic material consumption for all EU countries since the year 2000, disaggregating the total flows into fossil fuels, biomass (crops and forest products), metal ores, and nonmetallic minerals (European Commission, 2001; Eurostat, 2013). Eurostat's methodological guides for economy-wide material flow accounts offer detailed procedures for the inclusion of biomass (food, feed, fodder crops, grazed phytomass, wood, fish, hunting, and gathering activities), metal ores and nonmetallic minerals, and for all forms of fossil fuels as well as for all dissipative uses of products, including organic and mineral fertilizers, sewage sludge, compost, pesticides, seeds, road salt, and solvents (Eurostat, 2009; Schoer *et al.*, 2012). Eurostat aggregates also include unused materials (mining overburden, losses accompanying phytomass production, soil excavation, dredging, and marine by-catch), and quantify emissions (CO_2, water disposal, and landfilled wastes) but leave out oxygen and water.

In 1882, the US Congress mandated the annual collection of statistics for mineral commodities produced and used in the country. The US Geological Survey was responsible for this work, then the US Bureau of Mines, and since 1995 the task has reverted to the USGS. These statistics were the basis for preparing the first summary of America's material flows aggregated by major categories and covering the period between 1900 and 1995 (Matos and Wagner, 1998). An updated inventory, with data for aggregate categories extending until 2006, was published in 2009 (Matos, 2009) and data on individual elements, compounds, and materials are updated annually (USGS, 2013).

The USGS choice of items included in its national material accounts is based on concentrating only on the third class of the material triad; leaving out food and fuel and aggregating only the materials that are used domestically in all branches of the economy. The series offers annual totals for domestic production, exports, imports, and domestic consumption; it excludes water, oxygen, hidden material flows, and all fossil fuels and it includes all raw materials produced by agricultural activities (cotton, seeds yielding industrial oil, wool, fur, leather hides, silk, and tobacco), materials originating in forestry (all kinds of wood, plywood, paper, and paperboard), metals (from aluminum to zinc), an exhaustive array of nonmetallic minerals (be they extracted in their natural form – such as gypsum, graphite, or peat – or processed before further use – such as crushed stone or cement – or synthesized, such as ammonia) and nonrenewable organics derived from fossil fuels (asphalt, road oil, waxes, oils, and lubricants and any variety of solid, liquid, or gaseous fossil fuel used as feedstock in chemical syntheses).

Very few of these inputs are used in their raw, natural form as virtually all of them undergo processing (cotton spinning, wood pulping, ore smelting, stone crushing, or cutting and polishing) and, in turn, most of these processed materials become inputs into the manufacturing of semi-finished and finished products (cotton turned into apparel, pulp into paper, smelted metals into machine parts, crushed stone mixed with sand and cement to make concrete). This compilation of agriculture- and forestry-derived products, metals, industrial minerals, and nonrenewable organics gives a fairly accurate account of annual levels and long-term changes in the country's material flows. While all imports and exports of raw materials are accounted for, the series does not include materials contained in traded finished goods: given their mass and variety their tracking would be very difficult.

Where does this leave us? Those material flow studies that conceive their subject truly *sensu lato* (that is as virtually any substance used by humans) include everything with the notable exception of water; that is not only biomaterials used in production of goods, all metals, nonmetallic minerals, and organic feedstocks, but also all agricultural phytomass (harvested food and feed crops, their residues, forages, and grazed plants), and all (biomass and fossil) fuels and oxygen needed for combustion. Slightly more restrictive studies exclude oxygen and all food and feed crops, and consider only those agricultural raw materials that undergo further processing into goods, but include all phytomass and fossil fuels. In contrast, the USGS series exemplifies a *sensu stricto* approach as it includes only raw biomaterials used for further processing and excludes oxygen, water, all fuels (phytomass and fossil), and all hidden (and always tricky to estimate) material flows. My preferences for setting the analytical boundaries are almost perfectly reflected by the USGS selection, but instead of simply relying on that authority I will briefly explain the reasons behind my exclusions.

Leaving out oxygen required for combustion of fuels is a choice that is easily defensible on the basis of free supply of a virtually inexhaustible atmospheric constituent. Claims about the danger of serious O_2 depletion through combustion were refuted a long time ago (Broecker, 1970). Complete combustion of 1 kg of coal carbon consumes 2.67 kg of oxygen, and burning of 1 kg of hydrocarbons requires 4 kg of O_2. Global combustion of about 8 Gt of fossil carbon in 2010 thus claimed about 21 Gt of O_2 or about 0.0014% of the atmosphere content of 1.5 Pt of O_2 – and even a complete combustion (a clear impossibility) of the generously estimated global resources of fossil fuels would lower the atmospheric O_2 content by no more than 2%.

There is thus no danger of any worrisome diminution of supply (to say nothing of exhaustion) of the element, and yet, once the choice is made to include it in material flow accounts, it will dominate the national and global aggregates. For example, as calculated by the comparative WRI study, oxygen was 61% of the direct US processed material output in 1996, and in Japan in the same year the element's share was 65% (Matthews *et al.*, 2000). Consequently, magnitudes of national material flows that incorporate oxygen needs would be nothing but rough proxies for the extent of fossil fuel combustion in particular economies.

The reasons for excluding hidden flows from the accounts of national material flows are no less compelling: after excluding oxygen they would dominate total domestic material output in all countries that have either large mineral extractive industries (especially surface coal and ore mining) or large areas of cropland subject to heavy erosion. Not surprisingly (after excluding oxygen), in the WRI analysis these hidden flows account for 86% of the total domestic material output in both the USA and Germany, but with much less mining and with limited crop cultivation the rate was lower (71%) in Japan (Matthews *et al.*, 2000). The undesirable environmental impacts of these associated flows should not be ignored when analyzing particular extractive or cropping activities, but the flows cannot be quantified with high accuracy. They are dominated by unusable excavated earth and rocks, mine spoils, processing wastes, and eroded soil; and earth and rocks moved around as a part of construction activities will make up a comparatively small share.

But the principal problem with the inclusion of hidden flows is not their unsurprising dominance of domestic output of materials in all large, diversified economies, but the indiscriminate addition of several qualitatively incomparable flows. An unusable mass of stone left in a quarry after it ceases its operation may be no environmental burden, even no eyesore, and once the site is flooded to create an artificial lake that hidden material flow may be truly hidden as part of a new, and pleasing, landscape. On the other hand, bauxite processing to extract alumina (to give one of many possible common examples) leaves behind toxic waste (containing heavy metals) that is also often slightly radioactive waste and is very caustic (high pH).

And no less fundamental is the difference between *in situ* hidden flows generated by mineral extraction (abandoned stone, gravel and sand quarries, and coal and ore mines with heaps, piles, layers, or deep holes or gashes full of unusable minerals or processing waste) and by rain- and wind-driven soil erosion that transports valuable topsoil not just tens or hundreds but as much as thousands of kilometers downstream or downwind. The first kind of hidden flow may be unsightly but not necessarily toxic, and its overall environmental impact beyond its immediate vicinity may be negligible or nonexistent, but erosion is a globally important, often regionally highly worrisome, and locally devastating process that reduces (or destroys) the productivity of crop fields, silts streams, contributes to eutrophication of fresh and coastal waters, and creates lasting ecosystemic degradation and substantial economic losses.

My reasons for excluding water are based on several considerations that make this indispensable input better suited for separate treatment rather than for inclusion into total material requirements of modern economies. The first, obvious, reason is, once again, quantitative: with the exception of desert countries, water's inclusion would dominate virtually all national material flow accounts and would misleadingly diminish the importance of many inputs whose annual flows are a small fraction of water withdrawals but whose qualitative contribution is indispensable. For example, in 2005 the total water withdrawals in the USA were just over 5 Gt (Kenny *et al.*, 2009), while all materials directly used by the country's economy (the total dominated by sand, gravel, and stone used in construction) added up to less than 3.8 Gt (USGS, 2013).

Moreover, there are fundamental qualitative differences between these two measures. The most voluminous water withdrawal (accounting for nearly 60% of the total), that of cooling water for thermal electricity-generating stations, is not a consumptive use because all but a small (evaporated) fraction of that water becomes available almost instantly for further downstream uses. In contrast, materials that become embedded in long-lasting structures and products are either never reused or are partially recycled only after long period of being out of circulation. And the majority of the second most voluminous water use, about 30% of the US 2005 total used for irrigation, is also nonconsumptive: all but a tiny fraction of the irrigation water is evapotranspired by growing plants, re-enters the atmosphere, and eventually undergoes condensation again and is precipitated. And if the inclusion of water were driven by resource scarcity concerns, then a critical distinction should be made between water supplied by abundant precipitation and water withdrawn at a high cost from deep and diminishing aquifers that cannot be replenished on a civilizational timescale.

At this point, it might be useful to call attention to yet another (comparatively minor) problem with aggregate measures of material flows that, to the best of my knowledge, has not been raised by any assembler of national and global accounts: that of the water content of sand and of harvested biomass. Even when looking just at those biomaterials that are used as industrial inputs, their water content ranges from less than 15% for raw wool to more than 50% for freshly cut tree logs (the range is wider for food crops, ranging from only about 5% for dry seeds to more than 90% for fresh vegetables).

Freshly excavated sand can contain more than 30% water, purified sands contain 15–25%, storage in drainage bins reduces that level to about 6%, and drying in rotary bins or in fluidized bed dryers expels all but about 0.5% of moisture for sands used in such processes as steel castings or hydraulic fractioning under pressure. Obviously, the best solution would be to report the masses of any moisture-containing materials in terms of absolutely dry weight in order to make their flows comparable to those of materials that contain no moisture. This is not the case in practice, and hence all national material aggregates contain far from negligible shares of water.

Foodstuffs and fuels are obviously indispensable for the survival of any civilization, and their flows have been particularly copious in modern high-energy societies enjoying rich and varied diets, while traditional biofuels remain important in many low-income countries. Moreover, unlike water or oxygen, their inclusion would not dwarf all other material flows combined: for example, even in the fuel-rich USA the mass of annually consumed coal, crude oil, and natural gas is equal to about 50% of all nonenergy minerals. So why leave them out? Exclusion of food and fuel is justified not only because these two large consumption categories have been traditionally studied in separation (resulting in a rich literature on achievements and prospects) but also because they simply are not *sensu stricto* materials, substances repeatedly used in their raw state or transformed into more or less durable finished products.

Unlike raw biomaterials (wood, wool, cotton, leather, silk), metals, nonmetallic minerals, and nonrenewable organics (asphalt, lubricants, waxes, hydrocarbon feed-stocks) foodstuffs and fuels are not used to build long-lasting structures and are not converted or incorporated into the still increasing array of ephemeral as well as durable industrial, transportation, and consumer items. Foods are rapidly metabolized to yield energy and nutrients for human growth and activity; fuels are rapidly oxidized (burned) to yield, directly and indirectly, various forms of useful energy (heat, motion, light): in neither case do they increase the material stock of modern societies.

Finally, I must defend a conceptual change that concerns the handling of materials placed into the category of dissipative flows by the EU's material balances. According to the EU definition, the eight categories of dissipative losses are a collection of disparate residuals: some of them add up to small total flows (think about solvents escaping from dry cleaning or about rubber tires wearing away on roads), others are more substantial (manures, sewage sludge, and composts applied to cropland) but dissipative losses contributed by both of these material categories are not monitored and are very difficult to quantify. The USGS approach accounts for the largest flows in this category (salt and other thawing materials – including sand and grit – spread on winter roads, nitrogenous and phosphatic fertilizers, and potash applied to crops and lawns) by including them in the industrial minerals group.

While salt and sand are abundant materials whose production is not energy-intensive, inorganic fertilizers are critical material inputs in all modern societies that cannot be ignored and that will receive a closer look when I examine advances in the production of synthetic materials. But I would argue that most of the remaining dissipative flows add up to relatively small amounts whose inherently inaccurate quantification appears to outweigh any benefits of including them in any grand total of consumed materials. And while manures and sludges represent relatively large volumes to be disposed of, they are more about recycling water than biomass: sludge contains least 80% water, fresh manures 70–85%; moreover, in many (perhaps most) instances, sewage sludge should not be recycled as it contains heavy metals, pathogens, pesticide and drug residues, steroids, and hormones.

This leaves me with an argument for a single addition to the USGS list, for the inclusion of industrial gases. Although air (oxygen) is needed for combustion of fossil fuels, the dominant energizer of modern civilization, adding air to the total material input would have (as I have already explained) a skewing and confusing effect similar to that of counting all uses of water; but assessing the use of gases separated from the air in order to enable many industrial processes is another matter. In simple quantitative mass terms the global use of oxygen, hydrogen, nitrogen, and rare gases such as argon or xenon constitutes only a minor item, but in qualitative terms their use is indispensable in industries ranging from steelmaking (basic oxygen furnaces are now the principal means of producing the metal) to synthesis of ammonia (using nitrogen separated from air and hydrogen liberated from methane) and efficient lighting.

2

How We Got Here

The Earth's biosphere teems with organisms that use materials for more than just their metabolism; moreover, in aggregate mass terms the material flows commanded by humanity do not appear to be exceptionally high when compared with the work of marine biomineralizers. But it is the combination of the overall extent, specific qualities, and increasing complexity of material uses (extraction, processing, and transformation to particular inputs destined for infrastructures and myriads of products) that is a uniquely human attribute. To set it into a wider evolutionary perspective, I will first note some of the most remarkable material uses by organisms ranging from marine phytoplankton to primates, those distinguished either by the magnitude of their overall fluxes or by their unique qualities.

Afterwards I will proceed with concise chronological surveys of human use of materials, focusing first on the milestones in our prehistory, above all on those still poorly explained feats of megalithic construction that required quarrying, transportation, and often remarkably accurate placement of massive stones. Then I will review and quantify some notable deployments of traditional materials (stone and wood) during antiquity, the Middle Ages, and the early modern era (1500–1800), concentrating on the advances in building roads, aqueducts, ceremonial and religious structures, and ships; on the origins and developments in metallurgy; and on materials used by households.

I will close the chapter with two closely related sections that will describe the creation of modern material civilization during the nineteenth century and its post-1900 spatial expansion and growth in complexity. I will focus on key quantitative and qualitative advances in the use of materials that laid the foundations for twentieth societies as they supported fossil fuel extraction, industrialization, urbanization, and evolution of modern transportation modes on land, water, and in the air. These developments were based on materials whose production required high energy inputs and whose introduction and use have been dynamically linked with enormous advances in scientific and technical capabilities. In turn, new materials have been the principal drivers of increased food production and improvements in sanitation that led to unprecedented gains in quality

Making the Modern World: Materials and Dematerialization, First Edition. Vaclav Smil.
© 2014 John Wiley & Sons, Ltd. Published 2014 by John Wiley & Sons, Ltd.

of life. They also expanded capabilities for mechanized and automated production and for long-distance travel, information sharing, and telecommunication.

2.1   Materials Used by Organisms

Inevitably, all organisms use materials: that is the essence of metabolism. Global photosynthesis, the foundation of life in the biosphere, creates new biomass by incorporating annually more than 60 Gt of carbon, absorbed by leaves as CO_2 from the atmosphere (Smil, 2013), and millions of tons of the three key macronutrients (nitrogen, phosphorus, and potassium, absorbed by roots) that are incorporated into complex compounds forming plant tissues and organs. But these metabolic necessities – mirrored by the nutritional requirements of heterotrophs, be they herbivorous, carnivorous, or omnivores organisms – are not usually included in the category of material uses that is reserved for active, extrasomatic processes.

In terms of the initial acquisition, these material uses fall into five major categories. The rarest, and in aggregate material terms quite inconsequential, category is the use of collected natural materials as tools. The second category with limited aggregate impact is the use of secreted materials to build protective or prey-catching structures (as is done, often spectacularly, by spiders). The next one is the removal of biomass (and now also manmade) materials and their purposeful emplacement to create often remarkably designed structures (ranging from beaver dams to intricate nests); then comes the removal and repositioning of soils and clays (termite mounds, intricate rodent burrows); and, finally, the most massive endeavor is the extraction of minerals from water, mostly to build exoskeletons; the process dominated by marine biomineralizers including phytoplankton, corals, and mollusks.

Tool-using activities have been well documented in species as diverse as otters and seagulls and elephants and finches (Shumaker *et al.*, 2011), but they have reached their greatest complexity, and have gone as far as resulting in specific cultures, among chimpanzees who use blades of grass or twigs to collect termites or small stones and stone anvils to crack open nuts (Wrangham *et al.*, 1996; Boesch and Tomasello, 1998; Whiten *et al.*, 1999). Spider silk is certainly the most remarkable secreted material, with a tensile strength similar to that of good-quality steel (Brunetta and Craig, 2010). At the other end of secretion spectrum might be the frothy nests excreted by spittle bugs.

Use of collected materials is quite widespread among heterotrophs. Even some single-cell amoebas can built portable, intricate, ornate sand-grain houses whose diameter is a mere 150 μm (Hansell, 2007). Perhaps the most remarkable collecting activity among insects is that of leafcutter ants (genus *Atta*) as they harvest leaves, dragging them underground into elaborately excavated nests in whose chambers they cultivate fungus (Hölldobler and Wilson, 1990). And beavers are active harvesters of wood used to build their dams. But birds' nests offer the most varied and sometimes spectacular examples of construction using collected materials; they range from simple and rather haphazard assemblies of twigs or stems to intricate constructs produced by *Ploceidae*, the family of tropical weaver birds, and they may use a single kind of a collected material or an assortment of tissues (Gould *et al.*, 2007; Burke, 2012).

Birds use not only a wide range of collected plant tissues (slender blades of grass to the heavy twigs used by storks and eagles) but also feathers of other species and spider silk (most passerine birds), and some nests may contain thousands of individual pieces. Use of mud (by swallows) is not that common, but many ground-nesting birds (including penguins) collect small stones, while the elaborate structures prepared by some bower birds of Australia and New Guinea to attract females may include not only such colorful natural objects as shells, berries, leaves, and flowers but also discarded bits of plastic, metal, or glass, and some species even make courts creating forced visual perspective for the courted females (Endler *et al.*, 2010). Some insect species also use collected material to build their nests: paper wasps cut tiny pieces of wood and mix them with their salivary secretions; mud wasps shape mud into cylindrical nests. In contrast, primates, our closest animal predecessors, use branches and leaves to build only simple, temporary structures on the ground or in the trees.

Soil-displacing species engage mostly in digging tunnels, burrows, and nests but also use soils and clay to build above-ground structures. The earliest burrow constructs date to the pre-Cambrian (650–700 million years ago) oceans, coinciding with the emergence of macropredation (Turner, 2000). As demonstrated by Darwin in his last published book, earthworms are capable of such prodigious efforts of earth displacement (passing the particles through their guts and excreting the worm casts on the surface) that they can bury monuments of human activity in remarkably brief periods of time (Darwin, 1881). Rodents are diligent builders of often extensive subterranean networks of tunnels and nests that may also help with temperature control and ventilation and that facilitate escape.

Termites are the greatest aggregate movers and users of soils in subtropical and tropical environments as they construct often impressively tall and voluminous mounds that not only shelter the massive colonies, but also provide induced ventilation driven by pressure differences (Turner, 2000).

Biomass densities of these abundant warm-climate insects range from $2\,g/m^2$ in the Amazonian rainforest (Barros *et al.*, 2002) to around $5\,g/m^2$ in Australia's Queensland (Holt and Easy, 1993); to $10\,g/m^2$ in the Atlantic forests in northeastern Brazil in Sao Paulo, as well as in dry evergreen forest of northeast Thailand (Vasconcellos, 2010); and in African savannas their total fresh-weight biomass can be more than twice the biomass of elephants (Inoue *et al.*, 2001). Species belonging to the genus *Macrotermes* select clay particles to build conical mounds that are usually 2–3 m tall but can reach 9 m, with a typical basal diameter of 2–3 m, although much wider mounds are not uncommon.

The typical mass of mounds (wall and nest body) is between 4 and 7 t, but spatial density of mounds varies widely, with as few as 1 and 2 and as many as 10/ha (Fleming and Loveridge, 2003; Abe *et al.*, 2011; Tilahun *et al.*, 2012). As a result, the total mass of termite mounds varies between just 4 and 8 t/ha to as much as 15–60 t/ha. A very conservative estimate of the clay mass used to build termite mounds would be 5 Gt (average 5 t/ha, an area of about 10 million km² of tropical and subtropical grasslands inhabited by mound-building insects) but the actual total may be several times larger. In any case, this means that the annual use of materials by these tiny heterotrophs would be of the same order of magnitude as our civilization's global extraction of metallic ores and other nonfuel minerals at the beginning of the twenty-first century.

In aggregate terms, both the mass of materials collected by vertebrate animals to build structures and the mass of soils displaced by burrowing heterotrophs, earthworms, and termites are negligible compared to the mass of compounds excreted by species capable of biomineralization, above all by phytoplankton, protists, and invertebrates. More than 30 biogenic minerals (two-thirds of them being carbonates) are produced by a small number of vascular plants (belonging to Bryophyta and Trachaeophyta), animal species ranging from Porifera to Chordata, some fungi, many protoctists, and some Monera (Lowenstam, 1981; Boskey, 2003). Some biomineralizers deposit the minerals on organic matrices, but most of them produce extracellular crystals similar to those precipitated from inorganic solutions.

In mass terms, by far the largest users of natural materials are the marine biomineralizers that are able to secrete the inorganic compounds they produce from chemicals absorbed from water. Marine biomineralizers use dissolved $CaCO_3$ to form calcite or aragonite shells, two almost identical minerals that differ only in their crystal structure ($2\,HCO_3^- + Ca^{2+} = CaCO_3 + H_2O$). Reef-building corals (Anthozoa belonging to the phylum Cnidaria) are the most spectacular communal biomineralizers, while coccolithophores (calcareous marine nanoplankton belonging to the phylum Prymnesiophycae) encase themselves in elaborate calcitic microstructures (smaller than $20\,\mu m$), and foraminifera (amoeboid protists of the eponymous phylum) create pore-studded micro shells (tests). Unicellular coccolithophores are abundant throughout the photic zone in nearly all marine environments of the Northern hemisphere and up to about $50°S$ in the Southern Ocean (O'Brien *et al.*, 2012). They also form massive ocean blooms that last for weeks, cover commonly $10^5\,km^2$ of the ocean surface, and are easily identifiable on satellite images.

Coccolithophores cover themselves in coccoliths that are formed inside the cell and are extruded to form a protective armor; many coccoliths detached from cells also float freely in water. Coastal blooms have a coccolith to coccolithophore cell ratio of 200–400, but the ratios for open waters are much lower, between 20 and 40. *Syracosphaera*, *Umbellosphaera*, and *Gephyrocapsa* are common genera but *Emiliania huxleyi* is the biosphere's leading calcite producer (Stanley *et al.*, 2005; Boeckel and Baumann, 2008). The species is also unusual because it produces fairly large coccoliths at a fast rate, sheds about half of them and, unlike many other planktonic species, is a relative newcomer that originated only about 270 000 years ago.

The photic zone can extend from just a few meters to about 200 m; densities of coccolithophores can be as low as a few thousand cells per liter, while in blooms they surpass 100 000/l; daily calcification rates range from less than 10 to 80 pg of calcite per cell per day; the largest blooms can cover $10^5\,km^2$ for periods of weeks and their annual extent adds up to about 1.5 million km^2 (Lampert *et al.*, 2002; Boeckel and Baumann, 2008). Given these natural variabilities, it is impossible to offer any reliable estimate of coccolithophore-mediated annual global calcification in the ocean – but a conservative set of assumptions should yield at least the actual order of magnitude.

Assuming continuous coccolithophore production in 60% of the world's ocean at depths only up to 50 m, with an average concentration of just 25 000 cells/l and a calcification rate of just 10 pg of calcite per cell per day, would result in annual global sequestration of about 900 Mt of calcite. More liberal assumptions (50 000 cells/l, 20 pg/cell a day) would yield an annual withdrawal of roughly 3.7 Gt of calcite. In comparison to this ocean-wide process the periodic blooms, no matter how spectacular, make a minor contribution. A bloom covering $250\,000\,km^2$ in the northeastern Atlantic in June 1991 had calcification rates up to $1.5\,mg\ C/m^3/h$, and in less than a month it had sequestered about 1 Mt of carbon in calcite (Fernández *et al.*, 1993). That would be just over 8 Mt of calcite and, if a similar rate were to apply to the about 1.5 million km^2 coccolithophore blooms that cover the ocean every year, the total sequestration would be just on the order of 50 Mt of calcite.

The same order of magnitude is obtained by assuming 50 m of highly productive photic zone, 150 000 cells/l, daily calcification at 70 pg/cell, and an average bloom duration of 30 days: that combination yields annual sequestration of about 25 Mt of calcite in coc-colithophoride blooms. The best conservative estimate is thus annual coccolithophoride-mediated calcification on the order of a few gigatonnes per year, a low rate from a geological perspective because the high Mg/Ca ratio and low absolute concentration of calcium in the modern ocean limit the production by most extant species (Stanley *et al.*, 2005). The most obvious testimony to high productivities of coccolithophores in the past are the immense sequestrations in Cretaceous and Tertiary chalk deposits (including the white cliffs of Dover).

Silicon is the other mineral that is massively assimilated by marine microorganisms; above all diatoms, silicoflagellates, and radiolarians. They use silicic acid ($Si(OH)_4$) to create their elaborate opal (hydrated, amorphous biogenic silica, $SiO_2 \cdot 0.4H_2O$) structures. Tréguer *et al.* (1995) estimated the rate of that uptake to be about 7 Gt Si a year. This means that the total mass of calcareous and siliceous materials sequestered annually by marine phytoplankton is on the order of 10 Gt/year, larger than the total extraction of all metallic ores and roughly the same as annual production of all fossil fuels (coal, crude oil, and natural gas).

In comparison to marine processors of calcium and silica, aggregate use of these elements by other organisms is orders of magnitude smaller; but its forms are often not only elaborate but also quite beautiful. This is true for many mollusk shells: some of them are quite simple but others show remarkable geometric properties (Abbott and Dance, 2000). Carbonates are also bioprecipitated by reptiles and birds to build their eggs, and by snails to form their shells. Curiously, a leading student of structures built by animals deliberately leaves out such activities from his surveys: Hansell (2007) acknowledges that the Great Barrier Reef may be (as its common description claims) the world's largest structure built by living animals, but in his books he focuses on building that requires behavior, an attribute absent in coral polyps or coccolithophores that just secrete their skeletons.

2.2 Materials in Prehistory

Evolution of hominins (the human clade that diverged from chimpanzees more than 5 million years ago and that had eventually produced our species) should be more accu-rately seen as a dynamic co-evolution of several traits that have made us human: upright walking, endurance running, cooperative hunting, eating meat, symbolic language, and tool making, using natural materials to fashion objects that provide simple but practical extensions and multipliers of human physical capacities. That is why the archeology of hominin evolution traces the developments in making stone tools nearly as much as it does the changes in bone structure, diets, and socialization.

The oldest stone tools of the lower Paleolithic, discovered in East Africa (the Olduvai Gorge in Tanzania, Afar triangle, Koobi Fora in East Turkana) were dated to about 2.6 million years ago, but even earlier origins cannot be excluded (Davidson and McGraw, 2005). For millions of years the aggregate mass of stones used to make Paleolithic tools (the era that ended about 10 000 years ago) and then Neolithic artifacts remained minuscule, as the hominin population counted in mere tens or hundreds of thousands (it was no higher than 125 000, 500 000 years ago). The global total of modern humans (with *Homo sapiens* clearly identifiable in the fossil record from about 190 000 years ago) was reduced to perhaps less than 10 000 as a result of Toba's megaeruption 74 000 years ago (Harpending *et al.*, 1998; Ambrose, 1998); and it then rose to just a few million people by the time of the first sedentary crop-cultivating societies nearly 10 000 years ago.

Given these low population totals, it is obvious that using stones for tools was not a matter of quantity and mass supply but of specific quality and manufacturing skills to produce desired shapes and edges. Obsidian (volcanic glass formed by rapid cooling of magma) and flint (crystalline quartz) were the best materials to produce sharp cutting edges and piercing points by knapping; tools for pounding and pulverizing were made by grinding from basalt, rhyolite, and greenstone. Perhaps the most notable innovation in making stone tools was to modify stone properties by heating: there is clear evidence that 164 000 years ago our species began to use fire as an engineering tool to heat-treat stones in order to improve their flaking properties (Brown *et al.*, 2009).

As is attested by numerous finds in both the Old and New World, the combination of expert stone selection and appropriate manufacturing methods (heat treatment and skilled flaking, knapping, or grinding) produced adzes, axes, hammers, awls, arrows, tips, and knives that were ingeniously designed, esthetically pleasing, admirably ergonomic, and practical. In contrast, wooden tools have been preserved only in rare instances, particularly when buried in anoxic layers. There is no doubt that hominins used wooden sticks to dig up roots and wooden clubs to kill small animals, but carefully crafted wooden weapons arrived relatively late. Although some massive herbivores could be killed without any tools – carefully planned stampeding of buffalo herds over cliffs, whose best known example is Alberta's Head-Smashed-In Buffalo Jump near Fort Mcleod in Alberta (Frison, 1987) – killing of megafauna (from mammoths to elands) generally required projectile weapons.

The earliest, and surprisingly well-preserved, throwing spears were found in Germany in 1996: six 2.25 m long spruce-wood spears that were dated to between 380 000 and 400 000 years ago (Thieme, 1997); that is nearly 200 000 years before the appearance of our species (Trinkhaus, 2005). And new evidence from the site of Kathy Pan 1 in South Africa (fracture types, modification near the base, distribution of edge damage) indicates that stone points could have been hafted to spear tips about 500 000 years ago, pushing the time of the first hafted multicomponent tools to about 200 000 years earlier than previously thought (Wilkins *et al.*, 2012).

Other designs of food-procuring wooden tools are much more recent. Dating of the first bows (from the most suitable species of wood, above all yew, white ash, black locust, osage orange) and arrows (merely sharpened wood or stone-tipped) remains elusive, but they were common in the late Paleolithic. Hunting sticks made of wood or mammoth tusks were used in Euroasia long before the invention of the Australian boomerang (returning stick) some 10 000 years ago. The earliest use of fishing nets is impossible to estimate, as they were undoubtedly woven from flexible plant stems or boughs that decayed rapidly; the oldest known net, made from slim willow branches and found in Finnish Karelia, is about 10 000 years old (Miettinen *et al.*, 2008).

Foragers living without permanent abodes left no traces of their temporary shelters built from branches, grasses, reeds, or palm fronds. The oldest preserved components of shelters are, naturally, stones arranged for protection (walls, roofs), and mammoth bones and tusks made into walls, timbers, and hides. In contrast to the vanished living spaces of foragers (save for the caves, some adorned with remarkable paintings of animals made by the Neolithic hunters), archeologists have found thousands of foundations belonging to the houses, sheds, and store rooms of many pre-agricultural or early agricultural societies. Perfect reconstructions of most of these Neolithic structures are usually impossible, and we can only speculate about the actual use of wood (slender trunks of small trees, branches), reeds, straw, or clay in their walls and ceilings. Neolithic foragers also built the first wooden vessels: the oldest excavated examples of the simplest design, canoes dug out from a single tree, are nearly 10 000 years old.

Although prehistoric societies managed without any metals, some of them succeeded in changing the properties of minerals by producing fired pottery. Excavations of some of the earliest permanently inhabited sites indicate that they used quicklime (burnt lime), a material whose production required a processing almost as sophisticated as the smelting of ores. In order to extract quicklime (CaO) from limestone ($CaCO_3$) the rocks had to be first crushed by hand to a fairly uniform small size (but tiny pieces and dust had to be avoided in order to keep the charge porous); the earliest firing took place in pits filled with layers of firewood and crushed limestone, later small stone structures (kilns) were built to confine the process and to concentrate the heat needed for calcination.

After prolonged firing came the most hazardous operation, removal of the highly caustic quicklime whose reaction with water (that is with moist skin, eyes, or lungs) produces severe irritation and will eventually burn, while ingestion of the dust brings abdominal pain and vomiting. Controlled addition of water to quicklime produces hydrated (slaked) lime ($Ca(OH)_2$, calcium hydroxide), the key ingredient of whitewash, mortar, and plaster. Producing lime is clearly a complex process that requires a considerable amount of planning (collecting sufficient firewood, fashioning a suitable pit, or a simple kiln) and management (the temperature inside a kiln must reach at least 825 °C and it must be maintained for a prolonged period) – but lime products dating to 9600 BCE were excavated at Göbekli Tepe (Courland, 2011). Kilning limestone was thus the first successful industrial process dependent on a chemical reaction – and one whose fundamentals remained nearly identical until the nineteenth century.

Many of these wood-and-stone gatherer-hunter societies also mastered the firing of shaped clays to produce pottery, the first process used by humans to convert commonly available minerals into utilitarian or ornamental objects (Cooper, 2000). Firing removes water from shaped clays, sets their shape, and strengthens their final form. Firing clay to produce a small object requires relatively low temperatures of 500–600 °C, and the oldest such piece of ceramics is the famous Venus of Dolní Věstonice in Moravia, an 11.1-cm tall statue of a nude female that was made by Paleolithic foragers between

25 000 and 29 500 years ago (Vandiver *et al.*, 1989). To make larger, utilitarian pieces of simple earthenware, temperatures must reach about 1000 °C, and while this could be achieved (for brief periods of time) in covered fire pits or mounds, it was much better done in stone kilns. The earliest pottery pieces were undoubtedly pit-fired, but the dates of the oldest fired objects have been receding with new finds, particularly in Asia.

According to the latest findings from Xianrendong Cave (in northern Jiangxi province), Upper Paleolithic hunters may have been firing simple small vessels in China as far as 19 000–20 000 years ago (Wu *et al.*, 2012). Japan's Jōmon (cord-marked) pottery (first various round-bottom bowls up to 50 cm tall, then larger flat-bottom vessels for food storage and cooking) is up to 12 000 years old (Habu, 2004). After 6000 BCE ceramic objects – vessels, cups, vases, chalices, plates, figurines, many of them finely ornamented, others fired at higher temperatures to produce more durable stoneware – became common throughout Europe (as attested by numerous finds of post-5500 BCE *Bandkeramik* in Germany, the Czech lands, and Austria), the Middle East (Egyptian ceramic was particularly diverse, both in terms of raw materials used and shapes produced), and East Asia. This was also the time when the potter's wheel (allowing, in skilled hands, a superior execution of the intended shape) came into widespread use.

Pre-agricultural and proto-agricultural societies have also left behind an almost indestructible record of megalithic structures, beginning with remarkable stone circles with carved animal relief at Göbekli Tepe. Western Europe (from the Iberian peninsula through France and Britain to northern Germany) has a particularly high density of megalithic monuments. They include menhirs (single-standing stones, usually upright, some also horizontal); grouped standing stones, such as the famous parallel arrays at Carnac in southern Brittany, and Stonehenge and nearby Avebury (construction began in the late Neolithic, about 2800 BCE); and, the most common category (about 5000 in all), burial chambers with corbeled roofs or vaults, often containing large capstones (Daniel, 1980).

Despite centuries of speculation and decades of interdisciplinary scientific studies, we still cannot offer any definitive explanation concerning the specific methods of quarrying these massive stones, transporting them across often rough terrain, and erecting them in a predetermined fashion, sometimes in remarkably accurate alignments with periodical configurations of celestial bodies. Transportation arrangements remain particularly unclear, as both the unit masses and the straight-line distances were often considerable. The inner circle at Stonehenge was made of 80 4-t bluestones that had to brought from Prescelly Mountains in southwestern Wales about 380 km from the Salisbury plain (and hence their move almost certainly involved shipments in coastal vessels), and although the outer circle came from Marlborough Downs only 32 km north of the site, moving 50-t stones was an even greater logistic challenge (Thorpe and Williams-Thorpe, 1991).

Finally, there is no doubt that the first textiles are of prehistoric origin, but the poor preservation of natural materials used for clothing makes it impossible to offer any reliable chronology (Ginsburg, 1991). Bone sewing needles used to make the warm clothes needed to survive Ice Age temperatures became common during the Solutréan phase of the Upper Paleolithic in Europe (21 000–17 000 years ago). The first deliberately cut, processed, and shaped body covers made of plant material (leaves, bark fibers) and animal skins have long decayed, precluding any dating. Skin tanning, needed to keep hides supple in low temperatures, is definitely of prehistoric origin, but identification of specific processes (using plant materials or minerals) becomes possible only during the Mesopotamian and Egyptian antiquity. The quality and functionality of prehistoric clothes made by Neolithic foragers is best demonstrated by *annuraangit*, Inuit skin-and-fur garments that provided excellent protection in the Arctic environment (Oakes, 2005).

Flax weaves were produced even before the plant was domesticated in the Middle East more than 10 000 years ago, and seed size indicates that by the third millennium BCE different forms of flax (for oil and for fiber) were cultivated in Europe north of the Alps (Zeist and Bakker-Heeres, 1975; Karg, 2011). Cotton cultivation began in Asia about 7000 years ago, much later in Mesoamerica, and phylogenetic analysis shows that the African-Asian species and the New World species stem independently from wild progenitors and converged morphologically under domestication (Wendel *et al.*, 1999). Primitive looms used to make rough linen fabric came around 6000 years ago, rough wool cloth was spun and woven more than 10 000 years ago; woolly sheep were domesticated about 8000 years ago but it took at least another 2000 years before the appearance of first woven wool fabric (Broudy, 2000).

2.3   Ancient and Medieval Materials

The material world of the oldest settled societies – whose crop productivity and population growth allowed the establishment and expansion of cities – was determined by their immediate environment. As a result, living quarters included caves excavated in silt (in China's Shaanxi) or in limestone (in central Anatolia), light wooden frames with walls of bamboo and clay (in Japan), unbaked brick structures (in sub-Saharan Africa), painstakingly assembled stone houses (in the Mediterranean and high-lying parts of Asia), massive log houses (in Scandinavia and Russia), and sturdily built structures utilizing fired bricks or mortared stones (throughout continental Atlantic Europe).

Environment made yet another, strategic, difference. As Adshead (1997) has pointed out, a key reason for China's (and more generally East Asia's) choice of ephemeral wooden housing, in contrast to Europe's preference for sturdy construction using stone and bricks, was the high frequency of destructive earthquakes; and I would also add of annual typhoons and major floods. The other reasons were the Chinese preference for low initial capital outlay and high cost of maintenance, the reverse of the European approach. Historical realities also mattered, with medieval and Renaissance Europe admiring and emulating many examples of monumental Roman stone structures.

Some of the earliest sedentary societies soon mastered the construction of monumental structures whose erection required not only ingenious quarrying and transport of stones to building sites but also, unlike with pre-historic megaliths, careful (and often detailed and extremely accurate) cutting and intricate assembly of massive structures, or production and emplacement of prodigious quantities of bricks. Some ancient societies also developed remarkably ingenious infrastructures designed to serve their growing cities, with the Roman aqueducts being perhaps the best illustration of these capabilities. The first sedentary societies of the Middle East, Mediterranean, and East Asia were also the first complex civilizations to smelt metals, initially only copper and tin on limited scales (for ornamental or limited military uses), later mastering also the production of zinc, lead, iron, mercury, silver, and gold.

In the Middle East, India, and Europe, iron became the dominant metal after roughly 1200 BCE, and its artisanal production was eventually on a scale sufficient to supply relatively inexpensive nails for house and ship construction, shoes for horses, a variety of hand tools and machine components, and a wide range of weapons. These metallurgical advances led Christian Thomsen (1836) to divide material evolution into the Stone, Bronze, and Iron Ages, but this was not a universal sequence as some societies (perhaps the best example is Egypt before 2000 BCE) had a pure copper era, while others skipped the Bronze Age and moved directly from stone to iron artifacts. In any case, in terms of total material flows, all pre-industrial societies remained in the wooden age: in all forested environments, no materials were as ubiquitous as timber for buildings and wood for tools, utensils, implements, and machines.

The first woody phytomass used in construction consisted of small trunks, tree branches, and bushes, and only the availability of stone adzes (and later good metal axes) and the acquisition of the requisite carpentry skills made it possible to build the first massive log houses made of shaped and joined timbers; later still came metal saws that could be used to produce long boards and other precisely cut components. Most pre-industrial wooden houses were poorly built and had short lifespans, but properly constructed wooden structures were surprisingly durable, even in monsoon climates where they have been exposed to annually heavy rains: wood for Japan's Hōryūji pagoda in Nara, the world's oldest extant wood building, was cut 1400 years ago.

Remarkably, these Japanese (and Korean and Chinese) buildings did not use any fasteners, instead their frames, beams, and roofs relied on a variety of precise mortise-and-tenon joints that not only allowed the wood to expand and contract in reaction to changing atmospheric humidity but also to withstand frequent earthquakes. Moreover, tall pagodas were essentially earthquake-proof because of the *shinbashira*, a central pole that does not support the structure but acts as a massive stationary pendulum (Atsushi, 1995). Its action is equivalent to tuned mass damping in modern skyscrapers, with their large concrete or steel blocks suspended at the tops of tall buildings.

Wood was also the only material used during antiquity and the Middle Ages for the hulls and masts of ocean-going vessels. Small river boats were made of a single tree trunk, but Homer credited Odysseus with cutting down 20 trees to build his ship (*Odyssey* 5:23): that would translate to mass of 10–12 t. Egyptian boats used to transport stone obelisks were much heavier, but the ships that made the first Atlantic crossings were remarkably light: a Viking ship (based on a well-preserved Gokstad vessel built around 890 CE) required the wood of 74 oaks (including 16 pairs of oars). Typical Mediterranean vessels also remained small and hardly increased in size during the millennium between the demise of the Roman Empire and the first great intercontinental voyages of the late fifteenth century.

Stone's durability made it the preferred material for utilitarian structures built to last (aqueducts, roads), and for funerary (pyramids, tombs) and religious or ceremonial (temples, gates, palaces, towers, obelisks, statues) monuments. Many stone structures impress due to their mass, but among these the Egyptian pyramids at Giza are unique: Khufu's pyramid not only remains the largest stone structure ever built (195 m high, it required 2.5 million stones whose average weight was 2.5 t) but this mass of more than 6 Mt of stone was also laid down with an admirable degree of accuracy. In comparison, the great Mesoamerican pyramids (Teotihuacan and Cholula) are not as imposing, not only because Teotihuacan's flat-topped Pyramid of the Sun rises only to just over 70 m (including the temple) but also because its core is a mass of soil, rubble, and adobe bricks, and only its exterior was clad with cut stone and plastered with lime mortar.

Other stone structures are distinguished by their ubiquity, others yet by a complexity and lightness that belies the heaviness of the material (medieval cathedrals), and some display an almost incredibly tight fit and accurate emplacement of stone. During the Roman Empire the largest mass of stones went into well-engineered aqueducts (including elevated bridges, inverted siphons, and distribution networks within cities) as well as into their extensive system of roads (*cursus publicus*), but most of their monumental structures (triumphal arches being a major exception) were of marble-clad bricks. Intricately designed stone monuments dominated in South and Southeast Asia (including Angkor Wat, the world's largest Hindu temple in Cambodia, and Borobodur, the world's largest Buddhist monument in Java) as well as in Mesoamerica and South America (the Maya, Aztec, and Inca empires, including the precision-cut stones of Pumapunku near Tiahuanaco in Bolivia). And Africa's most massive stone structure, the 240-m long oval enclosure of Great Zimbabwe whose construction began after 1100 CE, required an estimated 1 million stone blocks (Ndoro, 1997).

Medieval Europe brought stone architecture to unprecedented heights (literally and figuratively) with its bold and intricate designs of large cathedrals with arched vaults, flying buttresses, and tall spires (Fitchen, 1961; Schutz, 2002). Prak (2011) showed that this was achieved by relying on surprisingly simple underlying principles of modular design (transferred on a personal basis by designers/builders) augmented with on-site experimentation. The most obvious commonalities of stone construction are elaborate quarrying (the task being all the more remarkable during the earliest period when, as in ancient Egypt, the masons had only copper chisels and dolerite mallets), often highly accurate dressing, frequent needs for long-distance land- or water-borne transport to often distant building sites, and ingenious ways of emplacing massive components. Even today we are impressed by the inventiveness and skill used in the quarrying, transporting, and shaping of these massive stones, and in the design and organization of their final emplacement, often within admirably tight tolerances.

This is particularly true in case of monoliths (used for statues or as steles or obelisks: Rome has eight of them imported from ancient Egypt) from traditional civilizations; we still can only speculate about the many specific artisanal techniques and the construction management required to build the pyramids on the Giza plateau or to fit the polygonal stones of massive Inca walls at Sacsayhuamán and Ollantaytanbo with incredible tightness. Nor do we know exactly how the ancient Egyptians orchestrated the transport of a 1000-t stone statue, by a Nile ship, 270 km from Aswan to Thebes. Not surprisingly, the specific modalities of building many of these structures remain elusive: the only certainty is that their construction required careful planning and extraordinary supply and worksite management, particularly for those structures that were completed in what appear to be incredibly short periods of time.

Egypt's largest pyramid took only 20 years to complete (2703–2683 BCE). The Parthenon, a simple post-and-lintel building but one possessing perhaps the best-ever proportions, required only 15 years (447–432 BCE). And the astonishing Hagia Sophia in Constantinople, a relatively light structure deriving its impact from enormous vaults, did not take even five years to complete (527–532). In contrast, centuries could elapse between the laying of the foundation stone and the consecration of many European cathedrals: construction of Notre Dame took nearly two centuries, from 1163 to 1345; St. Vitus, dominating the Prague Castle, was begun in 1344 and was completed only in 1929.

Nor can we, obviously, quantify the mass or volume of construction stone used annually in antique or medieval societies, but revealing calculations can be made for the requirements of some large projects. The correct order of magnitude of materials deployed by the Romans to build their roads can be estimated by assuming an average width of 4 m (surviving stretches of Italian *via* have widths between 2.4 and 7.5 m) and depths of at least 1 m (with paving stones topping layers of gravel, pebbles, clays, and sand): leaving aside subsequent repairs, the initial construction of 85 000 km of major roads would have required 425 Mm^3 of aggregate (sand and gravel) and quarried stone.

In alluvial plains, clays transformed into bricks became the dominant material of construction: Mesopotamian kingdoms built their ziggurats, palaces, and walls from bricks (both adobe and better quality made from fired clays). Bricks were also indispensable for the building boom in imperial Rome, which needed relatively inexpensive and rapidly (and locally) produced material in order to build massive buildings whose exteriors were then clad in marble. The first bricks were made with puddled clay (just mixing clay and water with additions of sand, chopped straw, or dung), these mixtures were shaped in wooden molds and the bricks were left to dry in the sun.

Early Mesopotamian cultures (Sumer, Babylon, Assyria) produced prodigious amounts of these bricks (the Babylonian standard was a square-shaped prism, $40 \times 40 \times 10$ cm) which they used to build their houses, palaces, and towers. The poor quality of such bricks turned even the most massive structures into mounds of clay: a few exceptions aside – above all Chogha Zanbil, a partially preserved Elamite structure in Iran's Khuzestan with three of the five storeys still standing (Porada, 1965) – Mesopotamia's stepped flat-topped temple towers (ziggurats) built after 2200 BCE are now nothing but hillocks on an arid plain. Bricks burnt in kilns fueled with wood or charcoal were used first in ancient Mesopotamia, they were common in both European and Asian antiquity and remained the most important quotidian construction material in many Old World cultures until the arrival of concrete and structural steel.

Slim, oblong Roman bricks ($45 \times 30 \times 3.75$ cm) were clad in marble in all monumental buildings, but even after that protective layer was stripped away they survived fairly well in the dry Mediterranean climate where their poor heat conductivity helped to keep the structures cool in summer. Romans also used bricks to construct remarkably expansive vaulted ceilings. There were many different methods of laying down the bricks and combining them with stones and wood (common Roman choices were *opus reticulatum*, *vittatum*, *mixtum*, and *testaceum*), all of them seen in the ruins stripped of their marble cladding.

Structures built with fired bricks proved more durable (naturally, with some repairs) even in Asia's tropical climates: many *stūpas* (meaning, literally, heaps in Sanskrit), huge mound-like constructs built to house Buddhist relics, have survived for more than a millennium (Longhurst, 1979). The Mahabodhi stūpa (Bodhgaya temple), built on the site of Buddha's enlightenment 2500 years ago, the great sandstone-clad Sāñcī stūpa of Bhopāl (36 m base diameter, 15 m tall), the Phra Pathom Chedi stūpa in Thailand, and the 122-m tall Jetavanaramaya in Anurādhapura (Sri Lanka) built some 1700 years ago, are among the best examples of these massive fired-brick structures (Bandaranayake, 1974; Pant, 1976). The number of bricks (made of 60% fine sand and 35% clay) required to build Jetavanaramaya was put at between 93 and 200 million (Leach, 1959).

Fired earthenware and stoneware (the latter material having much lower water absorption after firing) were also used to make roof, floor, and decorative tiles. Ceramic roof tiles were commonly used around the Mediterranean from antiquity, while the first examples of glazed bricks and decorative tiles can be found in even older Mesopotamian structures. Floor and wall tiles were eventually produced in a great variety of finishes (unglazed, glazed, mosaic, hand-painted), and particularly intricate tiling designs became one of the signature marks of medieval Muslim builders who excelled in matching colors and shapes; a less exuberant ancient tiling tradition also evolved in a number of European countries, especially in Italy, Spain, and Portugal.

Clay amphoras were the most common containers of sea-borne Mediterranean commerce (Twede, 2002). These ceramic jars were used to transport wine, olive oil, and also such processed foodstuffs as *garum*, Roman fish sauce. Romans also used wooden barrels made by skilled craftsmen (coopers) from hardwood staves kept tight by iron hoops, with a typical barrel consuming 50–60 kg of wood (Twede, 2005). During the Middle Ages, these containers dominated European trade in liquids as well as in pickled and salted foods. Porcelain, the highest-quality material made of fired clay (specifically kaolin, soft white kaolinite clay indispensable for making fine ceramics) was invented around the sixteenth century BCE (during the Shang Dynasty) in China and was soon made in Korea and Japan; this East Asian monopoly was broken only during the early eighteenth century with the production of true porcelain in Europe, first in Germany and France, and soon afterwards in England (Atterbury, 1982).

Durable construction using bricks (and tiles) requires good bonding material, but the first clay and mud mortars were weak; the calcination of limestone ($CaCO_3$) resulting in quicklime (CaO) whose hydration produced $Ca(OH)_2$, slaked lime, made a superior material. As already noted, the kilning of limestone predates the era of the oldest centralized empires and slaked lime was used routinely in ancient Egypt. Davidovits (2002) caused a controversy by claiming that the blocks of the Giza pyramids are not quarried stones but mixtures of granular limestone aggregate and alkali alumino-silicate-based binder that were cast *in situ*. Although this hypothesis was rejected by most Egyptologists, it continues to have adherents even among material scientists.

Romans are credited with the invention of concrete, but this is an inaccurate attribution. Concrete is a mixture of cement, aggregates (sand, pebbles), and water and cement is a finely ground mixture of lime, clay, and metallic oxides fired in kilns at a high temperature. There was no cement in Roman *opus cementitium* and hence this sturdy mixture, strong enough to build large vaults and domed structures, was not the material now known as concrete. *Opus cementitium* contained aggregates (sand, gravel, stones, broken bricks, or tiles) and water but its bonding agent was lime mortar (Adam, 1994). The combination of slaked lime and volcanic sand from the vicinity of Puteoli near Mount Vesuvius (*pulvere puteolano*, later known as pozzolana), produced a superior mixture that could harden even under water and that could be used to build not only massive and durable walls but also spectacular vaults. The coffered ceiling of the Pantheon (118–126 CE), with a dome span of 43.2 m, remained unsurpassed for nearly two millennia (although San Pietro's dome, designed my Michelangelo and completed in 1590, came close with 41.75 m).

The last ancient material invention that required heat processing was glass (Macfarlane and Martin, 2002). The earliest small Mesopotamian pieces are about 5000 years old, the art of making glass objects was advanced during the New Kingdom era of the ancient Egypt and glass finds are not uncommon at Roman sites – but it became an important material only during the Middle Ages, both in construction (as attested by intricately designed multicolored cathedral windows) and in everyday use (as shown by the elaborate glass goblets of Venetian and Bohemian glass). Crown glass production was the only practical way to produce pans of limited sizes until the mid-nineteenth century. Molten material was mouth-blown to create a large bubble and then spun into a disc (with a circle in the middle, up to 1 m in diameter) that was, after cooling, cut into panes. This glass was too expensive for the glazing of ordinary homes, and glass windows became common only during the early modern era.

But the most consequential material development in antiquity was not the routine use of a wide variety of biomaterials (wood, bones, hides, plant-, and animal-textiles) and common construction (stones, clays, sand, concrete) and ornamental (tiles, glass) materials, but the ability to smelt and to shape a growing array of metals. Ore mining and metal smelting resulted in an epochal advance that began with the use of copper and its alloys and was followed by smelting of iron ores, making iron the dominant metal of the ancient Greece and Rome, the two great Mediterranean civilization whose accomplishments influenced so much of subsequent European, and global, history.

Copper, soft and malleable when pure and easy to alloy, was the first metal used by Stone Age societies as early as almost 10 000 years ago, initially without any smelting as the bits of outcropping pure metal were shaped cold or subjected to repeated heating and hammering (annealing). Copper smelting and casting took off more than 6000 years ago, first in Mesopotamia (Gilgamesh, the great Sumerian epos (about 2500 BCE) refers to a copper box and a bronze bolt and excavations found copper knives, axes, and spears in the region's ancient empires), a millennium later in Egypt, and about 3500 years ago it was common in China. Relative abundance of copper sulfide ores made the element the dominant metal between 3000 and 1000 BCE.

Ore reduction (Cu has a fairly high melting point (1083 °C)) was carried out with wood or charcoal, first in clay-lined pits, later in small clay furnaces, and metal was purified by further heating. Producing copper from abundant sulfide ores (chalcopyrites) was more complicated: crushed ore (by hand, later by harnessed horses or waterwheel-driven hammers) had to be first roasted to remove sulfur and various associated metals (As, Fe, Pb, Sn, Zn). The roasted ore was smelted in shaft furnaces and then smelted once again to yield 95–97% pure metal. All of this devastated local and regional wood resources, and copper smelting was a leading cause of Mediterranean deforestation, particularly in Spain and Cyprus.

But annealed copper was a soft metal and had low tensile strength. That is why the much stronger and harder bronze became the first practical alloy in history and why Christian Thomsen chose it for his now classic periodization of the Stone, Bronze, and Iron Ages (Thomsen, 1836). Bronze, an alloy of copper with 5–30% of tin (10% being typical), has a tensile strength nearly four times that of annealed copper and it is nearly six times harder, good enough for durable knives, swords, axes, and medals (as well as for bells and musical instruments). Brass is an alloy of copper (ranging from less than 50 to about 85%) and zinc; its smelting dates to the first century BCE, but it became common only during the high Middle Ages. Zinc increases the alloy's tensile strength and hardness to about 1.7 times that of cold-drawn copper, without reducing malleability and resistance to corrosion. Pewter is an alloy dominated by tin (about 90%) with added antimony and copper.

Besides copper, tin, bronze, and brass, the other color (nonferrous) metals whose production was mastered by ancient metallurgists included zinc, lead, mercury, silver, and gold. Analysis of lead in an ice core from Greenland made it possible to reconstruct the world's longest series of metal production. Hong *et al.* (1994) estimated that large lead–silver smelting by Greeks and Romans increased total lead output from about 250 t/year in 750 BCE to nearly 80 000 t by 50 CE. Lead's low melting point and easy malleability made it an early candidate for making water pipes in Roman cities, but the greatest demand for the metal was for inverted siphons (U-shaped pipes connecting a header tank on one side with a lower-lying receiving tank on the opposite bank), the preferred way of Roman engineers to cross valleys where stone bridges would have to be taller than 50–60 m. Post-Roman decline and stagnation kept lead output at only about 12 500 t 1000 years later, and global output surpassed the highest antique production only by the middle of the eighteenth century.

Gold and silver were used in antiquity for ornaments (Tutankhamun's burial mask, from about 1320 BCE, is perhaps the most famous golden object of the era) and jewelry in both the Old and the New World, and they were also used to mint coins, often debased by the addition of cheaper metals: gradual debasement of the Roman *denarius* is perhaps the most notorious example of this process (Salmon, 1999). Patterson (1972) made the following rough estimates of annual rates of Greco–Roman silver production: 25 t between 350 and 250 BCE, 200 t between 50 BCE and 100 CE, and, again, 25 t during the fourth century CE. Mercury was incorporated into ointments and cosmetics (just about the worst possible use imaginable), and later became a key element in experimental alchemy.

Iron ores are widely distributed, many early finds were rich in metal (pure magnetite contains 72% Fe) and at many sites they could be mined without sinking shafts, but smelting the metal requires higher temperatures than copper or lead production. The first iron artifacts date to around 5000 BCE, but the metal came into common use only less than 3000 years ago, both in ancient Egypt and in Asia. Cast iron, an alloy of the metal with carbon, was the easiest ferrous metal to produce, but its tensile strength was no higher than that of cold-drawn copper and its hardness was similar, or only slightly superior, to that of bronze. But the greater abundance of iron ores and slow advances in smelting the metal eventually made iron the most important metal of antiquity and, in aggregate mass terms, it has kept its dominance ever since: as I will show in detail in the third chapter, ours is still the Iron Age, with the total consumption of other metals adding up to a small fraction of iron use.

Due to its relatively high carbon content (between 2 and 4.3%) cast iron (or pig iron, the term chosen because of the resemblance of cast forms to a small pig) has low tensile strength (lower than bronze or brass), low impact resistance, and also very low ductility. But the alloy is strong under compression, making it good enough for a large variety of tools, utensils, and objects (ranging from nails to horseshoes and from pots to fireplace grates), and weapons (including heavy guns and cannon balls), but precluding its use wherever the metal was subject to greater tension: iron columns are fine, iron beams could never support tall buildings. Iron smelting began in simple, partially enclosed hearths, producing small (around 50 kg) slag-contaminated masses of the metal.

Bloomery hearths were gradually transformed into small shaft furnaces whose development led to the first blast furnaces (during the fourteenth century in the

Rhine–Meuse region) in whose columnar bodies smelting takes place, as upward movement of hot CO-rich gases reduces ferrous oxides into iron at temperatures reaching more than 1600 °C. For the next five centuries, blast furnace sizes were limited because charcoal could not support the heavier charges of iron ore and limestone (used to remove impurities). Fundamental innovation that allowed mass production of inexpensive iron came only during the late eighteenth century with the replacement of charcoal by metallurgical coke.

The oldest way to make steel is by carburizing (cementation), that is by adding carbon to practically carbon-free sponge iron (small porous masses of iron and slag from bloomery furnaces) by prolonged heating in charcoal; this produces an outside layer of hardened steel and uniform carbon distribution (needed for steel destined for weapons) and is achieved only by repeated forging (Birch, 1968). Decarburizing (removal of carbon by oxygenation) from cast iron was first done in Han China, producing an alloy good enough to make chains for suspension bridges to span deep gorges. But such uses were exceptions, as steel remained expensive and hence in limited use until the nineteenth century.

2.4   Materials in the Early Modern Era

I prefer a simple delimitation of the early modern era as the period of the three centuries between 1500 and 1800; if I had to choose more specific years I would opt (for obvious reasons) for 1492 and 1789. These were fascinating centuries, an amalgam of many persistent, old, medieval realities and new (as is clear in retrospect) fundamental departures that, undoubtedly, laid the foundation of the modern world. In material terms the era was characterized by some new qualities but even more so by much increased quantities. No new materials were introduced either in construction or in (still overwhelmingly) artisanal manufacturing, but the growing sizes of cities and ocean-going ships, more reliance on water wheels and windmills, larger fortresses, ingenious construction of new canals, ports, and roads, and (first in England and Wales) the expansion in mining of coal led to higher demand for the two basic high-quality construction materials – wood and iron – and required greater removal and emplacement of larger volumes of soil, sand, gravel, and stone.

The early modern era was a time of higher rates of population growth, incipient urbanization, and proto-industrialization and these processes stimulated changes in material consumption. The best available reconstructions indicate that the global population increased by less than 60% during the 500 years between 1000 and 1500 but then more than doubled (from about 460 million to nearly a billion) by 1800 – but remained overwhelmingly rural, with cities accounting for less than 5% of all humanity (Klein Goldewijk *et al.*, 2010). Proto-industrialization relied on cheap rural cottage and urban workshop labor, and its products reached wider national and even international markets.

In Europe the regions affected by this process included parts of the British Isles (Cotswolds, Ulster), France (Picardy), and Germany (Westphalia, Saxony, and Silesia); in Asia large-scale artisanal production was concentrated in the coastal areas of Qing China, in Mughal India, and in the cities of Tokugawa Japan, and an increasing share of manufactures from all of these regions found its way to other Asian and

European markets. Larger-scale artisanal production implies incipient mass consumption and, as Mukerji (1981) showed, materialism was not a consequence of industrialization because in parts of the Atlantic Europe it was quite evident during the sixteenth century.

Paintings of the Dutch artists active during the Golden Age (1581–1701, or essentially the entire seventeenth century) show many neat spacious houses in Amsterdam, Haarlem, or Delft, with small but neat backyards, clean tiled floors, large glass windows, walls adorned with pictures or maps, musical instruments, and no shortage of well-built furniture and bedding. Interiors painted by Jan Molenaar, Pieter de Hooch, and Jan Vermeer leave us with unmistakable impressions of material comforts and the incipient affluence enjoyed by many Dutch burghers long before European industrialization ushered in the successive waves of modern mass consumption. Moreover, owners of these houses were eager buyers of an expanding range of consumer goods, ranging from better cookware to finer clothes and from engravings to porcelain imported from China and Japan (the Dutch having the trading monopoly with the Tokugawa shogunate).

The greatest source of information about the accomplishments of the early modern era – the world's first encyclopedia published between 1751 and 1777 under the editorship of Denis Diderot and Jean le Rond D'Alembert – is filled with descriptions and engravings of a large variety of machines, including many complex and intricate designs whose adoption presaged even greater advances of the nineteenth-century industrialization (Diderot and D'Alembert, 1751–77). At the same time, housing remained commonly primitive and inadequate: even in France during the sixteenth century most village dwellings were just mud huts covered with straw or rushes, and in cities living quarters were often combined with (or were adjacent to) workshops or stores (such as in Kyōtō's dark, long *machiya* houses). Buildings were also poorly heated and badly lit: the dominant fireplaces were inconvenient and inefficient, and more efficient stoves (most notably German *Kacheloffen* and Dutch and Scandinavian tiled designs) diffused only slowly from their areas of origin.

As a result, fuel-wasting fireplaces and braziers resulted in a huge demand for fuelwood and charcoal to heat the expanding cities of the pre-coal era. In Paris, the demand rose from more than 400 000 loads of wood in 1735 to more than 750 000 loads in 1789 (about $1.6\,Mm^3$) and the same amount of charcoal, prorating to more than a ton of fuel per capita (Roche, 2000). Cramped rooms in rural houses were also littered with tools, and even in urban dwellings furnishing was often minimal. The chair, absent in most houses during the Middle Ages (people sat on the ground, on benches, ledges, or cushions), became a common piece of furniture, but a good bed remained very expensive: in France, just before 1700, its value was at least 25% of the total for all furniture in low-income families, and nearly 40% of that for servants (Roche, 2000).

Even during the seventeenth century it was still common to cook food and eat it from the same pots. Louis XIV ate with his fingers early in his reign (during the 1660s) but a century later even middle-class urbanites had on their tables an increasing number of objects devoted to specific uses, ranging from egg cups to teapots. This advance was accompanied by a retreat of noble materials (fewer objects made from silver or crystal glass) and a rise of cheaper objects made from common metals, china and blown glass, a shift that marked the onset of ephemeral consumption and rapidly changing fashions. Poverty was still widespread, but for increasing numbers of people daily living had gone beyond necessities As Roche (2000, p. 77) put it, "a new pattern of cultural behavior, made up of aspiration to well-being and dignity, asserted itself."

More substantial rural housing began to appear in some countries, but its wood requirements continued to reflect the environment: light structures in earthquake-prone regions, solid log houses in parts of the Atlantic Europe and eastern North America. Japan's *minka* (people's houses) used post and beam construction, clay or bamboo walls, sliding doors, paper partitions, and earthen floors. As a result a 100-m^2 *minka* often required no more than 8 m^3 of pine, cedar, or cypress wood (Kawashima, 1986) – while the equally small Scandinavian log house (*stock hus*) needed 100 m^3 of wood for walls, doors, ceilings, and roofs, and large farmhouses in Germany or Switzerland required commonly 1000 m^3 of timber (Mitscherlich, 1963). Frequent repairs and rebuilding added to this wood demand.

Wood remained indispensable not only for building houses and transportation equipment (carts, wagons, coaches, boats, ships) but also – as iron smelting rose in parts of Europe – for charcoal production for blast furnaces (substitution by coke began only during the latter half of the eighteenth century and was limited to the UK). And as Europe's maritime powers (Spain, Portugal, England, France, and Holland) competed in building large ocean-going vessels – both commercial and naval – the increasing number of such ships and their larger sizes brought unprecedented demand for the high-quality timber needed to build hulls, decks, and masts. At the very beginning of the early modern era, Columbus's *Santa Maria* displaced about 110 t, while Magellan's *Victoria*, the first ship to circumnavigate the world, displaced 85 t. With wooden hulls, masts, and spars being as much as 70% of the total mass (the remainder was divided among ballast, supplies, sails, armaments, and crew) these pioneering vessels contained 60–75 t of sawn timber (Fernández-González, 2006).

By the end of the eighteenth century large two-decked naval battleships (originally of French design) were about 54 m long, and carried 74 guns and crews of up to 750 men (Watts, 1905). Such a ship needed about 3700 loads (equal to 1.4 m^3) of oak timber, or about (with density of 650 kg/m^3) 3400 t of wood, roughly 50 times the mass used for the first intercontinental sailing vessels three centuries earlier. But because as much as 60% of all timber bought by the builders disappeared – as smaller pieces of wood were commonly taken from shipyards by workers for fuel, for making simple furniture, or to be sold (Linebaugh, 1993) – the actual mass removed from a forest was typically more than 5000 t per ship.

Naval ships of the early modern era also provide an excellent illustration of the increasingly massive production of armaments. Ships of the early sixteenth century had typically fewer than 10 guns; in 1588 the English vessels that defeated the Spanish Armada averaged only 12 guns; before the end of the seventeenth century the large men-of-war carried up to 100 guns; and during the battle of La Hogue in 1692 the British and Dutch captains deployed a total of 6756 guns (Anderson and Anderson, 1926). And iron for naval armaments was not the only reason for the metal's rising consumption. Higher demand also came from the expansion of mining, and the growing production of nails, wire, horseshoes, and weapons for land armies. English statistics show pig iron output rising from only about 10 000 t in 1700 to 26 000 t by 1750 and to 156 000 t by 1800 (Bell, 1884).

Iron production in small blast furnaces required enormous quantities of charcoal and combined with inefficient wood-to-charcoal conversion this led to widespread deforestation in iron-smelting regions: by 1700 a typical English furnace consumed 12 000 t of wood a year (Hyde, 1977). Reprieve came only with the slow adoption of metallurgical coke: it was first used in a blast furnace in 1709, but even in the UK it became dominant only a century later. Steel remained a commodity of restricted supply even in the mid-eighteenth century. During the late 1740s, Benjamin Huntsman (1704–76) began producing his crucible cast steel by carburizing wrought iron, but that metal was used only for such specialized, limited-volume applications such as expensive weapons (famous Damascene or Japanese swords), razors, cutlery, watch springs, and engineering (above all metal-cutting) tools whose quality justified higher prices (Bell, 1884). Mining for all kinds of ores expanded in Europe, with particularly notable innovations introduced in Germany, France, and Italy.

But the greatest change came from the exploitation of the New World's metals, thanks to an unprecedented expansion of gold and silver mining. The first questions put by Columbus to the natives after his landing in the New World were about gold, but the Spanish fortunes changed only with the discoveries of both gold and silver in Mexico (Zacatecas), Peru, and Bolivia (Potosí, 1545). During the next 250 years American silver enriched the monarchy through direct transfers, but the trans-Pacific transfer of the metal to the Philippines and then to China created a true world system of exchange (Frank, 1998).

Barrett's (1990) estimates show total annual silver flows to Europe rising from 40 t during the early 1500s, to 600 t during the last four decades of the eighteenth century. Total annual silver flows to Asia (through the Dutch and English companies and through the Levant trade, and directly from the Americas by trans-Pacific sailings) rose from 75 t during the first decade of the seventeenth century to nearly 170 t by the mid-eighteenth century. And Flynn and Giráldez (1995) argue that founding of Manila in 1571 (in order to link, for the first time in history, Asia and America with "substantial, direct, and continuous trade") was the birthday of true world merchandise exchange.

The eighteenth century also saw a notable expansion of textile-making, because the first category of consumer goods that saw an obvious and widespread increase of expenditure was clothing, starting in cities and spreading to the countryside, eventually creating what Roche (2000) called "the unification of sartorial habits." Vertical looms had been used since antiquity, and many traditional societies in Asia and America used portable body-tensioned looms. In either case weavers could produce cloth only as wide as their arm span and wider cloth required the cooperation of two weavers. This changed in 1733 with John Kay's invention of a flying shuttle that could be sent by flicks of the wrist from one end of the loom to another; the most important innovation allowing cheaper, mass-scale production of cloth came in 1785 with Edmund Cartwright's power loom (initially powered by steam).

A no less important innovation was the response to imports of Indian printed cotton fabrics (calico, produced in Calicut since the eleventh century) to Europe. Adoption of Indian techniques by European craftsmen began in France before 1650, before 1700 both French and English workshops were able to produce lasting colors and the practice spread to the Netherlands, Germany, Switzerland, and Austria. A century later the

Lancashire calico industry was gaining a clear competitive advantage over the India product, especially after Thomas Bell's invention of printing by copper rollers was adopted in 1785 (Jenkins, 2003). Soon the trade was radically reversed and British statistics show that by 1835 the total number of cloth pieces (plain and dyed) sent from Britain to India was 169 times that imported to Britain (National Archives, 2012).

Finally, a few paragraphs about stone and other construction materials in the early modern era. Cut stone remained the dominant material for monumental buildings, opulent private structures, and religious architecture. The splendors of the late Renaissance and baroque Rome are perhaps the best examples of this deployment in palaces, basilicas, churches, and colonnades designed by such masters as Michelangelo Buonarotti, Carlo Maderno, Gian Lorenzo Bernini, Francesco Borromini, and Girolamo Rainaldi. New markets for cut stone opened up with the expansion of urban housing. Perhaps the most notable example of this development is the expansion of Paris. For centuries the Lutetian cut limestone (*pierre de taille*) was supplied by subterranean quarries on the city's outskirts, but the city's expansion above the mined galleries led to collapsing tunnels and surface subsidence (Blanc *et al.*, 1998).

A commission set up by Jean-Baptiste Colbert, France's minister of finance under Louis XIV (between 1665 and 1683), identified limestone from Saint-Maximin quarries in the Val d'Oise only about 40 km north of the city's center as a nearly perfect color match for the city's major stone monuments (Destination Oise, 2013). This soft stone is easy to cut and yet it is fairly resistant to weathering, and since the late seventeenth century massive blocks for weight-bearing foundation floors and slimmer cuts for façades have given Paris its inimitable look in a few subtle shadings ranging from whites to faint yellows. The first great wave of this construction took place between 1715 and 1752 when 22 000 substantial new houses were added, nearly every fifth one with a *port cochère* (Brice, 1752). This remarkable pace of construction was surpassed only a century later with some 40 000 houses added during Haussmann's reshaping of Paris between 1853 and 1870 (Des Cars, 1988).

But in parts of Europe, stone gave way to brickwork in the new, and more material-intensive, types of fortresses that emerged in response to the greater capability of long-distance artillery. While typical medieval fortifications relied on compact (and very often elevated) sites and were built with high and thick stone walls, the new designs were radically different: they were variations of star-shaped polygons (hexagons, octagons) adapted to local (and often entirely flat) terrain, but always with relatively low masonry walls backed by massive earthen embankments that were to absorb artillery shelling, to conceal and protect, and, at the same time, to establish clear lines of defensive fire.

Sébastien Le Prestre Vauban, military engineer and eventually Marshal of France, was the most prominent architect of these fortifications that required an unprecedented mass of bulk construction materials. During the 40 years between 1667 and 1707 he upgraded the fortifications of some 300 cities and built 37 new fortresses along the western, northern, and eastern borders of France (they are now on UNESCO's list of World Heritage sites), including such spectacular projects as the citadels of Le Palais and Mount Louis, the city of Besançon, and an island redoubt of Saint Martin-de-Ré (Duffy, 1985; Hebbert, 1990). Lille fortress required 60 million bricks, and his largest project, the fortifications of Longwy in northeastern France, consisted of some 640 000 m^3 of rock and earth and 120 000 m^3 of brick masonry (Anderson, 1988).

2.5　Creating Modern Material Civilization

When judged in basic existential terms – that is looking at prevailing diets, longevities, typical living conditions, fuel consumption, prime movers, and everyday choices, uses, and abundance of materials – even advanced pre-industrial societies of the late eighteenth century (China during Kangxi's rule, late Tokugawa Japan, pre-revolutionary France, Russia under Catherine the Great) were not radically different from their late medieval precursors. Famines were infrequent, but prevailing diets were barely adequate, monotonous, and overwhelmingly cereal-based; high infant mortality reduced the average life expectancy at birth to less than 40 years; housing was crowded, unhygienic, and uncomfortable; wood (and charcoal made from it) and crop residues (mostly cereal straws) were the dominant fuels; human and animal muscle were the most important prime movers; and wood, stone, and clay (shaped and fired as bricks) furnished the basic building materials, while possession of objects made from metals and alloys (iron, copper, bronze, brass) was uncommon.

Only in England did coal displace wood as the dominant fuel. James Watt's (still quite inefficient) steam engines began to offer the first competition to windmills and waterwheels as reliable, and increasingly powerful, prime movers, while replacement of charcoal by coal made it possible to build larger blast furnaces, to produce more iron, and to lower the cost of many common iron objects. But even in the UK diets were often marginal, life expectancies were short, and the material possessions of most people were limited. The nineteenth century changed all that, first in the UK, parts of Western Europe, and eastern USA, then in most European countries and across North America, while Japan became the first Asian country to begin its quest for modernity after 1870. In material terms modernization – driven by industrialization and urbanization – was marked, above all, by two processes: a greatly expanded extraction of traditional construction materials, and a rapidly increasing consumption of metals.

The first kind of material expansion is usually neglected, as historians of industrialization focus on the consumption of fossil fuels and the production of metals and machines; but unprecedented quantities of stone had to be cut, blasted, crushed, and shaped, and even greater volumes of soil, sand, and clay had to be moved, emplaced, or incorporated in bricks and concrete in just a few decades in order to span countries with railways and better roads, to house the millions of former peasants who moved every year to cities, and to build the productive infrastructure of the modern economies (mines, ports, and factories) that enabled countries to move from agrarian societies to economies dominated by manufacturing in just two or three generations.

Perhaps the most esthetically appealing paragon of a much expanded demand for the cut stone would be the modernization of Paris that began during the Second Empire (1852–70). Reshaping of Paris' streets and housing, directed by Georges Eugène Haussmann between 1853 and 1870 (Carmona and Camiller, 2002), and the city's subsequent growth that nearly tripled its 1850 population by 1900, created a new wave of demand for the characteristic creamy stone from Saint-Maximin quarries that had been (as already explained) chosen by Colbert in the seventeenth century to replace locally quarried limestone. Haussmann used the stones for the typical solid-looking five-story buildings with angled mansard roofs that lined his straight, wide boulevards. An order of magnitude estimate could be made by assuming that 40 000 new buildings erected during Haussmann's renewal averaged 350 t of stone for the foundation and 250 t for the façade: that would imply about 25 Mt of stone and quarrying, transportation, cutting, and manipulation waste could easily double that total.

Urbanization, with its rising demand for windows, had, finally, brought new methods of mass-scale plate-glass production: in 1848 Henry Bessemer (later to become much more famous thanks to his new and inexpensive method of steel production) patented the production of flat glass by drawing a slightly cooled material upwards through a flat nozzle between asbestos rollers, a superior solution to the centuries-old (and severely size-limited) crown glass and to the still size-limited cylinder glass of the early nineteenth century. But it was only during the mid 1950s that Alastair Pilkington introduced the molten tin bath that allowed production of very large pieces of near-perfect uniformity (Pilkington, 1969).

But perhaps the best illustration of the much expanded removal and emplacement of bulk construction materials – including soil, sand, and gravel – is the growth of railroads. Their construction began in 1830 (56 km between Liverpool and Manchester), 30 years later their worldwide length topped 100 000 km, and by 1900 the railroads on five continents added up to 775 000 km, with about 250 000 km in Europe, more than 190 000 km in the USA, 53 000 km in Russia, and 30 000 km in the UK (Williams, 2006). Given the wide range of terrains covered by railtracks it is impossible to estimate a typical volume of bulk construction materials – earth displaced and replaced to create cuts or embankments, stone cut to create tunnels or incision in mountainsides, and stone quarried to produce gravel for access roads and railbeds – that had to be handled for an average kilometer of new track.

Even a highly conservative assumption of 3000 m^3/km would result in nearly 2.5 Gm^3 of bulk materials associated with the global railway construction of the second half of the nineteenth century. A similarly conservative assumption of at least 2000 t of ballast (crushed stones packed underneath and around ties) per kilometer would translate to at least 1.5, and more likely to 2 Gt, of coarse gravel applied to hold in place the tracks built between 1830 and 1900. Mineral aggregates were also needed in unprecedented volumes for the building of new factories, for the expansion of ports, and for the construction of hard-top roads. But the era of asphalted roads came only after 1900, and so the most important new material in the construction of the nineteenth century was concrete produced by mixing cement with an aggregate and water. Aggregate's coarseness ranges from sand to various sizes of gravel, and hydration (an exothermic reaction) of cement hardens the mixture even under water.

A suitable cement to produce good-quality concrete became available after 1824 when Joseph Aspdin, an English bricklayer, received a patent for a hydraulic mortar/stucco made by firing limestone and clay at high temperatures. This process vitrifies the alumina and silica materials and produces a glassy clinker whose grinding yields Portland cement, the name chosen by Aspdin because, once set, the color of the material resembles the limestone from the Isle of Portland (Shaeffer, 1992). Hydration of this cement (its reaction with water) produces material that could be readily form-shaped and which is strong under compression but weak in tension. This weakness can be overcome by reinforcing the material with iron, the combination made possible by the fact that concrete and iron form a solid bond and because hydraulic cement actually protects iron from rust.

The development and commercialization of reinforced concrete was a gradual process, with contributions made by French inventors (François Coignet in the early 1860s and Jacques Monier, who patented first a reinforced concrete beam and then in 1878 a general system for reinforced structures), and during the 1870s by William Ward in the UK and Thaddeus Hyatt in the USA (Newby, 2001). The true beginning of modern rebar came only in 1884 when Ernest Ransome patented his system of concrete with reinforcing bars, which became widely franchised, particularly for building new industrial structures (Newby, 2001). In 1886, Carl Dochring patented an ingenious idea of pre-stressing the reinforcing bars inserted in concrete: they are stretched when set in the wet material and release this tension once the material sets (Abeles, 1949). The first concrete skyscraper, Chicago's Monadnock Building, was completed in 1891, and during the 1890s the introduction of modern rotary kilns (with temperatures reaching up to $1500\,^{\circ}C$) made it possible to produce low-cost, high-quality cement. All the prerequisites of the concrete era were thus in place before 1900.

Even as the demand for wood was falling due to the displacement of fuelwood and charcoal by fossil fuels and coke (in France, coal began to supply more than half of all energy by the mid 1870s; in the USA the tipping point between fuelwood and coal and oil came in 1884–85) and owing to the shift of ship construction from timber to steel, new markets for sawnwood were created by the large-scale expansion of coal mines and by construction of railways. Demand for timber props in underground mining varied with the depth and thickness of seams, but typical requirements in Europe and in the USA were between 0.02 and $0.03\,m^3/t$ of coal and my best estimate is that the global need for mining timber surpassed $20\,Mm^3$ (about 15 Mt) by 1900 (Smil, 2013).

All ties (sleepers) installed during the nineteenth century were wooden; concrete sleepers were introduced only around 1900 but remained uncommon until after World War II. Standard construction practice requires the placement of about 1900 sleepers per km of railroad track, and with a single tie weighing between roughly 70 kg (pine) and 100 kg (oak) every kilometer needed approximately 130–190 t of sawn (and preferably creosote-treated) wood. My calculations show that the railtracks laid worldwide during the nineteenth century required at least 100 Mt of sawnwood for original construction and at least 60 Mt of additional timber for track repairs and replacements (Smil, 2013).

Multiplied demand for metallic elements and their alloys included those that had been known for long time but were used previously in only limited quantities (cast iron, steel, and copper are the best examples in this category) as well as those that had only been recently isolated during the nineteenth century and had swiftly found new industrial markets (aluminum). Improvements in iron smelting (larger blast furnaces, more efficient use of coke) reduced the energy requirements of pig iron from almost 300 GJ/t in 1800 to less than 100 GJ/t by 1850 and to only about 50 GJ/t by 1900 and made the metal much cheaper (Heal, 1975). Cast iron was used in some pre-1850 buildings in the UK, and afterwards, once it became inexpensive, its applications became more common in the USA – particularly in the southern states – not just for columns but also for bridges. Cast iron columns work well because the metal is strong under compression, but its weakness in tension limited its structural use, or led to catastrophic failures.

But the best choice to meet all those high-tension demands was steel, also an iron–carbon alloy but one whose carbon content is limited to just 0.05–1.5% and that can be tailored to many different requirements by adding variable shares (less than 2% to more than 10%) of other metals (including Co, Cr, Mn, Mo, Ni, Ti, V, and W, sometimes in combination). Specialized steels have uses ranging from sheets for car bodies and cutting tools for machining, to stainless steels used in medical devices, chemical syntheses, and food processing. The best varieties (tool steels) have a tensile strength an order of magnitude greater than cast iron (1600–2500 MPa compared to 150–400 MPa), and steel is the strongest and the hardest of all common metals: its typical tensile strength is roughly seven times that of Al and nearly four times that of Cu, and its hardness is four times that of Al and eight times that of Cu. Steel's impact resistance can be more than six times that of cast iron (130 vs. less than 20 J) and the highest temperature it can withstand before losing its structural integrity is as high as 750 °C compared to only about 350 °C for cast iron.

As the advancing industrialization needed more tensile metal, particularly for the fast growing railways, wrought iron filled that need. Low-carbon wrought iron was produced by reheating and puddling, manual turning, and pushing heavy chunks (nearly 200 kg) of pig iron in order to expose it to oxygen in shallow hearths (decarburize it) and produce a nearly pure metal (with less than 0.1% C). Wrought iron was then reheated and shaped (rolled, hammered) into final forms: all early rails laid before the late 1850s were made in this way, and so was the Eiffel tower. Inexpensive steel became a possibility in 1856–57 when Henry Bessemer (in England) and William Kelly (in the USA) patented their decarburization process, during which molten pig iron in a tilting vessel with a refractory lining was subjected to 15–30 min blasts of cold air that decarburized the metal (Bessemer, 1905).

But the process did not remove the phosphorus present in many iron ores, and the solution to this problem came only during the late 1870s when Sidney Gilchrist Thomas and Percy Carlyle Gilchrist introduced basic (limestone) refractory linings and added lime to the charge: this made it possible to remove phosphorus in slag (Almond, 1981). By the late 1880s the basic Bessemer process dominated both European and American steel production – in the USA its importance peaked in 1890 at 86% of the total output (Hogan, 1971) – and it produced most of the world's steel output until 1910. Its brief dominance ended with a rapid diffusion of open-hearth steelmaking furnaces built according to an 1866 patent shared by William Siemens and Emile Martin.

After a slow start, the deployment of open-hearth furnaces lined with basic refractories began to soar during the late 1880s (Almond, 1981) and pushed up the alloy's share in the final output quite rapidly. In the antebellum USA only about 1% of the US pig iron was converted to steel, by 1900 the share was almost 75%, and by 1906 the conversion became virtually total (Hogan, 1971). The 1880s also saw the introduction of new steels developed by Robert Mushet (self-hardening steel containing tungsten and manganese) and Robert Abbot Hadfield (with 13% Mn) that made it possible to make superior metal-cutting tools as well as durable ball-bearings. And by the decade's end American steel output surpassed British production: it rose from about 200 000 t in 1800 to roughly 2.5 Mt by 1850; then it doubled in 20 years to almost 6 Mt in 1870 and it reached 9.5 Mt by 1900. Global production of pig iron rose from about 5 Mt in 1850 to more than 30 Mt by 1900, while steel output rose from just half a million tons in 1870 to 28 Mt by 1900 (Smil, 2005).

For the first time in history inexpensive steel could be used to build structures, devices, and machines that were previously made of wrought iron or wood, and steel products conquered new markets thanks to their superior structural attributes and durability. In agriculture (still the dominant economic activity in 1850) steel went first into moldboard ploughs (first patented in 1868 by John Lane Jr.) whose mass deployment opened up America's Great Plains and Canada's Prairies for cultivation, then into grain reapers, other implements (harrows, seeders), and the first horse-drawn combines (the first self-propelled combine was introduced in 1911). New markets for steel were created in during the late 1880s with the invention of seamless steel pipes by Reinhard and Max Mannesmann. Introduction of better rifles, the inventions of the machine gun and modern explosives, and the production of grenades and bombs and (starting in the 1880s) the building of large and heavily armored battleships (though the first *Dreadnought* was built only in 1906) were other major sources of new demand.

New modes of transportation became the fastest rising markets for steel. Rapid development of railways created a huge demand for steel used in locomotives, passenger and freight cars, and for rails. Rails used during the nineteenth century weighed between 20 and 30 kg/m and, assuming an average of 25 kg/m, the railway construction between 1850 and 1900 would have required about 20 Mt of steel, while replacement would have more than doubled that total. Steel became the favorite material for railway bridges: its use was pioneered by the cantilevered 2529 m-long Firth of Forth crossing in Scotland, designed John Fowler and Benjamin Baker and built between 1883 and 1890 with 51 000 t of the metal (Forth Bridges, 2013).

The second large-scale global transportation market was created once Lloyd's Register of Shipping, in 1877, accepted steel as an insurable material for ship construction: within a generation the metal conquered the shipping market and the age of commercial wooden freight ships was over. By 1900 shipyards in the UK, Germany, and France were routinely launching passenger ships that needed more than 10 000 t steel to make. The third transportation segment that eventually created enormous demand for steel also originated before 1900 – but the age of mass car ownership in the USA only began with Ford's Model T in 1908.

Late-nineteenth century industrialization also created enormous new markets for steel in the steelmaking industry itself (due to its massive expansion that required greater numbers of larger blast furnaces and steel mills), in the new electrical industry (starting in the early 1880s, and requiring heavy machinery including boilers and steam turbogenerators, as well as steel for transformers, transmission towers, and electrical wires), in oil and gas extraction and transportation (steel for drilling pipes, drill-bits, well-casings, pipelines and vessels, pipes and storage tanks in refineries), as well as in traditional textile and food-processing industries, where advancing mechanization resulted in the adoption of a greater array of steel-based machines and other processing and storage equipment. These demands led, in turn, to a higher steel requirement in manufacturing of various industrial machines, tools, and components.

Inexpensive and highly tensile steel made it possible to design and to build inhabited structures of unprecedented height and functionality. Structural steel (more specifically long I-beams riveted from smaller pieces) made skyscrapers possible by doing away with thick load-bearing walls. Chicago's Home Insurance Building (designed by William Le

Baron Jenney, completed in 1885, in 1931 replaced by the Field Building) was the first tall (10 story, 42 m high) structure where steel columns and beams carried the weight, cut the total mass of the structure by two-thirds, and allowed for more floor space and for the installation of larger windows.

Five years later the World Building in New York went up to 20 stories and 94 m, and just before the century's end Henry Grey's invention of the universal beam mill, capable of rolling long H-beams, made steel construction easier by eliminating most of the need for laborious riveting. Skyscrapers required elevators (Otis, still the leader today, installed its first electric elevator in 1889) and hence more steel for machinery, cables, and cabins. The other, hidden, use of steel was its use in reinforced concrete. Joseph Monier, a Parisian gardener, patented first a version using simple metal netting in 1878, and the 1880s saw the first common applications in France, Germany, and Austria, especially in new industrial buildings.

Copper's ancient use for coins and alloys (brass, bronze) continued during the era of industrialization, but two new applications emerged to dominate the market. The first was the general adoption of indoor plumbing in growing cities, with copper as the standard choice first for water pipes, and later also for heating and cooling systems. The second was the invention of large-scale electricity generation during the early 1880s, followed by a rapid increase in demand for new power plant capacities and transmission lines. Copper is an excellent conductor of electricity and it is the preferred metal for wires and cables (with applications ranging from massive bundled units in data centers to the circuitry in microprocessors) and in electric motors; other uses include connectors, bearings, brakes, and radiators in motor vehicles, as a roofing material and (because of the metal's antimicrobial properties) in frequently touched public surfaces.

The lightest of all commonly used metals was also the last to be produced in commercial quantities. Aluminum was discovered by Hans Christian Oersted in 1825, and for the next 60 years it was produced only in minuscule amounts to make novelty jewelry; as late as 1884 its single largest application was a 2.85 kg pyramidal cap topping the newly built Washington monument (Binczewski, 1995). Commercial aluminum production was sought not only because of the metal's very low density (at 2.7 g/cm^3 only a third of iron's 7.9 g/cm^3) but also because of its excellent conductivity (surpassed only by Ag, Au, and Cu), and its extraordinary combination of high malleability (making it easy to roll, extrude, or stamp), high tensile strength (surpassed only by special steels and enhanced by alloying with Cu, Si, and Zn), and resistance to corrosion. Moreover, composite materials of aluminum and ceramics have unusual stiffness and durability, the nontoxic metal is suitable for uses where it comes in contact with food and drink, and discarded items can be easily compressed for highly energy-rewarding recycling.

Henri Saint-Claire Deville was the first experimenter to produce the metal by electrolytic means during the 1850s, but none of the subsequent variants of his process opened the door to mass production. Discovery of practical aluminum smelting in 1886 is among the most famous cases of a virtually identical, but independent, solution of a technical challenge: both inventors, Charles Martin Hall in the USA and Paul Louis Toussaint Héroult in France, were just 23 years old, and both moved rapidly to commercialize their process, with the first two plants in operation by 1888 (Borchers, 1904). The Hall–Héroult process still remains the only way to produce large quantities of the metal.

The process was made easier by Karl Joseph Bayer's 1888 invention of efficient alumina (Al_2O_3) from bauxite, a reddish compound particularly common in the tropics (lateritic bauxite) and named after Les Baux in southern France where Pierre Berthier first discovered it in 1821. Al_2O_3 (a nonconducting compound) is dissolved in molten, and highly conductive, cryolite (Na_3AlF_6, sodium hexafluoroaluminate), a perfect combination for efficient electrochemical separation in large electrolytic vessels; between 4 and 5 t of bauxite are needed to produce 2 t of alumina, and reduction of that compound yields 1 t of pure Al. By 1900 the global aluminum output was only about 8000 t, but the price fell rapidly to just 5% of the 1888 level (Borchers, 1904).

And yet another material should be singled out besides construction aggregates and metals: paper. Its large-scale production was revolutionized at the beginning of the nineteenth century with the introduction of a continuous paper-making machine, patented first in 1799 by Louis-Nicolas Robert in France but commercialized after 1801 in England by Henry Fourdrinier and subsequently called by his name. This immense machine consists of series of large cylinders used to form, dewater, and dry large continuous sheets of paper (Nuttall, 1967). The pulp is laid on a continuous wire mesh at the machine's wet end, most of water is then expelled in the felt press section, and the paper-making process is completed by conveying paper over a set of heated cylinders. But the raw material remained the same: until the 1870s paper was made from recycled rags and this obviously limited the produced mass and kept the price high.

This is why nineteenth-century innovators experimented with scores of natural fibrous materials, ranging from acacia and agave to thistles and yucca, but – based both on the available biomass and its properties – wood provided the best solution (Smith, 1970). During the 1870s paper production from wood began with mechanical pulp, but its high lignin content yields inferior paper (prone to yellowing) and it is now restricted for uses that include cheap newsprint, some toilet papers, and, above all, cardboard and building board (Smil, 2005). The first commercial operations using chemical pulp relied on the alkaline (soda) process but were soon displaced by the acid (sulfite) method that yielded a stronger pulp but a brittle paper: all printed matter using such acidic paper will eventually disintegrate. But the paper was cheap and the effects on consumption were obvious: for example, in 1872 Montgomery Ward's catalogue had a single page, two decades later it had more than 600 pages (Montgomery Ward & Company, 1895). The superior solution (pulping using the sulfate process) was invented in 1879 by a Swedish chemist, Carl F. Dahl, but it was commercialized widely only after 1900.

As for textiles, the nineteenth-century revolution was mostly quantitative, as mechanized weaving, based on Carthwight's power loom and Watt's more efficient steam engine, opened the way to the mass production of fabric and as higher incomes in urbanizing and industrializing societies created new markets for all kinds of textile products by expanding the ownership of clothes beyond what was commonly just a single set (or two) of outer garments. Other technical innovations helped the industry's expansion: factories using water power benefited from higher-capacity water wheels and (later) turbines, replacement of traditional wooden shafts by iron, and the introduction of iron to iron-frame looms lowered operating costs and improved productivity. British numbers illustrate the industry's progress: in 1800 the country had only about 2000 looms, by the early 1830s their total surpassed 100 000, and 250 000 were operating by 1857 (Hills, 1993).

Finally, scientific and engineering advances during the second half of the nineteenth century led to the first commercial extraction of gases from the atmosphere. In 1852, Thomas Joule and William Thompson (Lord Kelvin) found that highly compressed air forced through a porous nozzle cools slightly, about 0.25 °C for every 100 kPa pressure drop (Almqvist, 2003). Repetition of this expansion results in a cooling cascade that can gradually bring a gas to its liquefaction point; the first practical application of Thomson–Joule effect was patented in 1895 by Carl von Linde (Linde, 1916). More than a century later the eponymous company he founded remains the world's leader in gas liquefaction.

2.6   Materials in the Twentieth Century

I have argued that the modern, high-energy society of the twentieth century was created by the unprecedented combination of technical, scientific, and managerial advances that took place between 1865 and 1913 – and also that the transformations of these advances (together with some new notable innovations) resulted not only in new quantitative, but also in new economic and social, arrangements and in higher quality of life (Smil, 2005, 2006). In the last section of this chapter I will offer brief reviews of how some of the most important shifts in the production and use of materials (ranging from traditional biomass to modern electronics) that took place during the twentieth century helped to create these new realities.

Because of their renewability, annually harvested crop residues used to be indispensable materials in all traditional agricultural societies. In many deforested regions they were the only source of household fuel, straw–clay mixtures were made into bricks and straw bundles were used for roof thatching, in some countries peasants wore straw sandals and coats, and cereal straws were used as both feed and bedding for domesticated ruminants (only they can digest the cellulose that makes up as much as half of some straws). During the twentieth century, thatched roofs and straw sandals almost completely disappeared, in affluent countries straw is now mostly (more than 60% of annual yield) recycled to improve soil quality, but in many countries crop residues have retained their importance as fuel and feed.

My reconstruction of the twentieth century global crop residue production shows that, in 1900, they accounted for nearly 75% of the total harvest of 1.5 Gt and that the share was still about 70% in 1950 (Smil, 2013). Subsequent diffusion of modern cultivars and the rise of average harvest ratios (also known as grain:straw ratios, with cereal crops now yielding as much grain as straw) reduced the share to 65% by 1975 and to only about 58% by the year 2000. I calculate that in that year the global mass of food and feed crop residues was at least 3.75 Gt of dry weight. For comparison, Wirsenius (2000) put the global total of crop residues at 3.46 Gt for the years 1992–94 and Haberl *et al.* (2006) at just 2.71 Gt for 2000. There are no reliable data about the final fate of crop residues: in many agroecosystems they should be directly recycled to maintain soil organic matter and to prevent erosion, but often their mass is judged to be excessive and they are simply burned in fields. This undesirable practice is particularly common in rice-growing regions of Asia.

Straw continues to be burned even in some affluent countries, most notably in Denmark where about 1.4 Mt of wheat straw (nearly a quarter of the total harvest) is used for house heating or even in centralized district heating and electricity generation (Stenkjaer, 2009). Construction boards are made from shredded and compressed (or fused with resins) straw by such companies as Stramit International and Agriboard Industries, and some green architects promote straw-bale buildings (StrawBale, 2013). Residues can be also used as a source of organic compounds: furfural (a selective solvent in crude oil refining and in phenolic products) can be derived from corn stover and cereal straws (Di Blasi *et al.*, 2010). Straw can be also used as a substrate to cultivate mushrooms: wheat straw for white button mushrooms, rice straw for straw mushrooms (De Carvalho *et al.*, 2010).

The twentieth century was unmistakably the era of metals and plastics and their novelty and ubiquity made wood an underappreciated material: true, in all affluent countries its per capita consumption has declined, but aggregate demand has risen substantially as wood retained and actually expanded all but a few of its traditional markets. The only category of wood consumption that became relatively unimportant during the twentieth century was shipbuilding: wood is now limited to a niche market of small vessels (boats and yachts) and for interiors of more costly ships. The material's two leading uses, as lumber and pulp for making paper, will be reviewed in detail in the next chapter; here I will note the continuing importance of fuelwood and wood for railroads and coal mining.

All totals of global fuelwood harvests are aggregates based on estimates and assumptions whose large uncertainty (on the order of 50%) is no smaller for the year 2000 than it was for 1950 (Fernandes *et al.*, 2007). The FAO put global fuelwood combustion in the year 2000 at 1.825 Gm^3, and added 75 Mm^3 of wood residues and 49.2 Mt of charcoal – whose production (using an average multiplier of 6.0) required nearly 300 Mm^3 wood and raised the grand total of solid biomass fuels to about 2.2 Gm^3 of wood (FAO, 2013). With about 0.65 t/m³ and 15 GJ/t of air-dry wood, that mass equals 21.5 EJ or 1.43 Gt. But because in many countries most wood does not come from forests, FAO's totals are obvious underestimates of the real demand.

My best estimate is about 2.5 Gt of air-dry wood or 2 Gt of absolutely dry matter containing about 35 EJ, and burning of crop residues by rural households adds at least 10 EJ for a total of 45 EJ in the year 2000 and an equivalent of 3 Gt of air-dry wood (Smil, 2010), an excellent agreement with the 45 ± 10 EJ total calculated by Turkenburg (2000) and only about 20% higher than the total of 2.457 Gt (37 EJ) of solid phytomass fuels (75% as wood, 20% crop residues) published by Fernandes *et al.* (2007). A global aggregate of around 40 EJ in 2000 is thus a good consensus value and implies a nearly 70% increase in biomass fuel demand between 1950 and 2000 and a doubling of wood and crop residue harvests during the twentieth century.

But the intervening high population growth greatly reduced the average per capita consumption and the huge expansion of fossil fuel extraction cut the biofuel share from 50% in 1900, to less than 10% of global primary energy supply in the year 2000, and (because of inferior efficiencies of wood and straw combustion) to less than 5% in terms of useful final energy supply. Among the major economies, wood has the highest national share of primary energy supplies in Brazil, at about 10%, while its share in affluent nations ranges from negligible values (just 1% in the UK and Spain) to about 20% in Sweden and Finland, with the US share falling from about 4.5% in 1950 to just 2% in 2010 (Eurostat, 2012; USEIA, 2013).

Wooden railway ties, that quintessential nineteenth-century innovation, maintained their high share of the global market throughout the twentieth century. During the 1990s, 94% of America's ties were wooden – and even with a relatively high proportion of concrete ties in parts of Europe and Asia, only 15% of the world's sleepers were made of materials other than wood (Sommath *et al.*, 1995). Better treatment of ties prolonged their average lifespan from about 35 years in 1940 to 40–50 years by the year 2000 (James, 2001). European and North American tie markets have been basically limited to replacements, but Asia's railway's networks keep on expanding, although the continent's most impressive new link, the Tibet railway from Golmud to Lhasa, uses only concrete sleepers.

The worldwide market for underground mine props kept on growing until after World War II, but the subsequent decline of coalmining in Europe and the shift of extraction to surface mines in the USA and Australia weakened that demand. I calculate that the volume of timber used in coal mines surpassed $20\,Mm^3$ (about 15 Mt) in 1900, and $40\,Mm^3$ in 1950 – when it accounted for about 2% of the world's roundwood harvest (Smil, 2013). Reduced reliance on coal in Europe and Japan, and expansion of surface mining and widespread adoption of long-wall mining using movable steel roof supports, reduced the post-World War II demand for pit props throughout the Western world.

By contrast, China's coal extraction increased about 75-fold between 1950 and 2010 (from 43 to 3235 Mt), but in that deforested country specific consumption of timber has been always much lower than the rate of around $0.025\,m^3/t$ that was typical in the West. China's low specific use of mining timber (now just $0.005\,m^3/t$ of coal) means that, even with the country's large increase of raw coal extraction, the roundwood demand in Chinese coal mining is less than 10% of the country's total wood consumption (SFA, 2009). However, due to its massive exports of manufactured products, China has become a leading user of wood, plywood, cardboard, and paper required for packaging.

As I have already noted, all the key technical breakthroughs needed to make concrete – the world's most common building material – were in place by 1900, but the first major initiative to make concrete the popular choice was a peculiar experiment directed by Thomas Edison, who became engaged in a futile experiment to design and build cast-in-place concrete homes (Courland, 2011). The inventor launched the project in 1906 (after failing to develop a vastly improved battery) and five years later, as the project was failing, he tried to revive it by also promising cheap concrete furniture, including entire bedroom sets, and even a concrete phonograph. Edison's dreams of inexpensive concrete housing failed, but some architects began to choose the material for the signature buildings of their careers.

In France, Auguste Perret had already designed elegant apartments and the Theatre des Champs-Elysees before World War I, and in the USA Frank Lloyd Wright was the material's great American proponent. After designing smaller pre-World War I structures in the USA, he built Tokyo's Imperial hotel, completed just before the great earthquake destroyed the city in 1923; the hotel survived with only minor damage and was demolished in 1968. Wright's other famous pioneering concrete structures included the Johnson Wax Headquarters in Racine, WI (in 1939), the Fallingwater House in Pennsylvania (in 1935), and the Guggenheim Museum in New York (in 1959).

Other famous post-World War II structures made possible by reinforced concrete include Jørn Utzon's Sydney Opera House (noted for its elegance and completed after many delays in 1973) and Burj Khalifa Tower in Dubai, the world's tallest building, finished in 2010. Many new, bold-looking designs took advantage of the fact that pre-stressed structures require much less steel and concrete to bear the same loads (reductions are, respectively, about 70 and 40%) and can be made much more slender. Eugène Freyssinet pioneered this type of construction, and he also introduced post-stressing that uses tensioning wires threaded through ducts in pre-cast concrete (Grotte and Marrey, 2000).

During the first half of the twentieth century, reinforced concrete also became the most important building material for new bridges and dams, new offshore structures, and the foundation of modern transportation infrastructures. The Swiss architect, Robert Maillart, had already pioneered the design of graceful concrete bridges before World War I (Billington, 1989), and the world's longest bridge, the 164.8-km Danyang-Kunshan Grand Bridge in China (completed in 2010) is also a reinforced concrete structure. America's largest dams, built during the 1930s (Hoover Dam on the Colorado and Grand Coulee on the Columbia River) became the precursors of even larger structures built around the world after World War II; the largest of them, at 185-m tall and 2.3-km long, China's Sanxia (Three Gorges) dam on the Yangzi, contains almost 28 Mm^3 of concrete and 500 000 t of reinforcing steel.

In 1982, the *Statfjord B* oil drilling platform became the heaviest object ever moved, as 816 000 t (mostly in reinforced concrete in the structure's four massive concrete columns and storage tanks) were towed into position in the North Sea (Aker Solution, 2013). Modern commercial flight depends on the reinforced concrete that forms the runways, approaches, and aprons of airports: they must withstand the repeated traffic of airplanes with a mass commonly between 150 t (Boeing 737) and 277 t (Airbus 380); runway concrete is up to 1.5 m thick and the longest runways are 3600–4000 m long. But most reinforced concrete has not gone into iconic structures but into ever-increasing numbers of nondescript or outright ugly (or brutal looking) apartment buildings, high rises, factories, garages, roads, overpass bridges, and parking lots.

Pilkington's float-glass process made it possible to produce windows of unprecedented size, and many architects were quick to incorporate them into their building designs. By 2010, the global market for flat glass was about 56 Mt (and growing at 4–5%/year), with about 33 Mt of high-quality float glass, 1 Mt of sheet glass, and 2 Mt of rolled glass, with the remainder of roughly 20 Mt being low-quality float glass made mostly in China (NSG, 2011). Advances have also been made in the purity of the material, transforming glass into a conduit of information. By 1900 the best optical glass was about 10 000 times more transparent than the earliest glass produced in Egypt some 5000 years ago, and after 1970 the use of high-purity SiO_2 had improved that transparency by four more orders of magnitude, making it possible to use optical fibers for long-distance, mass-dispatching of voices and data (Agrawal, 2010).

Steel's dominance of the global metal market, firmly established by 1900, greatly increased during the twentieth century thanks to nearly continuous advances in iron smelting and steelmaking and to the emergence of new mass markets for the metal. Open-hearth furnaces remained the dominant producers of world steel for more than two-thirds of the twentieth century. The principle of slow blowing in open hearths remained unchanged, but the unit sizes and productivities grew substantially: just before 1900 the largest US steel furnaces had an area of about 30 m^2, before World War I the maximum size grew to 55 m^2, and by 1945 it was nearly 85 m^2, while the typical heat (batch) capacities grew from roughly 40 t in 1900 to 200 t during World War II

(King, 1948). Additionally, electric arc furnaces, first introduced in 1902, began to be used more frequently to convert the growing stocks of scrap metal into high-quality steel.

The first cars of the 1890s created only a small demand for steel because they had simple wooden bodies and were made in small numbers: only once mass-production took hold (with Ford's Model T introduced in 1908) did the auto industry become the leading consumer of steel, with its demand rising from about 70 000 t in 1910 to 1 Mt by 1920 (Hogan, 1971). For many decades the idea of producing noncorroding steel on a mass scale was rejected by nearly all metallurgical experts. Harry Brearley is usually credited with inventing commercial stainless steel (containing nearly 13% chromium) in 1913 (an innovation readily embraced by Sheffield's famous steel cutlery makers) but contemporaneous advances were made in Germany and in the USA (Cobb, 2010).

Besides cutlery, stainless steel was soon used in surgical implants and cookware, and also in many bulk food-handling applications in brewing, winemaking, bakeries, and the meat industry. The 1930s saw the first deployment of stainless steel in construction: in 1930 the 55-m tall tower and spire of New York's Chrysler Building was the first eye-catching application. A year later, the Empire State Building got stainless steel window trims and pilasters. Rapid passenger trains with aerodynamic steel bodies (starting with the Burlington Zephyr in 1934) were also popular, before they were displaced by lighter aluminum structures.

Steel output fell sharply during the Great Depression (in the USA by 75% between 1929 and 1932, down to just 12.4 Mt) but then rose to new record highs in order to support the unprecedented wartime demand for the metal (King, 1948). Inevitably, these increases brought many problems and new solutions. Higher blast rates led to faster deterioration of brick linings – and to the introduction of water cooling. Large increases in the volume of ore, coke, and limestone inputs could not be supported by manual handling – hence the mechanized skip hoists and automated dumping. The requirement for clean gas to run blowing engines led to better gas-cleaning equipment, including Frederick Cottrell's electrostatic precipitators, first introduced in 1919.

The metal needed for the armaments required to win World War II reinvigorated America's steel industry, and after 1950 it pioneered the use of coke-saving smelting under pressure, use of beneficiated ores, injection of gaseous or liquid fuels, enrichment of blast air by oxygen, and automated operation controls (Gold *et al.*, 1984). But, by 1960, innovative leadership in ferrous metallurgy shifted to Japan (the country became the world's second largest producer after the USSR) and to Europe. Larger, but more efficient, blast furnaces to smelt iron, basic oxygen furnaces to produce steel, and continuous casting of steel products made up the trio of technical innovations that transformed the global steel industry after World War II, and I will cover all of them in some detail in the next chapter when dealing with advances in metal production.

Mass production of inexpensive, high-quality structural steel enabled the construction of downtown high rises in the largest cities of all inhabited continents, and many of these buildings also used steel on the outside. In 1954, New York's Socony-Mobil building was the first skyscraper completely clad in stainless steel (and cleaned for the first time only in 1995). Among the numerous steel-clad skyscrapers that have followed (usually in the form of curtain walls) are the US Steel Tower in Pittsburgh and the Petronas Twin Towers in Kuala Lumpur (temporarily the world's tallest building). Burj Khalifa uses reflective glazing and textured stainless steel spandrels.

But much more steel (in the form of sheets and rods) has gone into cars and trucks and new transportation infrastructures on land (ranging from multi-lane highways and bridges to new airports) and into the construction of large oil tankers, bulk carriers (transporting anything from grain to ores), and, starting in the 1960s, container ships and ports. Steel allows particularly captivating design of long suspension bridges with woven cables supporting lengthy road spans: Japan's Akashi Kaikyō that links Honshū and Shikoku is nearly 2 km (1199 m) long. The transportation sector also became the leading user of aluminum: the combination of light weight and durability made the metal, and its alloys, an ideal choice for applications ranging from cooking pots to rapid train cars, and it is indispensable in the airline industry.

Between 1900 and 1943, its output, driven primarily by an unprecedented demand for the metal in aircraft construction, expanded nearly 300 times to almost 2 Mt, with 45% of the peak demand coming from the USA that was engaged in the largest-ever effort to build record numbers of military planes. Many new producers added to the global output after 1950, by the year 2000 electrolytic production of the metal was close to 25 Mt, and in 2008 it reached a new record of nearly 40 Mt, with China accounting for most of the post-2000 increase in demand (USEIA, 2013; IAI, 2013). Structural aluminum spread beyond airplanes to bridge decks or superstructures (starting in the 1940s) and vessels (ranging from all-aluminum small boats to structures on offshore drilling rigs, and from large cruise ships to expensive yachts) and railcars.

New, major post-1950 markets have also included irrigation pipes (particularly for the center-pivot systems now so common in many arid regions), heat exchangers (many applications in air conditioning, medical equipment, electronics), and even sheet piling. At the same time, aluminum has lost some of its traditional markets at both ends of its value range: in many demanding aerospace applications (especially for supersonic airplanes) it has been displaced by titanium, while in many simple manufactures it has been replaced by less energy-intensive and less expensive plastics. Titanium production requires about twice as much energy (around 400 GJ/t) as the chain of operations needed to produce aluminum (for details see the next chapter) but the metal's melting point (1667 °C) is 2.5 times higher than for aluminum (660 °C), making it well suited to the cladding of supersonic aircraft.

Post-World War II demand for copper has been dominated by five major final markets: copper in construction goes into electrical wiring, plumbing, refrigeration, and air conditioning conduits, and also has visible uses (copper sheathing and roofing); industrial machinery, fittings and wiring, and heat exchangers; every category of transportation machinery; industrial electrical and electronic products, above all telecommunication and lighting; and a wide range of consumer products dominated by electronic gadgets and electrical cords and, in many countries, coins. Copper has maintained its third ranking among the twentieth century metals as its global consumption rose from less than 500 000 t in 1900 to more than 13 Mt in the year 2000.

The fourth most important metal has been zinc, with a consumption of 12.6 Mt in 2010; but the steadily rising demand for lead has brought this formerly more distant number five close to the zinc total: in 2011 the global refined lead supply surpassed, 10 Mt for the first time, to reach 10.6 Mt, with about 45% being primary metal and the rest coming from recycled material (ILZSG, 2013). Although the use of tetraethyl-lead as the leading automotive additive (to prevent knocking in gasoline engines) was outlawed in Japan in 1986, in the USA in 1995, and in the EU and China in the year 2000, the expansion of the global car and truck market (and hence the annual installation of tens of millions of new and replacement lead acid batteries) has led to record demand. With a total of just over 1 billion cars and light and heavy trucks, and with an average mass of 10 kg Pb in automobiles and 13 kg Pb in truck batteries, there was nearly 11 Mt of lead on the world's roads in 2010. In addition, this heavy metal also remains an indispensable solder in electronics.

Use of silicon, a light ($2.3 \, g/cm^3$) metalloid, has become synonymous with modern electronics, but the element was not present either at the industry's birth or during its pre-1950 development. Electronics' theoretical beginnings go back to the 1860s (James Clerk Maxwell and Rudolf Hertz), and practical applications got underway during the 1890s when several researchers and engineers tried to create wireless telegraphy, that is to send signals through the atmosphere. The group's most prominent members were Nikola Tesla, David Hughes, Alexander Stepanovich Popov, William Crookes, Edouard Branly, Oliver Joseph Lodge, and Guglielmo Marconi.

And it was Marconi who took the first commercial steps without relying on any new materials: his first broadcasting towers in Newfoundland were wooden, his spark generators and wire antennas were made of common metals. Later models of high-frequency alternators (built by Reginald Aubrey Fessenden and by Ernst Frederik Werner Alexanderson, and able to produce a continuous wave signal) were remarkable examples of precision engineering, but superior devices that opened the door to mass produced electronics came only with Fleming's invention of the diode (essentially just a light bulb with an added electrode). Its progeny – triodes, tetrodes, and pentodes – required careful manufacturing procedures, but in material terms these vacuum tubes amounted to just lots of glass and hot (mostly tungsten) filaments.

Radios, first introduced in the early 1920s, became inexpensive and widely owned by the 1930s, and television was the first expensive electronic device to achieve mass ownership. The first broadcasts took place in the UK and the USA before World War II, by 1948 fewer than 200 000 families had a bulky set with a small black-and-white screen, but by 1960 TVs were in 90% of all US homes. The first color broadcasts came in 1954, but affordable sets came only during the late 1960s and by 1975 two-thirds of American families had a color TV (Abramson, 2003). By that time, the conversion to solid state electronics based on silicon was virtually complete.

Silicon makes nearly 28% of the Earth's crust, and while it is abundantly present as SiO_2 (silica) in sand, sandstone, and quartz and in many silicates ranging from hard feldspars (rock-forming minerals) to soft kaolinite (a layered clay mineral), it is never found in pure, unbound elementary form. But the purest crystalline silicon is the material foundation of modern electronics: intricate webs of semiconductors and connections are etched into thin wafers of ultra-pure Si; optical fibers (made by fusing SiO_2 and GeO_2, their deposition inside a tube to form glass, and subsequent fiber-drawing in a tower) carry communications between countries and continents; and photovoltaic cells power communication satellites that relay data and messages between the continents (and are increasingly used for land-based electricity generation). Silicon's importance deserves extended coverage.

Plastics have been widely seen as the quintessential twentieth century material, with a particularly rapid post-World War II diffusion as they replaced wood, metals, and glass in many household, industrial, and transportation products (Strom and Rasmussen, 2011). Although the first synthetic materials go back to the 1870s, when John Wesley Hyatt patented his celluloid process and when chemists began to study reactions between phenol and formaldehyde, the real progress came only in 1907 when Leo Hendrik Baekeland, a Belgian chemist working in New York, prepared the world's first thermoset plastic formed between 150 and 160 °C (Baekeland, 1909). His General Bakelite Company, established in 1910, was the first large-scale commercial producer of a plastic material, and bakelite was soon turned into products ranging from telephones to electric insulators and from door knobs to parts of light weapons.

In 1912 came Jacques E. Brandenberger's cellophane (regenerated cellulose, first in France, since 1924 in the USA), styrene (its polymer has eventually become a leading insulator and packaging material), and cellulose acetate (Brydson, 1975). But the1930s became the still unsurpassed era of major plastic discoveries as a result of systematic institutionalized research by large chemical firms, above all the US DuPont, Germany's IG Farben, and the UK's Imperial Chemical Industries. First, in 1930, Wallace Hume Carothers led a Du Pont team that produced neoprene (synthetic rubber), and IG Farben synthesized polystyrene. In the early 1930s the ICI began to research organic reactions under very high pressure, and by 1935 this led to polyethylene.

The company also began its synthesis of methyl methacrylate in 1933, the year that Ralph Wiley at Dow Chemical made the accidental discovery of polyvinylidene chloride (Saran wrap for cling-packaging of food). In 1935 came the first Plexiglas, a year later polyurethanes (Otto Bayer at IG Farben), and in 1937 Carothers got the patent for his polymer 66, commercially known as Nylon (Carothers, 1937; Hermes, 1996). Toothbrush bristles were the first nylon products in 1938, soon followed by stockings. Also in 1938, DuPont's Roy Plunkett made the serendipitous discovery of polytetrafluoroethylene, branded as Teflon. This was followed by alkyd polyesters and polyethylene terephthalate (PET) during the 1940s. PET has been made into fibers (Terylene and Dacron), a film (Mylar), and since 1973 it has been the leading plastic for bottling water, plant oils, and salad dressings. These ubiquitous PET bottles and containers are also among the most recycled plastics and are transformed into polyester carpets, fiberfill for sleeping bags and coats, and car bumpers and door panels.

The greatest discoveries of the 1950s include polyimides (for bearings and washers and for heat- and chemicals-resistant applications in electronics) and polycarbonates (for optical lenses and windows, later for CD covers), but the most consequential advance was Karl Ziegler's method of synthesizing polyethylene at normal temperatures and low pressures by using new organometallic catalysts. Post-1960 additions have included polysulfone (a flame retardant), polybutylene (a flexible polyolefin used for pipes and plastic packaging), liquid crystal polymers (aromatic polyesters widely used in electronics), and Du Pont's plastics marketed under such well-known names as Lycra (Spandex, for athletic wear and clothes), Kevlar (bullet-proof para-aramid used for body armor), flame-resistant Nomex (used in firefighting equipment and pilot suits), and Tyvek (a form of high-density polyethylene used to wrap houses in an insulating but water vapor-permeable barrier).

Global production of all plastics remained below 50 000 t until the early 1930s, did not reach 1 Mt until 1949, and surpassed 6 Mt in 1960. Then, with a greater availability of hydrocarbon feedstocks, worldwide synthesis rose rapidly by an order of magnitude, reaching 100 Mt in 1989, 200 Mt in 2002, and 265 Mt in 2010: the last total was more than six times the worldwide output of aluminum and equal to about 18% of steel production in that year. But if the comparison is made in terms of volume, then plastic production surpassed the output of steel: assuming an average density of $1 g/cm^3$ it amounted to $265 Mm^3$ compared to $181 Mm^3$ of steel (averaging $7.8 g/cm^3$).

The increased availability of plastic waste and its many undesirable environmental impacts have led to intensifying efforts to recycle at least several major plastic varieties, and concerns about the long-term supply of hydrocarbon feedstocks have rekindled interest in renewable (plant) sources of the raw materials needed for plastic syntheses. The plastics industry now offers more than 50 kinds of materials, but only a small number of them account for most of the global output. I will take a closer look at the three dominant products – polyethylene, polypropylene, and polyvinylchloride – in the next chapter.

I saved fertilizer advances for the last entry in this brief review of material innovations of the twentieth century – but if the order of presentation were to be determined by the existential importance for the survival of our species, then Haber–Bosch ammonia synthesis should have come first. After all, we could have had (and until the 1950s we did) a prosperous civilization without any Si-based electronics, and we could have done (with more expense and sometimes with less comfort) with much less steel and without scores of plastics and metal alloys – but we could not have supported the twentieth century global increment of 4.5 billion people consuming increasingly better diets without a huge increase in nitrogen applications.

Availability of nitrogen was a key factor limiting the yields of traditional cultivars; assiduous recycling of organic matter (crop residues, animal manures, human urine and feces), rotations including leguminous species capable of fixing atmospheric nitrogen (alfalfa, clover, vetch), and regular fallowing were the only options to maintain soil fertility. In many affluent countries, higher yields were the only way to expand the necessary food supply and higher applications of nitrogen were the key (the other two macronutrients, P and K, are more readily secured by extracting P- and K-containing minerals). In 1900 there were only three limited options to obtain fertilizer nitrogen: declining imports of guano, nitrates from Chile, and ammonia as a by-product from coke ovens. Global consumption from these three sources amounted to less than 350 000 t N, only about 2% of N needed by that year's crops.

Continuing reliance on the recycling of organic matter and planting N-fixing legumes could not guarantee the high yields required to provide better nutrition to the greater urban population with higher disposable incomes. The solution came in 1909 when Fritz Haber, chemistry professor at the University of Karlsruhe, demonstrated the practicality of catalytic synthesis of ammonia from nitrogen and hydrogen; a no less remarkable achievement was that of Carl Bosch, at the BASF corporation in Ludwigshafen, whose team was able to transform Haber's laboratory demonstration into full-scale commercial synthesis in just four years (Smil, 2001). Most of Germany's ammonia synthesis was then diverted to make munitions and explosives for World War I, and the interwar years saw only limited progress in increased nitrogen fertilization. But the post-1950 change was rapid, and after 1970 most of the gains came from nitrogen applications in populous, low-income nations.

Global output of synthetic fertilizers (in terms of pure N) rose from just 150 000 t in 1920 to 3.7 Mt in 1950 and 85.13 Mt in and 2000, an increase of two orders of magnitude (roughly 570 times) in 80 years. Remarkably, that was not even an exceptionally large gain, as the global production of other new materials saw even greater increases over the course of the twentieth century: three orders of magnitude for aluminum (roughly 3600 times, from just 6800 t in 1900 to 24.3 Mt in 2000) and nearly four orders of magnitude for plastics (from about 20 000 t in 1925 to 150 Mt in 2000). And even long-established materials had very large global production multiples, roughly 30 times for paper, 30 times for steel (from 28.3 to 850 Mt), and 27 times for copper (from 495 000 t to 13.2 Mt). In comparison, the global population increased 3.8 times between 1900 and 2000, and the gross world product (in constant monies) rose about 20-fold, which means that during the twentieth century material use had greatly intensified both in per capita terms and per unit of economic output.

Before examining in detail some long-term trends in material consumption, presenting a systematic review of their energy requirements (energy costs), highlighting some of the environmental burdens resulting from their production and use, and noting some successes and challenges of material recycling (all in the Chapter 4), I will take closer, systematic looks at the most important components of the modern world's material consumption – ranging from bulk construction materials to extremely pure silicon – and I will do so in all cases at the generic and global levels, in many instances also focusing on specific national developments.

3

What Matters Most

Given the enormous variety of materials used in modern civilization, any systematic, but still far from comprehensive, review of their production, their energy costs, their most common uses, and their associated environmental, social, and economic impacts would require either multi-volume coverage or would fit into a mid-size book only when restricted, encyclopedically, to a sequence of very brief entries. I am adopting what is perhaps the most efficacious solution to this challenge: in this chapter I will note the diversity of every major category of materials whose incessant supply enables the functioning of modern civilization – but I will take a closer look only at a limited number of products that are either quantitatively dominant or qualitatively indispensable. I will survey the basic production methods of each of these key materials and note some of the important industrial advances of recent decades.

My reviews start with the two leading biomaterials: lumber used in construction and furniture-making and wood pulped to make paper. Then I will turn to the world's leading construction materials, that is to all varieties of sand and stone (gravel) and then to the cement, a fairly energy-intensive product whose combination with aggregates and water yields the world's most important construction material: concrete. The modalities and consequences of iron and steel production will dominate the third section that will also take a closer look at the other leading metal, aluminum. Afterwards, I will deal with the most important plastic materials, with their derivatives and, necessarily, with the hydrocarbon feedstocks required to produce them, and the many worrisome environmental consequences of their use and challenges of their recycling.

This will be followed by a section devoted to three leading industrial gases whose commercial supply has become indispensable for industrial sectors ranging from high-quality steel to nitrogen fertilizers, and whose synthesis will be the main topic of the next section in which will also address less complicated efforts to produce the two key mineral fertilizers: phosphates and potash. As we have now entered the second century of electronics – I am dating the field's birth to Fleming's invention of diode in 1904 (Fleming, 1934) – I will take a closer look at the material that is at the very core of

Making the Modern World: Materials and Dematerialization, First Edition. Vaclav Smil.
© 2014 John Wiley & Sons, Ltd. Published 2014 by John Wiley & Sons, Ltd.

modern computing, as well as imaging and telecommunication. Hence the last section will explain the ways of producing polysilicon, the growing of ever-larger, ultra-pure silicon crystals and wafers, and their slicing to provide substrate for ever-more powerful microchips.

3.1   Biomaterials

The category of biomaterials is large and diverse: it includes products as different as construction lumber, straw, cotton, wool, beeswax, birds down, and animal hides – but in mass terms, as well as in the overall importance and indispensability of uses, it is dominated by wood. As already noted (in the previous chapter) crop residues, above all cereal straws, come next, and they remain an important source of fuel in the rural areas of many low-income countries. But, as most of them are either burned in fields (particularly rice straw in Asia), recycled to renew organic matter and prevent soil erosion, or removed as a feed or bedding for animals, their use as construction materials is now quite marginal (Barreveld, 1989; Unger, 1994; Smil, 1999).

In mass terms, the third largest category of biomaterials are fiber crops, whose aggregate harvest was about 26 Mt in 2010 (FAO, 2013). Cotton lint is dominant, with global harvests rising from 7 Mt in 1950 to more than 12 Mt in 1975, nearly 19 Mt in 2000 and 23.3 Mt in 2010. Coir (coconut husk fiber) is a distant second with about 1 Mt in 2010. Wild flax (*Linum usitatissimum*) fibers (dated to 30 000 years BCE and used to make cords) were found in a Paleolithic cave in Georgia (Kvavadze *et al.*, 2009), the plant is the oldest cultivated fiber crop, and its harvest surpassed 600 000 t in 2010 – followed by sisal (fiber from stiff *Agave sisalana* leaves, about 360 000 t in 2010) and ramie (fibers from a bushy perennial *Boehmeria nivea*, nearly 120 000 t in 2010). Worldwide harvests of each of three less important crops – kapok (fiber surrounding seed pods of a large tropical tree *Ceiba pentandra*), abaca (fibers from leaves of a banana species *Musa textilis*), and hemp (fibers of *Cannabis*) – come close to 100 000 t a year.

Two species of ancient plant fibers that used to be woven into fabrics – coarse hemp that was made into sailcloth, sack, and canvas, and the fine linen that furnished many societies with excellent cloth fabric and crisp handkerchiefs – have been in retreat, in many countries to the point of near disappearance. How many men now regularly wear puff-folded pocket squares in their jackets, and how many ships are tied to bollards with hempen ropes in a new age when cannabis is grown mostly as a drug? But cotton is more ubiquitous than ever as it clothes more people than any other natural fabric. The next most massive category of biomaterials is natural rubber, latex extracted from the Amazonia rubber tree *Hevea brasiliensis* (Sethuraj and Mathew, 1992). Transplanted throughout the tropics, the species now produces natural rubber mainly in Asia, with Thailand, Malaysia, Vietnam, and Indonesia being the largest exporters (IRA, 2013). Annual production was below 2 Mt until the early 1960s, it surpassed 3 Mt a decade later, reached nearly 7 Mt in 2000 and 10 Mt by 2010 (FAO, 2013), compared to about 15 Mt annually of synthetic rubber.

Animal hides and skins come next, and their production has risen along with the expanding demand for meat: the annual output of bovine (cattle and water buffalo) hides surpasses 6 Mt, that of sheep and lambskins over 400 000 t, and some 300 000 t of goat and kidskins are turned into leather product annually (FAO, 2011). Production of wool, the most important animal fiber, rose from about 960 000 t in 1950 to 2.9 Mt in 1970, fluctuated afterwards (peaking at 3.3 Mt in 1990), and declined to just below 2 Mt in 2011 (FAO, 2013). In contrast, production of silkworm cocoons has more than doubled during the past 50 years, to about 500 000 t in 2010. And, just for the sake of completeness, here are the recent annual outputs of minor biomaterials: about 60 000 t of beeswax; 20 000–30 000 t of lac – resinous secretions of insects belonging to the genus *Kerria*, with 70–80% of output coming from India (Singh, 2006); and between 30 000 and 50 000 t of gum arabic (exudates from *Acacia seyal* and *A. senegal*) mostly from Sudan (World Bank, 2007).

Consumption of many of these biomaterials has been in long-term decline and some of them will become even more marginalized; aggregate demand for other plant and animal products may diminish but they will remain highly valued by some buyers due to their intrinsic qualities or, at least, due to the perceived status conveyed by their acquisition (be they beeswax rather than tallow candles, or silk rather than synthetic saris or purses). And there can be no doubt that the demand for the three biomaterial leaders – lumber, paper made from wood pulp, and cotton clothes – will remain strong. The only major wood-based industry that has seen significant post-1950 changes and inroads made by substitutes is furniture-making.

Quality furniture has seen a great shift to veneer-covered particle board or plywood, and those two products can be sourced from many species; low-end furniture has been made using laminate materials (most commonly melamine-clad panels of particle board) and many smaller household items formerly made of wood have been replaced by plastics and metals. Solid wood pieces remain favorites only in some markets and for some specific applications. Similarly, wood use in wine-, spirit-, and beer-making has declined with the adoption of stainless steel tanks and aluminum kegs, but cumulatively significant and irreplaceable (or at least preferable) uses of wood remain: for many products oak barrels and casks will remain in demand, as will select woods for musical instruments and some sport implements.

Scores of tree varieties are still harvested (both for lumber and pulp), but the preference of modern, large-scale processing for fairly uniform inputs has led to an increasing dominance of a relatively small number of coniferous (softwood, above all pines, spruces, firs) and deciduous (hardwood, mainly ash, eucalyptus, poplar) species, with an increasing share of them harvested from newly planted and replanted monocultures: in parts of Europe, coniferous monocultures of firs and spruces are already in their third, or even fourth, generation. Some of these managed forests have also seen remarkable increases in productivity (Kauppi *et al.*, 2010).

In forestry statistics, industrial roundwood is the composite category that includes all sawlogs, veneer logs, and wood harvested for pulp to be made into paper. The reliability of wood harvest statistics ranges from excellent in Western countries, to chronically unreliable in many nations in Latin America, Africa, and Asia where illegal cutting has been common. FAO's data show the worldwide roundwood harvest at about 1.7 Gm^3 in 1950, 2.88 Gm^3 in 1975, 3.43 Gm^3 in the year 2000, and the basically identical total of 3.40 Gm^3 in 2010 (FAO, 2013). The 2010 total was split 55:45 between fuelwood and industrial roundwood, whose harvest was just a bit above $1.5 \, \text{Gm}^3$. Assuming an average dry weight of 420 per kg of green roundwood, the total 2010 harvest would have been about 650 Mt.

The actual annual cut of woody phytomass is considerably higher, not just because the standard statistics exclude illegal harvests but above all because they leave out all phytomass that does not become a part of commercial products. A very conservative adjustment for illegal cutting would raise the global industrial roundwood harvest by 15%, to about 750 Mt in 2010. A much larger adjustment must be made for the cut phytomass that is not commercially harvested that includes small trees (with a breast-height diameter of less than 12 cm) and nonmerchantable woody biomass, a category that includes trees of poor form, stumps, bark, branches, and tapering tree tops.

Birdsey (1996) and Penner *et al.* (1997) published formulas for converting merchantable stem volumes to the totals of cut above-ground phytomass. The US multiplier averaged about 2.1, with species-specific values ranging from about 1.7 (southeastern loblolly pine) to more than 2.5 (spruce and fir); the Canadian average is higher at 2.56 (Wood and Layzell, 2003). Consequently, even a very conservative assumption would double the harvested total to about 1.5 Gt, and adding at least 20% in order to include the root phytomass that remains in place after wood harvests would yield about 2 Gt, with half of this assignable to the category of hidden material flows. Is this an excessive harvest or could this extraction continue without further massive deforestation? To answer this question on the global level is complicated by unreliable forest inventories, but it is fairly easy for affluent nations with good forestry data.

According to the US inventory of timberlands, the tree growing stock in the year 2000 stood at $23.65 \, \text{Gm}^3$ and it produced an annual growth of $670 \, \text{Mm}^3$ (USDA, 2001). This productivity compares to the removal of about $490 \, \text{Mm}^3$ (USDA, 2003). The annual cut was thus only 2% of the standing stock and less than 75% of the annual growth of merchantable timber, and the shares are much smaller, respectively just 1 and 6%, when compared to the inventory of all standing phytomass in all forests, including those in protected areas (Joyce *et al.*, 1995). Clearly, modern management makes it possible to exploit forests in a nondestructive manner. Going back to the global roundwood harvest, that $1.5 \, \text{Gm}^3$ cut in 2010 ended up as about $390 \, \text{Mm}^3$ of sawnwood and 185 Mt of pulp for paper. I will follow each of these material streams: I will first describe the main handling and processing methods that lead to commercial lumber (sawn pieces) as well as other structural wood-based products, above all particle board and plywood, and then I will look at a variety of paper products and current changes in their consumption.

As I will show in the next chapter, substitutions by concrete, metals, and plastics have greatly reduced wood intensities of modern societies (that is mass of wood used per unit of GDP) but wood remains an important structural material and its annual production surpasses the combined output of all metals and plastics. Rising post-World War II wood demand has been due to great surges in wood-based house construction, the use of wood for concrete forms, packaging of goods for long-distance shipments and intercontinental trade, and use in paper-based information flows. Similarly, the progressive adoption of fossil fuels and hydroelectricity in the modernizing countries of Latin America, Africa, and Asia has reduced the average per capita demand for wood and charcoal, but rapid population growth on those three continents has resulted in large increases in aggregate demand for fuelwood (see the previous chapter).

Hundreds of tree species are used to make building materials, with their properties (density, resistance to moisture, durability, homogeneity, etc.) and cost determining many specific applications ranging from structural timber, cladding, internal paneling and flooring, to roofing, windows, doors, veneers, plywood, and engineered components. Wood's properties (chemical, mechanical), moisture relations, drying, durability, fire safety considerations, and construction methods increase the material's strength and lifespan have been studied for generations in great detail: the centennial edition of the *Wood Handbook* is perhaps the best single-volume source of information (Forest Products Laboratory, 2010).

Turning logs into sawnwood generates inevitable milling losses, although modern automated computer-guided operations make accurate cuts and minimize waste. A recent survey of European milling practices lists representative averages disaggregated for coniferous and nonconiferous species (UNECE, 2010). The overall means for all tree species are as follows: 54% of logs end up as sawnwood (the extremes are 45–62%), 29% as chips and slabs, 11% as sawdust, and 1% as shavings, with the shrinkage loss averaging 4%. None of this material is wasted: it is used for making particle board and pulp, for mulching, cellulose extraction, or to produce steam and electricity. Using a rounded average of 0.55 for logs-to-sawnwood conversion would mean that the global 2010 production of about 390 Mm^3 of sawnwood required about 710 Mm^3 of roundwood.

The durability of lumber is obviously a key consideration in all external wood uses, and for untreated wood this is determined by its location: extremes range from close to a millennium (elm, oak, and beech as well as pine and spruce can last 800 years, sometimes even longer, when always dry) to less than 5 years for wood in contact with earth (Berge, 2009). Treated timber can last significantly longer. Creosote derived from coal tar remains the first choice for treating railway ties and electricity poles, but various copper-based compounds and new nonmetallic preparations have replaced the chromated copper arsenate used to pressure-treat construction lumber (Richardson, 1978). Biodeterioration (wood attacked by fungi and insects) is a universal problem but is particularly rapid in the tropics; borates offer the best, and the safest, protection against termites, the main natural destroyers of wooden structures in warm climates.

New housing construction in industrializing economies of the twentieth century shows a clear divide among a small number of countries where wood has remained dominant (USA, Canada, Japan) and a majority where traditional brick structures have been increasingly augmented by buildings using concrete and some structural steel (above all EU and China). Thanks to North America's abundance of low-cost lumber, wood remains often the only structural component of single-family houses, and is used for framing studs, plywood for outer walls, wooden trusses for roofs, and wooden joists, subfloors (plywood), and often floors as well, staircases, doors, and door and window frames. A new North American house of 200 m^2 requires about 14 t of lumber (typically yellow pine) and another 14 t of panel products (mostly plywood) for a total mass of 28 t. The greater use of plywood and engineered lumber has reduced the relative demand for dimensional sawnwood, but in aggregate terms this trend has been countered by the steadily increasing size of an average American house (doubling from 110 m^2 in the early 1950s to 220 m^2 by 2005).

Wood-based composite products – including fiberboard, particle board, plywood (layered wood veneers whose commercial use began during the 1850s), laminated veneer lumber, glued-laminated lumber (Glu-lam), manufactured trusses, and finger-jointed lumber – have been displacing traditional sawn wood in many applications. Glu-lam is a particularly outstanding material (Glulam, 2013). Because it is made from thin flexible lamellas it is easy to machine and it can be used not only for columns, rafters, and trusses but can also be turned into optimal shapes including arches and domes for large clear spans and curved portals.

Glu-lam is lightweight (its density is $0.55–0.7\,g/cm^3$ compared to $7.8\,g/cm^3$ for steel) which means that even though it must be fastened with steel bolts or dowels and steel plates, the overall weights of structures made with it are much lighter than those made of steel. But the material is strong enough (more than twice that of sawn timber, per unit of mass comparable to structural steel) to be shaped into major load-bearing structures, including large bridges. Glu-lam is also fairly resistant to aggressive chemicals and it burns poorly, it insulates well and the same piece can be used both inside and outside a building without any cold bridging problems.

Industrial construction has also been a major wood consumer because of its need for concrete forms. Plywood makers produce framework panels with a moisture-resistant adhesive of specified strength and stiffness. Wood requirements vary with the thickness and shape of the poured concrete; forming a typical 20 cm-thick basement wall of an American house will require $10\,m^2$ of plywood (weighing about $20\,kg/m^2$) for every cubic meter of emplaced concrete. But the forms, particularly the standard-sized pieces, can be reused and those properly engineered (with smooth, abrasion-resistant surfaces) can be deployed up to 200 times (APA, 2013). And while in Western countries standard-sized tubes or pipes are now the norm for vertical scaffolding components, wooden boards still provide horizontal pieces and bamboo scaffoldings (often on buildings in excess of 30 stories, built with standard lengths of 6 m) with wooden boards continue to be common in East Asia.

In contrast to wood's stagnating or declining fortunes, paper is one of the signature materials of the modern world. By the late 1930s, Dahl's sulfate process, producing stronger (kraft) paper, became the leading chemical method in large-scale papermaking. Pulp treatment with inexpensive sodium sulfate and net energy production were its two main advantages, offensive emissions of hydrogen sulfide its leading drawback. By the year 2000, the sulfate process was producing roughly two-thirds of the world's pulp, semi-chemical and mechanical processes accounted for nearly all of the rest, with the sulfite method reduced to less than 2% of the total. The demand for raw material varies with the processing method and with the kinds of paper (Bajpai, 2010).

Mechanical pulping needs $2.5\,m^3$ of roundwood for a ton of pulp, and the rates are nearly $2.7\,m^3$ for semi-chemical process and almost $4.5\,m^3$ for the kraft process, with an overall global mean of about $3.8\,m^3/t$ (UNECE, 2010). This would mean that the 2010 global output of some 185 Mt of pulp required just over $700\,Mm^3$ of roundwood, slightly less than the total used for sawnwood. Analogical conversion rates (all in cubic meters of roundwood per ton of product) are about 2.9 for newsprint, 3.25 for packing paper, and 4.35 for sanitary and household paper, with the mean for all paper and paperboard products at about $3.6\,m^3/t$.

Worldwide expansion of paper-making during the second half of the twentieth century was driven by a combination of several trends: growing book, newspaper, and journal publishing; increasing demand for photographic paper (due to the expansion of both professional and amateur photography, the latter use promoted by inexpensive pocket cameras); a much higher demand for typing paper in the offices of the pre-computer era; a new demand created by the introduction of plain-paper copiers (first in the USA in 1959 by Xerox, soon a worldwide office necessity); the massive diffusion of office printers (first limited to line printers using perforated paper and serving large computers, since the late 1970s smaller machines attached to desk computers using standard-size office paper); the post-1980 acquisition of small printers attached to personal computers; and a much increased demand for paper used in industrial and retail packaging. Ironically, the rapid increase in the consumption of printing paper coincided with the diffusion of what was to be the paperless office of a new electronic age.

In the USA, additional paper demand came from the post-1950 housing expansion. By the mid 1950s, drywall had replaced traditional gypsum plaster in half of all new homes and a decade later it had all but a small share of the household market for internal partitions and ceilings (Dushack, 2013). American drywall is a thin (the standard thickness for interior partitions is half an inch or 1.27 cm) layer of gypsum that is sandwiched between two heavy liner paper boards; the mineral was formerly all mined, now nearly 50% of it originates as the by-product of flue-gas desulfurization in coal-fired power plants (Gypsum Association, 2013). The paper used in drywall production amounts to about $400 \, g/m^2$ (on two sides) of a panel and with an output of nearly $1.6 \, Gm^2$ the US gypsum board manufacturing industry used about 640 000 t of paper in 2010, virtually all of it being a secondary (recycled) material (Gypsum Association, 2011). Sheathing paper has been replaced by Tyvek, but tar paper is needed for the waterproofing of roofs as underlay for asphalt or wood shingles.

By the turn of the century, paper consumption had become saturated in many affluent countries, and wood pulp for domestic production began to decline because of more intensive paper recycling. In the USA, wood pulp production peaked during the mid-1990s (at just above 60 Mt/year) and since then it has been in decline, a trend roughly matched by the output of paper and paperboard whose decrease (more than 10% since 2000) accelerated after 2007 due to the economic downturn. This was accompanied by a massive decline in pulp, paper, and board mill employment (from 200 000 in 1999 to less than 120 000 a decade later) and a substantial rise in output per mill employee (from 450 t/year in 2000 to nearly 700 t/year in 2010). Similarly, disaggregated demand figures from Japan show that the consumption of "information-use" paper peaked during the late 1990s and has since declined by nearly 25%, that of plain copier paper began to fall after 2008, and demand for sanitary-use paper has been flat since 2005 (Japan Paper Association, 2013).

Among the trends that have contributed to the decline of paper consumption, none has been as relatively abrupt as the virtual demise of the market for photographic paper and the shift in disseminating news and images from printed to electronic forms. This formerly important paper-consuming sector was first affected by a widespread adoption of color slides, but its near demise came when printed photographs became gradually displaced by electronic images, first taken by digital cameras and viewed on screens, later by a mass ownership of camera-equipped cellphones with billions of images increasingly stored in the cloud. More importantly, the publishing sector has seen closures, reductions, and retrenchment as many newspapers (several hundred in the USA since 2008) and magazines have folded, including such well-known periodicals as *Gourmet* (1941–2009) and *Home* (1951–2008), or stopped printing and launched only e-editions (perhaps most notably America's second oldest weekly, *Newsweek*, published between 1933 and 2012).

But global consumption of paper keeps rising, driven by large increases in demand in populous Asian countries, above all in China. Worldwide production of paper and paperboard rose from 127 Mt in 1975 to nearly 325 Mt in the year 2000, and in 2011 it surpassed 400 Mt (FAO, 2013). China's 2011 production (103.1 Mt) was nearly three times the 2000 total and it accounted for just over 25% of the global supply, up from just 11% in 2000 and 3% in 1975. Not surprisingly, given China's limited forest resources, the country has become the world's largest importer of waste paper, buying 25 Mt in 2010 and 28 Mt in 2011, with the USA being its largest supplier (total US exports of recovered paper more than doubled from 10 Mt in 2000 to 21 Mt in 2011).

FAO(2013) data also show that packaging and paperboard accounted for about 64% of the output in 2010, followed by coated papers (11%, for writing and printing), newsprint (about 8%), and household and sanitary paper (about 7%). A report by the Bureau of International Recycling offers a different breakdown for 2010: 52% packaging and board, 36% graphic and printing, 5% tissues, and the rest special and other papers (Magnaghi, 2011). Per capita differences in average national consumption rates (calculated using FAO's production and trade figures) have reduced during the last two generations when the global mean rose from about 25 kg/year in 1960 to nearly 60 kg in 2010 (FAO, 2013). During the same period, the US rate rose from 195 to 240 kg and the Chinese rate soared from only about 3 to about 70 kg. Even so, large gaps persist, particularly in the consumption of household and sanitary paper, with rates ranging from nearly 20 kg in the USA to 11 kg in France, less than 5 kg in China, and less than 50 g (yes, 50 g) in India.

3.2   Construction Materials

Before I review the production methods and the uses of construction materials that dominate the world of the early twenty-first century, I must point out that hundreds of millions of people – Berge (2009) puts the total at more than 30% of humanity – continue to live in structures whose material, locally available clay, has not undergone any elaborate processing and that can be made without any modern energy inputs. Although the construction of earthen houses requires only a few simple tools it is rather labor intensive. Walls are usually built by ramming the manually dug-out earth between shuttering, or by pressing earth blocks and drying them in the sun before emplacement. These adobe bricks are made by adding fibrous plant materials: straw is used most often, but stalks of wild plants and cattle manure are also common ingredients.

Adobe buildings remain common in arid regions, and while some massive structures have proved surprisingly durable, many earthen houses have poor structural integrity and their collapsing has been a major reason for exceptionally high death tolls in areas that experience repeated earthquakes. In contrast, production of all durable soil- or earth-based materials requires firing in kilns, with temperatures ranging from less than 500 °C for low-quality bricks to as much as 1100 °C for ceramic tiles, 1300 °C for vitrified bricks, and 1400 °C for glass, while the pyroprocessing of Portland cement requires 1400–1450 °C (Berge, 2009). Damp clay used to make quarry tiles and terracotta is the same raw material that is used for bricks, and its firing in kilns sinters the minerals.

Dry pressed clay, often with added kaolin and finely ground waste glass, is used for vitrified ceramic tiles, and the firing produces smoother and finer pieces. The main problem with attractive glazing is that many of the pigments used (oxides of Cd, Co, Cr, Ni, V) are toxins that require careful disposal.

The classification of aggregates, bulk construction materials whose extraction dominates the mass of annually mined minerals, is based not on their chemical composition but on their basic physical attribute: grain size (Reeves *et al.*, 2006). Clays contain particles less than 0.002 mm in diameter that result from the disintegration of rocks; their composition is dominated by the two oxides most common in the Earth's crust, SiO_2 and Al_2O_3. Firing of clays produces hard and durable materials ranging from building bricks (with clays making up 20–30% of the initial mixture that contains 50–60% sand) to highly heat-resistant tiles. Silt – be it in such massive aeolian deposits as China's Loess Plateau or as river-borne material left behind floods – is coarser than clay, with particle diameter ranging from 0.004 up to 0.06 mm.

Construction aggregates include sand, gravel, and crushed stone. Sand grains have diameters up to 2 mm, gravel's irregular dimensions are between 2 and 6 cm, and pieces larger than that belong to various kinds of crushed stone. Sand and gravel are the unconsolidated results of rock disintegration, whose composition is dominated by siliceous and calcareous minerals. They are widely distributed in alluvial deposits and sedimentary beds and hence they can be usually produced from nearby deposits for local and regional consumption, but there is some longer-distance and even international trade. Their modern extraction is carried out either as pit excavation or by barge-based dredging of rivers, lakes, or along coastlines. These extractions produce raw, moist materials that can be used directly as a fill or a base layer, but most excavated sand and gravel is processed to meet market specifications.

Sequential washing, screening, crushing, and dewatering eliminate any organic matter and clay and produce a specific coarseness of material with low moisture. The best available estimates indicate that, in the USA, 41% of construction sand and gravel ends up as concrete aggregates, a quarter of the total is destined for road building, 13% for construction fill, and 12% for asphaltic concrete and similar mixtures (USGS, 2012). The small remainder is used for filtration, snow and ice control on roads (some municipalities also use salt), railroad ballast and golf courses, as well as for replenishment of eroding beaches (Hawaii's Waikiki being perhaps the most famous example, with sand imported from California's Manhattan Beach during the 1920s and 1930s, now dredged from the near offshore deposits). Sand and gravel production correlates highly with the fortunes of the construction industry, and during building slumps it can contract rapidly and significantly: in the USA it fell from 1.25 Gt in 2007 to just 760 Mt in 2010, a 40% decline in just three years (USGS, 2013).

Crushed stone used in construction (mostly in road building and maintenance and for cement manufacturing) is produced mostly from sedimentary rocks (above all limestone and dolomite), while granite is the most commonly quarried igneous rock. Dimension stone can be used in construction in both rough and dressed forms. Dressed stone is used in traditional ashlar buildings, for curbing, and thinner pieces for flagging and cladding, and because durability is important, most of it is granite, harder varieties of limestone (including different marbles), and slate (in addition to flooring and cladding also for roofing). Laborious stone cutting has been fully mechanized by using diamond blades on circular or frame saws, as well as water and flame jets. Finishing (grinding and polishing) can be done to exacting specifications.

Brick-making, one of the world's oldest industries, has reached new global highs thanks to China's extraordinary construction effort: in 2010 the country produced at least 900 billion (perhaps 1 trillion) bricks (about half of them were solid, a quarter of them hollow), nearly 60% of the worldwide output (Li, 2012a). If assuming, conservatively, an average mass of 2.8 kg/brick, the global output in 2010 had a total mass of close to 4.5 Gt, compared to 3.3 Gt of cement. In mass terms, sand, natural gravel, or crushed stone are the dominant materials in concrete, accounting for no less than 60% and as much 75% of the initial mix (PCA, 2011). Different types of these aggregates are needed for specific mixtures. Finer aggregates produce stronger concrete but require more water in the mix, coarser aggregates need less water but within the commonly used dimensions the size of aggregate matters less than its other qualities.

Water makes up anywhere between 15 and 20% of the initial mix and the rest, between 10 and 15%, is cement, a powdered mineral binder whose hydration produces durable solids even when used under water. Additives are also used in order to accelerate the rate of setting at a worksite, to delay setting during extended transport in mixer trucks, or to allow entrainment of a small volume of air and to reduce or repel water content. The resulting material has a density of $2.2–2.4 \, t/m^3$, poor tensile strength of just 2–5 MPa, but excellent compressive strength that has steadily increased. In 1900, the best mixtures had a compressive strength of about 25 MPa, by the 1970s 40 MPa was standard, and the subsequent development of high-performance mixtures for high-rise buildings and other special applications pushed the maximum strength to 150 MPa and for exceptional applications even to 200 MPa (Ulm, 2012).

Final disposition of cement varies by country: in the USA about 75% of the material goes to make ready-mixed concrete, less than 15% for the production of concrete masonry units and pre-cast concrete, in Germany the analogical ratios are about 55 and 30% (Sika, 2005). Because of its near-total dependence on construction, cement consumption fluctuates with economic booms and downturns. Construction of highways, dams, factories, and houses pushed US cement consumption by an order of magnitude in just 28 years: about 3 Mt of cement (87% of it produced domestically) was mixed with aggregates in 1900, and the total surpassed 30 Mt in 1928 (Kelly and Matos, 2013).

The Great Depression cut this to just less than 11 Mt in 1933; it then rose once again to 31 Mt in 1942, the peak year of wartime plant construction. After a brief post-war dip the domestic output, as well imports, kept growing almost without any interruption until 1973, when the consumption reached nearly 82 Mt. Construction of the US Interstate Highway System was a major component of this rising demand (USGS, 2006). About 60% of these multi-lane highways are paved in concrete whose standard thickness is 28 cm and hence 1 km of a four lane highway (each lane is 3.7 m wide) requires about $4150 \, m^3$. This adds up to roughly 10 000 t of concrete for every kilometer and the entire system of 73 000 km embodies about 730 Mt of concrete in driving lanes, with more emplaced in shoulders, medians, approaches, and overpasses.

The 1973 high was surpassed only in 1986, a new peak was reached in 2005 with 128.25 Mt, but five years later the total was down by 45% to just 71.2 Mt (reached for the first time in 1968). Until the mid 1980s, imports were less than 10% of domestic consumption, but then they rose to 22% of apparent consumption in the year 2010. Since 1986, global cement production has been dominated by China; during the 1980s its cement output rose by 20%, during the 1990s it more than doubled, during the first decade of the twenty-first century it more than tripled to nearly 1.88 Gt and in 2010 it accounted for about 55% of global output (USGS, 2013). In 2010, India, with 240 Mt, produced more than three times as much as the USA, with Brazil coming fourth.

The cement industry used to be a large source of uncontrolled particulate emissions (as much as 1 kg/t of the product) but modern fabric filters capture 99.6% of these particulates that are then returned into the kiln. This leaves relatively high CO_2 emissions as the industry's most serious environmental impact: in mass terms, cement production emits roughly a unit of gas for every unit of material, with half of this amount originating from the decomposition of $CaCO_3$, the rest from the heat and electricity required to supply the thermal and mechanical requirements of a rotating kiln. Emissions expressed per unit of standard concrete mixture (without any slag or fly ash) are roughly 480 kg CO_2/m^3 or 200 kg CO_2/t of concrete.

Global compilations of CO_2 emissions from the cement industry show its contribution rising from 1% of all carbon released from fossil fuels in 1950, to 2% in 1975, and almost 5% in 2010 (CDIAC, 2013). And in 2011 China's CO_2 emissions from cement production were only about 12% lower than Japan's total emissions of the gas. There has been a great deal of interest in improving the industry's energy efficiency and reducing its CO_2 emissions (USEPA, 2007; Worrell and Galitsky, 2008; Madlool *et al.*, 2011; Hasanbeigi *et al.*, 2012). Options include using cement and construction materials based on MgO, geopolymer cement, and cement using slag and fly ash. Use of fly ash is particularly appealing: not only does it reduce the energy cost of producing more clinker, but it also avoids landfilling the captured fly ash and extracting new raw minerals and it is the most effective way to curb CO_2 emissions.

In 2010, China's cement industry used all available slag (223 Mt) as well as 66.3% of 395 Mt of fly ash (Lei, 2011). And in the USA Ceratech, a cement company in Virginia, produces blends containing 95% fly ash and 5% liquid ingredients (compared to standard 15% fly ash blends), making a stronger concrete that also reduces the required mass of steel reinforcing bars (Amato, 2013). Recycling of concrete from demolished buildings or deteriorated roads and runways recovers the aggregate through crushing and sieving and the material can be used as a bulk fill (a cheaper option) or incorporated in new concrete (Wilburn and Goonan, 2013). Energy savings can be also realized through better ways of transporting and emplacing concrete, including improvements in pneumatic conveying, cold weather concreting, curing, and reusing returned material (Kermeli *et al.*, 2011).

The affordability of concrete and its ubiquity in modern civilization is not (even after leaving the contribution to anthropogenic CO_2 emissions aside) without major long-term costs. Some 60 Gt of cement were produced worldwide between 1945 and 2010, which means that about 500 Gt of concrete was emplaced in structures, with 60% of the total put in place in the two decades between 1990 and 2010, and 35% between 2000 and 2010. Concrete (particularly its reinforced form) is now by far the most important manmade material both in terms of global annual production and cumulatively emplaced mass. While this material provides shelter and enables transportation and energy and industrial production, its accumulation also presents considerable risks and immense future burdens.

These problems arise from the material's vulnerability to premature deterioration that results in unsightly appearance, loss of strength, and unsafe conditions that some-times lead to catastrophic failures, and whose prevention requires expensive periodic renovations and eventually costly dismantling. Concrete, both exposed and buried, is not a highly durable material and it deteriorates for many reasons (AWWS, 2004; Cwalina, 2008; Stuart, 2012). Exposed surfaces are attacked by moisture and freezing in cold climates, bacterial and algal growth in warm humid regions (biofouling recognizable by blackened surfaces), acid deposition in polluted (that is now in most) urban areas, and vibration. Buried concrete structures (water and sewage pipes, storage tanks, missile silos) are subjected to gradual or instant overloading that creates cracks, and to reactions with carbonates, chlorides, and sulfates filtering from above. Poor-quality concrete can show excessive wear and develop visible cracks and surficial staining due to efflorescence in a matter of months.

Alternations of freezing and thawing damage both the horizontal surfaces (roads, parking) that collect standing water, as well as vertical layers that collect water in pores and cracks. While concrete's high alkalinity (pH of about 12.5) limits the corrosion of the reinforcing steel embedded in the material, as soon as that cover is compromised (due to cracks or defoliation of external layers) the expansive corrosion process begins and tends to accelerate. Chloride attack (on structures submerged in seawater, from deicing of roads, in coastal areas from NaCl present in the air in much higher concentrations than inland) and damage by acid deposition (sulfate attack in polluted regions) are other common causes of deterioration, while some concretes exhibit alkali-silica and alkali-carbonate reactions that lead to cracking. Unsightly concrete blackened by growing algae embedded in the material's pores is a common sight in all humid (especially when also warm) environments. Given the unprecedented rate of post-1990 global concretization, it is inevitable that the post-2030 world will face an unprecedented burden of concrete deterioration.

This challenge will be particularly daunting in China, the country with by far the highest rate of new concrete emplacement, where the combination of poor concrete quality, damaging natural environment, intensive industrial pollutants, and heavy use of concrete structures will lead to premature deterioration of tens of billions of tons of the material that has been poured into buildings, roads, bridges, dams, ports, and other structures during the past generation. Because maintenance and repair of deteriorating concrete have been inadequate, the future replacement costs of the material will run into trillions of dollars. To this should be added the disposal costs of the removed concrete: some concrete structures have been recycled but the separation of the concrete and reinforcing metal is expensive.

The latest report card on the quality of American infrastructure gives poor to very poor grades to all sectors where concrete is the dominant structural material: bridges fared relatively well with C+, dams, schools, roads, aviation, public transit, and waste water treatment facilities got D, and levees and inland waterways received D-, for an overall average grade of D+, with an estimated investment of at least \$3.6 trillion needed by 2020 in order to prevent further deterioration (ASCE, 2013). Transposed to post-2030 China, this reality implies the need for an unprecedented rehabilitation and replacement of nearly 100 Gt of concrete emplaced during the first decade of the twenty-first century, at a cost of many tens of trillions of dollars.

Finally, here concrete is seen from two unusual global perspectives, one comparing areas, the other volumes. Elvidge *et al.* (2007) combined satellite observations of nighttime lights and population numbers to estimate the world's impervious surface area (built-up, paved) at about 580 000 km^2: that is less than 0.5% of ice-free surface, but an area equal to Kenya. In per capita terms, high-income countries in northern latitudes had the largest areas of impervious surfaces (Canada $350 \, m^2$, USA $300 \, m^2$, Sweden $220 \, m^2$) while in low-income countries the rates were below 100, or even below $50 \, m^2$ per capita. Of course, not all impervious surfaces are concrete but the material accounts for their largest share – and there are good reasons why there should be more of it on the ground.

Replacing mud floors with concrete floors in the hundreds of millions of the world's poorest dwellings would cut down the incidence of parasitic diseases by nearly 80%, while paving the streets boosts land and rental values, school enrollment, and overall economic activity and also improves the access to credit (Kenny, 2012). The case for more concrete is thus persuasive – and so is the argument, made first by J.R. Underwood in the early 1970s, that the enormous mass of concrete, bricks, tiles, and glass deserves to be put into a new category of anthropic rocks to complement the standard division into igneous, sedimentary, and metamorphic materials (Cathcart, 2011); three decades later he published the idea (Underwood, 2001) and more than a decade later further huge increases in concrete production support his suggestion.

At the same time, we should keep the mass of the anthropic rocks in perspective. In 2010, humanity put in place close to 40 Gt of them (dominated by 33 Gt of concrete and 4.5 Gt of bricks), an equivalent of at least $17 \, km^3$. For comparison, the volume of one of the world's best known mountains, Japan's Fuji, is about $400 \, km^3$ when approximated by a cone with the radius of about 13 km rising about 2800 m above the surrounding countryside (the mountain is 3776 m and the Lake Yamanka is 980 m above the sea level). So we are emplacing worldwide anthropic rocks equivalent to Fuji's volume roughly every two decades (in mass terms it would be more like every three decades: the basalt that forms the volcano, has a density of about $3 \, g/cm^3$), an achievement that is both impressive and modest: after all, the volume of the world's most massive volcano, Mauna Loa (including the mountain's entire 17 km elevation from its depressed sea floor foundation to its peak at 4170 m above the sea level), is, at about $75\,000 \, km^3$, two orders of magnitude larger (Kaye, 2002).

3.3   Metals

The combination of advancing industrialization, intensified private and public transportation, mechanization of agriculture, and the emergence of mass consumption created a rising demand for every metal during the twentieth century. In some affluent countries specific metal consumption (per capita, per unit of GDP) has leveled off but, as I will show later, there were few absolute declines. Iron, used overwhelmingly as steel, remained the dominant metal of the twentieth century, and by the year 2000 the global output of iron ore, pig iron, and steel had reached new global records: at 1 Gt/year iron ore extraction was surpassed only by the output of fossil fuels and bulk construction materials, pig (cast) iron production rose to nearly 600 Mt, and at roughly 850 Mt/year steel output was about 30 times higher than in 1900. That total was also almost 20 times larger than the aggregate smelting of aluminum, copper, zinc, lead, and tin, and in per capita terms it rose from less than 20 to about 140 kg/year. Demand for copper increased by a similar rate (27-fold, to 13.2 Mt) and zinc production rose almost 20-fold, from about 480 000 t to 8.77 Mt (Kelly and Matos, 2013).

Silver saw the smallest relative production increase (only about 3.4-fold, from 5400 t in 1900), followed by lead (4.3-fold, from 750 000 t to 3.2 Mt), while gold output rose nearly 7-fold, but in absolute terms it amounted only to about 2600 t in the year 2000, compared to 18 100 t for silver, and to millions of tons for Pb, Zn, and Cu. Aluminum emerged as the century's new signature metal, with global output rising from less than 7000 t to nearly 25 Mt, surpassing copper extraction nearly twofold. In demanding (high strength-to-weight) applications, the metal is often substituted by titanium: about two-thirds of the annual production of titanium sponge metal (roughly 140 000 t in 2010) is used by the aerospace industry, the rest goes on armaments and to the chemical industry, while TiO_2 is added as a whitener to paints, paper, and plastics (USGS, 2012).

Advances in the iron and steel industry changed every aspect of ferrous metallurgy during the twentieth century. Although most of the primary metal still comes from blast furnaces, their sizes, capacities, and energy efficiencies have been transformed by the continuous quest for lower cost and higher productivity (Kawaoka *et al.*, 2006). The largest furnaces now have volumes in excess of 5000 m^3: 5500 m^3 Shougang Jing Tang's furnace in Caofedian (new in 2009), 5513 m^3 Schwelgern 2 in Duisburg (ThyssenKrupp since 1993), and, at 5775 m^3 the world's largest, Japan's Ōita 2 (enlarged from 5245 m^3 in 2004) – with a hearth diameter of about 15 m and a maximum daily output of 12 000 t of hot metal a day (Hoffmann, 2012). And larger furnaces require less coke. In the USA, gradual decreases in coke charging and, later, the supplementary injection of pulverized coal, oil, or natural gas, reduced the specific consumption of coke in blast furnaces from about 1.3 t/t of hot metal at the beginning of the twentieth century to less than 0.5 t at its end (de Beer *et al.*, 1998).

In 1914 nearly 75% of US steel output came out of open hearths and nearly half a century later their share of steel production peaked at 88%. By that time, Japan and Europe were rapidly converting to top-blown basic oxygen furnaces (BOFs), Bessemer-like converters charged with molten pig iron and scrap that is subjected to blasts of supersonic oxygen. The first designs of these new furnaces were developed by Robert Durrer in Switzerland, and in 1948 he demonstrated the practicality of the process that used scrap as more than half of the total furnace charge (Durrer, 1948). The subsequent commercialization of the process was not done by a major established steelmaking company but by two Austrian steelmakers,*Vereinigte Österreichische Eisen- und Stahlwerke* (VÖEST) in Linz and Alpine Montan in Donawitz, with the first production starting before the end of 1952 (Starratt, 1960; Geschichte-Club VÖEST, 1991).

As a result, this new way of steelmaking became known worldwide as the Linz–Donawitz process and was almost instantly adopted in Japan, as the country was rebuilding its steel industry that had been destroyed in World War II. In contrast, major US steelmakers were reluctant innovators, adopting their first BOF only in 1964. By 1970 BOFs smelted half of the world's steel output, and 80% of Japanese production. Oxygen was initially blown only from the top, but later also from the furnace's bottom at rates of 50–60 m^3/t of hot metal. BOF capacities range mostly between 150 and 300 t per batch, can complete decarburization (to less than 0.1% C) in only 35–45 min (in contrast, in open-hearth furnaces it took 9–10 h), and boosted labor productivity 1000-fold (Berry *et al.*, 1999). By the century's end, BOFs produced slightly more than 70% of the world's steel, though this rate will not increase (it was just below 70% in 2011) as they face competition from electric arc furnaces (EAFs) using scrap metal.

EAF commercial use began with aluminum smelting in the 1880s, but their adoption by major steelmaking companies came only after World War II once electricity prices declined and there was an abundant supply of scrap metal. Their spreading adoption cut the traditional connection between iron- and steelmaking, as new smaller steel mills (minimills) could operate using only cold scrap metal as their raw material. As a result, the steel:pig iron ratio in the US output stood at 2.1 by the end of the twentieth century (USGS, 2013), and the country's steel output was split between EAFs and BOFs. A decade later, the US became an even better user of its obsolete metal: the steel:pig iron ratio rose further to 2.85 and, despite the fact that the country has been a major exporter of scrap, its EAFs produce 61% of all steel, compared to just 41% in the EU, 22% in Japan, and just over 10% in China.

Post-1950 changes have also transformed the processing of newly smelted metal. Standard practice was to cast steel into ingots (oblong chunks of 50–100 t) and reheat them in order to form semi-finished products (thick slabs, square-profile billets, rectangular blooms) that were eventually rolled into their final shapes as bars, beams, coils, plates, rods, rails, sheets, or wires. During the 1950s, these energy-inefficient practices began to be replaced by the continuous casting of hot metal, pioneered by a German metallurgist, Siegfried Junghans, and promoted by Irving Rossi, an American engineer (Morita and Emi, 2003; Tanner, 1998; Fruehan, 1998). Continuous casting speeds up production (from more than a day to less than an hour), cuts the metal losses (from about 10 to 1% or so), and saves 50–75% of energy compared to the traditional ingot–reheating–rolling sequence. Japanese steelmakers were early adopters of the process, by the century's end nearly 90% of global steel output was cast this way and the shares were up to 97% in the affluent countries, by 2010 even their global mean reached 95%, with China at 98% (WSA, 2013).

Fairly reliable worldwide statistics (BIR, 2012; WSA, 2013) make it possible to outline the global flow of materials in steelmaking in 2010. The dominant route (smelting iron ore to produce pig iron and then processing it in BOFs and a small number of remaining open hearth furnaces (OHFs)) starts with the mining of 1.59 Gt of iron-bearing ores that are charged into blast furnaces together with coke supplemented by pulverized coal injection (760 Mt of coal is used for these two inputs) and about 250 Mt of raw flux (limestone, dolomite) to produce 1.03 Gt of pig iron. This route depends on large-scale intercontinental shipments of iron ore: in 2009 nearly 60% of its production (924 Mt) was traded, with Australia and Brazil being the largest exporters and China the largest buyer, taking 65% of the world's exports and deriving nearly 70% of the ore it uses from imports (WSA, 2011). Larger ships have been used to carry this ore, with Brazil's Vale acquiring 35 362-m long Valemax carriers rated at 400 000 deadweight tons (Vale, 2013).

In 2010, the second route of steel production charged 430 Mt of scrap metal (190 Mt from the industry's operation, 340 Mt purchased, including 105 Mt from exports) into EAFs. The third route, the direct reduction of iron ores, has not diffused as rapidly as once believed and produced only 65 Mt of steel. The three routes produced 1.43 Gt of crude steel. Restated as specific rates (and with the charges expressed in net terms after handling and processing losses) the inputs are 1.4 t of iron ore, 770 kg coal, 150 kg of limestone, and 120 kg of recycled metal to produce 1 ton of crude steel; for the EAF method the charges are 880 kg of recycled metal, 150 kg coal, and 43 kg of limestone per ton (WSA, 2011). The twentieth century also saw major shifts in steel use (WSA, 2013). In 1900, rails were the leading product, now the output is dominated by hot-rolled flat pieces (to make vehicles and household goods, about 55% of the total output in 2011) and reinforcing bars (nearly 15%). At the same time, demand has increased for specialty steel, stainless steel, and high-performance alloys for the energy industry and for the construction of ever-taller buildings.

As already noted, global output of steel rose 30-fold between 1900 and 2000 (to nearly 850 Mt), it surpassed 1 Gt in 2004 and reached 1.43 Gt in 2010 and 1.52 Gt in 2011, with China producing 45% of the total and the EU about 12% (WSA, 2013). In per capita terms, the worldwide consumption of finished products prorated to about 220 kg, with national means of about 500 kg in Japan, 460 kg in China, 310 kg in the EU, 285 kg in the USA, and only 60 kg in India, with all Western rates much affected by the post-2008 economic downturn (in 2006 the US rate was at 400 kg). Exceptionally high per capita rates in South Korea and Taiwan (respectively 1160 and 780 kg) are not due to domestic consumption primarily, but instead due to large steel demand by ship-building industries producing vessels for export.

Fairly accurate output statistics make it possible to calculate that the global production of steel amounted to nearly 31 Gt during the twentieth century, with half of the mass produced after 1980. An obvious question to ask is: what happened to all of this metal? I estimate that at the beginning of the twentieth century the accumulated steel stock that could be potentially turned into new metal was about 15 Gt, or roughly 2.5 t/capita (Smil, 2005). My estimate is not that different from the most detailed recent attempt to quantify the global steel stock (Hatayama *et al.*, 2010). According to this analysis, for 42 countries the total reached 12.7 Gt in 2005 after doubling since 1980. Most of the metal is in structures (about 60%) and 10% is in vehicles; they also forecast that the structural and vehicular stock will reach 55 Gt by 2050, mostly due to a 10-fold increase of steel consumption in Asia. For comparison, Müller *et al.* (2009) estimate the total anthropogenic iron stocks at 25–30 Gt and the US iron stocks at about 3.2 Gt in 2000 – but the USGS put them at 4 Gt (Buckingham, 2006).

Global steelmaking has also seen major shifts in national dominance and in the structure of the industry. As steelmaking expanded and matured, both the primacy in production and the technical leadership shifted from the UK to the USA; but from the 1950s the center of innovation moved to Europe and above all to Japan, while the USSR became the world's largest, but technically inferior, steel producer. After 1974 (coincident with the first round of large oil price rises) the industry's almost uninterrupted post-World War II rise turned into a sequence of advances and retreats, with global production dipping in 10 of the next 25 years. The next dip came in 2008 and 2009 as a result of the global recession, but the output remained above 1 Gt/year (first surpassed in 2004) because of China's huge, and rising, demand.

In 1900, the US Steel Corporation produced nearly 30% of the world's steel and US output was 36% of the global total. At the end of World War II (with Japan and Germany in ruins) the US share was nearly 80% of the global output, but the post-1955 expansion of the Japanese and Soviet steelmaking began to reduce the industry's global share. By 1975, the USA still had three companies among the world's top 10 producers (US Steel as number 2), by 1990 US Steel was alone in the top group (number 5), and a year later USX, its parent company, was dumped from the Dow 30 to make place for Disney.

By the year 2000 there was no American company among the top 10: US Steel was placed 14th, and in 2011, when it was 13th, 6 out of the 10 largest steelmakers were in China, the USA made less than 6% of world steel and China produced 45% of the total.

During the twentieth century, aluminum became the second most important metal of modern civilization, but the fundamentals of its production remain as they were when invented in the 1880s, combining Bayer's production of alumina from bauxite (extraction with caustic soda followed by calcination) and Hall–Hérault's electrolysis of cryolite and alumina. But the entire process (from bauxite mining to electrolytic smelting and casting of ingots or rolling of sheets) consumes a great deal of raw materials and remains highly energy-intensive (IEA, 2007; IAI, 2007). Material requirements for the primary production of the metal include (per ton of cast ingot) about 5270 kg of bauxite (that will yield about 1925 kg of alumina) and 435 kg of petrol coke and pitch to make anodes.

Many industries, led by auto-making, have turned to aluminum in order to reduce the weight of their products without compromising structural integrity: as already noted, the metal's density is only a third of steel's but the tensile strength of many aluminum alloys is 400–500 MPa, overlapping with that of the most common structural steels whose tensile strength is 400–550 MPa (high-tensile steels rate well above 1 GPa). Aluminum can reduce the weight of many key car components by up to 40% compared to only 11% for high-strength steel, and that is why the US car industry now consumes about a third of the country's aluminum and why the use of the metal and its alloys reached a new high in 2012 models (nearly 156 kg/vehicle), making up 9% of the average car mass (but more than a third of the material scrap value), mostly in the engine and wheels, but increasingly in hoods, trunks, and door. It is also why the automakers, required to meet more stringent emission standards, intend to double the share of the metal in an average vehicle by 2025 (AA, 2011).

A new twentieth-century manufacturing sector has been particularly aluminum-dependent. The earliest airplanes had wooden and fabric covers, but aluminum bodies appeared by the late 1920s and monocoque construction was all-aluminum, with demand spurred above all by America's unprecedented program of new fighter and bomber construction during World War II, when the country's aluminum consumption more than quintupled in just four years (USGS, 2013). Military demand remained strong after World War II, and by the mid 1950s the era of commercial jetliners had arrived (starting with Boeing 707 in 1957) whose bodies are 70–80% high-strength aluminum alloys, with a single wide-bodied jet containing more than 100 t of the metal. Other post-World War II markets with a high reliance on aluminum alloys have ranged from house construction (window and door frames, siding, eavestroughs) to communication and spy satellites, cars, and beverage cans (Hosford and Duncan, 1994). Additionally, aluminum wires became the most important long-distance conductors of electricity.

Between the beginning of its commercial production during the 1880s and 2012 almost 1 Gt of the metal (956 Mt) was extracted from bauxite, and the best calculations show that nearly 75% of that mass is still in productive use, with about a third in buildings, close to 30% in transport, a similar share in electrical applications, and about 10% in machinery (IAI, 2012). This means that a significant share of the extracted metal has been recycled many times – but there is a limit to this reuse due to rising levels of other elements now added to make aluminum alloys (use of pure, 99.8%, aluminum has declined as alloys were selected to meet a variety of specific demands).

3.4 Plastics

Most of the now ubiquitous plastic materials belong to two categories of manmade polymers: thermoplastics account for nearly 80% of the total, thermosets for most of the rest. Thermoplastics are either linear or branched molecules that lack any chemical bonds and are softened by moderate heating and hence are easy to process; they harden again once cooled and retain fairly high impact strength (Ibeh, 2011). Polyethylene (PE) is by far the most important thermoplastic (it accounted for 29% of the world's aggregate plastic output, or roughly 77 Mt, in 2010), polypropylene (PP) comes next (with about 19% or 50 Mt in 2010), followed by polyvinyl chloride (PVC, about 12% or 32 Mt in 2010). In contrast to thermoplastics, thermoset plastics have bonds among molecules (and hence a greater dimensional stability) and cannot be softened by moderate heating (Goodman, 1998). Their output is dominated by polyurethanes and epoxy resins, followed by polyimides, melamines, and urea-formaldehyde.

As with most other materials, China is now the world's leading producer of plastics, delivering nearly a quarter of the total in 2010, followed by the EU and North America. Of course, in per capita terms China, with less than 50 kg/year (including a large share incorporated in exported consumer goods), is still far behind the EU and the USA where the average consumption rate of plastics was, respectively, about 115 and 125 kg in 2010. The final use of plastic products is well illustrated by the detailed breakdown available for the European market (Plastics Europe, 2011): in 2010 packaging consumed almost 40% of the total (mostly as various kinds of PE and PP), construction about 20% (mostly for plastic sheets used as vapor barriers in wall and ceiling insulation), the auto industry claimed nearly 8% (interior trim, exterior parts), and the electrical and electronic industry took about 6% (mostly for insulation of wires and cables).

The three dominant polymers that now make up about three-fifths of the global output of synthetics – PE, PP, and PVC – are structurally very similar: they all share the backbone of carbon atoms, but PE carries only hydrogen atoms, while in PP every other carbon atom carries a methyl (CH_3) group and in PVC chlorine alternates with hydrogen. There are other similarities: ethylene (IUPAC name ethane, C_2H_4) and propylene (the correct IUPAC name is propene, C_3H_6) are polymerized with the help of the same class of catalysts, they have many overlapping physical characteristics and they share many more everyday uses for packaging, construction, and a variety of household and industrial items. But there is a notable difference in terms of consumer perception and environmental impact: PE and PP do not get much negative coverage, but PVC has been widely portrayed as environmentally dangerous – its synthesis requires hazardous chemicals, it is not easy to recycle, and its disposal has undesirable consequences, no matter if the material gets landfilled or burned.

PE has been the world's most important plastic for three generations (Malpass, 2010). ICI patented high-volume polymerization of PE in 1936, and began its production of low-density polyethylene (LDPE, a branched ethylene homopolymer) using a chromium catalyst in September 1939, on the eve of Britain's entry into World War II. During the war, the new material was used as a cable insulator and to make light radar domes, but its production required very high pressure and high temperature (respectively, in excess of 100 MPa and 200 °C). This changed with the discovery of mixed organometallic catalysts, in 1953, by a group led by Karl Ziegler at the Max Planck Institute: they found that nickel in combination with triethylaluminum dimerized olefins (Ziegler, 1963). Further research discovered that other catalysts containing heavy metals, above all titanium in the form of $TiCl_4$, could polymerize ethylene to produce high-density polyethylene (HDPE, a linear, semicrystalline homopolymer of C_2H_4).

In the first trials production was done at a relatively moderate pressure of 20 MPa, later at ambient conditions (Ziegler, 1963). Italy's Giulio Natto used these catalysts to produce polymers from propylene, and in 1963 the two chemists shared the Nobel Prize in Chemistry. HDPE is a more resilient material: its density is only slightly higher than that of LDPE ($0.95-0.97$ g/cm^3 compared to $0.91-0.94$ g/cm^3), but its melting temperature is considerably higher ($135\,°C$ vs. $115\,°C$) and its tensile strength is superior, 32 MPa compared to 9 MPa for LDPE. LDPE is more flexible and more transparent and both polymers are inert. The propensity of early HDPE products to crack with aging was solved by synthesizing a material with a small share of branches in the linear chain.

Later, during the 1970s, the efficacy of titanium-based catalysts was greatly enhanced by the addition of MgCl, and this opened the way to the commercial synthesis of linear low-density polyethylene (LLDPE, a random copolymer of ethylene and α-olefins with a density between 0.915 and 0.925 g/cm^3). The material is stronger than LDPE and can be used to make thinner films. The range of ethylene-based materials now also includes ultra-high-molecular-weight polyethylene (UHMWPE, whose molecular weight is typically between 3.5 and 5.5 million and whose density ranges between 0.93 and 0.935 g/cm^3), medium-density polyethylene (MDPE, 0.94 g/cm^3), and cross-linked polyethylene (PEX or XLPE, an elastomer due to its cross-link bonds).

All of these products begin as ethane. In North America and the Middle East ethane is separated from natural gas, and low gas prices and abundant supply led to surplus production for export and favored further construction of new capacities: in 2012 Qatar launched the world's largest LDPE plant and, largely as a result of shale gas extraction, new ethylene capacities are planned in the USA (Stephan, 2012). The dominant feedstock for ethane in Europe, where prices of imported natural gas are high, is naphtha derived by the distillation of crude oil. Ethane is cracked to make ethane ($C_2H_6 \rightarrow C_2H_4 + H_2$) and the subsequent multi-stage ethylene purification yields a compound with 99.9% purity whose exothermic catalytic conversion (about 97% efficient) produces powdered PE.

Different kinds of PE are shaped into final products by techniques best suited to particular specifications: by extrusion and molding (through impact, blowing, or rotation) of objects and by casting or blowing of sheets (single layer or laminated) and insulation foams. Besides its low stiffness, PE's other disadvantages include flammability, poor weathering resistance, a limited range of temperature tolerance, and tendency to stress cracking. Final PE products are used for a still-expanding range of visible and hidden applications. Virtually ubiquitous are thin LLDPE, LDPE, and MDPE films made into transparent or opaque bags (sandwich, grocery, or garbage), sheets (for covering crops and temporary greenhouses), wraps (Saran, Cling), and squeeze bottles (for honey), HDPE garbage cans, containers (for milk, detergents, motor oil), and toys (including Lego bricks). Among a myriad of hidden PE applications are HDPE for house wraps (Tyvek) and water pipes; PEX for water pipes and as insulation for electrical cables; and UHMWPE for knee and hip replacements.

Flexibility, strength, durability, and resistance to UV radiation makes LLDPE an excellent choice for heavy-duty uses including pool liners and geomembranes used in construction, and the material is also preferred for ice and frozen foods bags. Size wise, PE applications range from massive LDPE water tanks (50 000 l or 50 m^3) to small machine parts (bearings, gears) made of wear-resistant UHMWPE. Another extreme range is indicated by soft LDPE cling and bubble wraps and UHMWPE (Spectra) used in bulletproof clothing as an alternative to aramid (Kevlar). PE products are increasingly recycled. HDPE (numbered 2 inside the recycling triangle molded on an item's base) returns again as garbage cans or detergent containers or is made into flower pots. LDPE (number 4) comes back as new bags.

Propene, a colorless but highly flammable and asphyxiating gas, is produced by hydrocarbon cracking and was first polymerized to make a crystalline isotactic compound (with all methyl groups attached to the same side of the carbon backbone) in 1954 by Giulio Natta in Italy and Karl Rehn in Germany. Much like PE, isotactic PP continues to be made by using metallic Ziegler–Natta catalysts while the production of syndiotactic PP (with regular alternation of opposing monomers) is catalyzed by soluble metallocenes, organo-metallic complexes whose deployment allows control of the molecular structure in order to produce polymers with specified qualities. Its commercial varieties include copolymers (homopolymer, block, and random) and rubber-modified blends. PP's disadvantages include flammability, degradation by UV radiation (both can be minimized by additives), susceptibility to chlorinated solvents, and poor impact strength at low temperature.

PP shares many market niches with PE – it is made into many kinds of food containers (yogurt, sour cream, mayonnaise) and their caps, crates, totes, garbage cans, storage bins, bottles, baskets, pails, and pipes – as well as with PVC (as a wire and cable insulator or a film). But its combination of low density (range 0.89–1.06, typical rate just 0.905 g/cm^3), substantial strength (25–40 MPa, more than the MPa for a film), high melting point (171 °C for perfectly isotactic polymer, 160–166 °C for commercial material), and resistance to acids and solvents makes it an ideal raw material for high-temperature uses (laboratory and hospital items requiring sterilization), heavy-duty applications (industrial pipes for hot or cold liquids, ropes, parts tolerating bending, such as hinges, and impact, such as container lids), nonwoven materials (including diapers and liquid and air filters), and fibers. The fibers range from indoor–outdoor carpeting to lightweight fabrics woven from PP yarn and used particularly for outdoor apparel, as the material insulates while staying dry.

PVC's history began unusually early for a plastic: vinyl chloride was first prepared by Henri Victor Regnault in 1835, and in 1872 Eugen Baumann succeeded in polymerizing the compound in the laboratory setting. Real commercial breakthrough came only in 1926, when Waldo L. Semon (working for BF Goodrich Company in Akron, Ohio) dissolved heated polymerized vinyl halide in a nonvolatile organic solvent and, after cooling, produced what he called a "stiff rubbery gel" (Semon, 1933). PVC production begins by combining ethylene and chlorine and converting ethylene dichloride into vinyl chloride that is polymerized (mostly in suspension but also in emulsion) inside reaction vessels to yield white PVC powder.

In order to illustrate the material's ubiquity in modern society, I wrote a brief narrative of a city woman's day pointing out PVC's presence in the objects she uses in the morning and as she leaves for work (ranging from insulated wires, water, and sewage pipes to food wraps and her car's interior and body undercoating) and in conclusion I noted that an even greater PVC presence would surround her if she had an accident and ended in a hospital:

> There she would be enveloped by objects made of different kinds of PVC: disposable and surgical gloves, flexible tubing for feeding, breathing and pressure monitoring, catheters, blood bags, IV containers, sterile packaging, trays, basins, bed pans and rails, thermal blankets, labware (Smil, 2006, p. 131)

PVC is also used in construction (house sidings, window frames), for outdoor furniture, water hoses, office gadgets, toys, and credit cards, but new plastic banknotes (now in circulation in Australia and Canada) are made from PP. However, there are concerns. Greenpeace calls the material "one of the most toxic substances saturating our planet and its inhabitants" and "the single most environmentally damaging of all plastics," one that contaminates people and the environment not only during its production but also during its use and incineration or landfilling, and calls for substitution by other materials in order to create a PVC-free world (Greenpeace, 2013). Plastic makers have rebutted these claims (Vinylfacts, 2013) and the compound has become so ubiquitous that creating a PVC-free world would be very challenging.

PVC production and incineration emit dioxins, and phthalates, plasticizers used to soften normally rigid PVC, are suspected carcinogens that enter the environment during the use and disposal of the polymer – but there are no easy, final judgments in these matters. On the one hand, Wilkinson and Lamb (1999) demonstrated that the levels of phthalates that could migrate from soft PVC toys to children during mouthing activities do not present a significant risk to children (being orders of magnitude lower than the no-observed-effect-level for chronic rodent exposure), and landfill simulation assays showed no degradation of PVC in landfilled waste and attributed released vinyl chloride to perchloroethylene rather than to PVC (Mersiowsky, 2002).

On the other hand Tickner *et al.* (2001) concluded that it is impossible to rule out carcinogenic responses to the leaching of phthalates from medical devices and argued for their assertive but careful substitution. Phthalate exposure and asthma in children (Bornehag and Nanberg, 2010) and possible health risks from the migration of chemicals from plastic piping into drinking water (Stern and Lagos, 2008) pose similar uncertainties. But problems with plastics go beyond any chemical effects: their sheer presence on land and in the ocean is no less worrisome. Plastics have a limited lifespan in terms of functional integrity: even materials that are not in contact with earth or water do not remain in excellent shape for decades. Service spans are no more than 2–15 years for PE, 3–8 years for PP, and 7–10 years for polyurethane; among the common plastics only PVC can last two or three decades and thick PVC cold water pipes can last even longer (Berge, 2009). But complete degradation may take decades or more than a century while torn films, broken pieces, and degraded bits of plastics accumulate in the environment.

3.5 Industrial Gases

When asked to make a short list of materials indispensable for the functioning of modern civilization, few people will include industrial gases; moreover, these gases are also usually excluded from the totals of material flows monitored in the EU and North America, and they are also commonly neglected by authors writing histories of modern inventions. Those readers who have not skipped any previous sections already appreciate that the three most important elements – oxygen, hydrogen, and nitrogen – deserve such ranking because without them we could not produce steel in the most efficient way, and could not have our modern petrochemical and nitrogen fertilizer industries. Other elements and compounds classified as industrial gases include acetylene, argon, carbon dioxide, helium, neon, and nitrous oxide.

As already noted, much of the early development in liquefying gases was carried out by Carl von Linde, who used combination of the Thomson–Joule effect and countercurrent heat exchange. In 1902, he patented separation of principal gases from liquefied air (at temperatures lower than $-140.7\,^{\circ}C$) through purification by rectification (countercurrent distillation), and by 1910 he had perfected this process for large-scale commercial applications by developing a double-column rectification that made it possible to produce pure oxygen and nitrogen simultaneously (Houseman, 1949; The Linde Group, 2012). Also in 1902 Georges Claude, after working with acetylene, introduced his variant of cryogenic air separation (Almqvist, 2003) and a century later his and Linde's designs remain the foundations of modern production of high-purity streams of nitrogen, oxygen, and argon from the atmosphere where their concentrations are, respectively, 78.08, 20.95, and 0.93%.

But Linde and Claude might be even more surprised to see how consequential their inventions have become. Most importantly, without the synthesis of ammonia (predicated on large-scale supply of pure nitrogen) we would not be able to feed billions of people, and without oxygen we could not produce most of the world's most important alloy. Ammonia synthesis is the world's largest consumer of nitrogen: in 2010 it required 130 Mt of the gas (about $112\,Gm^3$ of N_2). Nitrogen's other key uses as a feedstock include ammonia for the synthesis of nitric acid, hydrazines, and amines.

Liquid N_2 is used not only to preserve vaccines and tissues but also to freeze soils for easier drilling and tunneling, in tire and solvent recycling, for enhanced oil recovery and production of plastic moldings, and to achieve material shrinkage: nitrogen cooling of metal parts enables tight assembly fits and, in reverse, it allows the taking apart of closely-fitted parts. With the expansion of modern electronics, nitrogen found a new market in those instances (particularly during soldering) when it is necessary to reduce the presence of oxygen and to maintain a clean atmosphere (by 1985 this use claimed 15% of US consumption). And N_2 is also used as an inert blanket for flammable or explosive compounds and as protector of wine in opened bottles.

Ferrous metallurgy is by far the largest user of oxygen: the gas is blown into blast furnaces, EAFs, and BOFs need about $50\,m^3/t$ of hot metal. Chemical syntheses (above all ethylene oxidation) are the second largest market, and oxygen is also used in smelting color metals (lead, copper, and zinc furnaces), in the construction material industries (producing a more intense flame and reduced fuel use in the firing of glass, mineral wool, lime, and cement), and for delignification and bleaching (a superior choice to ClO_2) in the pulp and paper industry. Liquid O_2 is an excellent rocket propellant (combined with kerosene it powered the first stage of Saturn rockets used for all Apollo lunar missions); welding and cutting metal is done with oxygen and oxygen-acetylene flames; waste water treatment uses it for oxidation, incineration, and vitrification; incineration of hazardous wastes is another growing market; and aquacultural operations use it to aerate their ponds. The global market for liquid oxygen reached about $75 \, Gm^3$ in the year 2000 and it was more than $150 \, Gm^3$ in 2010 (Research and Markets, 2013).

Argon, the cheapest truly inert gas, goes into incandescent and fluorescent lights (the typical mixture is 93% Ar, 7% N_2) and between the panes of high-efficiency windows; it is used in stainless steel production as a shield gas in casting, and when blown into a converter it reduces Cr losses and prevents the formation of nitrides; it removes hydrogen and particulates from molten aluminum; it provides a shield for tungsten arc welding, as well as for the production of virtually pure silicon and germanium crystals and for the protection of wine in opened bottles; and it is used in micro-cryosurgery to destroy tiny bits of tissue.

There is too little hydrogen in the atmosphere (0.00005% compared to about 0.04% for CO_2) to separate it economically from air, and the standard pre-World War II choice was to produce it by steam reforming of coal and since the 1950s by steam reforming of hydrocarbons, with methane (CH_4) being obviously the best feedstock. During the first decade of the twenty-first century, the worldwide production of hydrogen (as an intentionally separated gas as well as a by-product gas generated by petrochemical industry for its own consumption) was about $50 \, Mt$ (roughly $550 \, Gm^3$) a year (Markets and Markets, 2011). Intentional gas production was about $450 \, Gm^3$ ($40 \, Mt$) in 2010, roughly split between steam reforming of natural gas and other fossil fuels, with electrolysis supplying less than 5% of the total. In 2010 global output of $159 \, Mt \, NH_3$ incorporated $28 \, Mt \, H_2$.

Crude oil refineries require hydrogen for cracking, dearomatization of and desulfurization treatment of oil (Chang *et al.*, 2012). Hydrotreating, hydrodesulfurization, and hydrocracking used to process roughly $3.7 \, Gt$ of oil in 2010 claimed (assuming that H_2 demand averaged 0.5% of the total crude input, or roughly $60 \, m^3/t$) about $20 \, Mt$ of the gas. The remainder was divided between other chemical syntheses (methanol, polymers, solvents, vitamins, pharmaceuticals), and industrial uses, including glass-making, semiconductors, food processing (to hydrogenate unsaturated fatty acids in order to produce solid fats), and rocket fuel.

Nitrous oxide (N_2O, prepared by heating of ammonium nitrate) is a common anesthetic and aerosol spray propellant. Acetylene (traditionally produced by hydrolyzing calcium carbide made by reducing lime by coke) was widely used for welding and cutting before it was largely displaced by the arc process. Helium (extracted from natural gas) is used mainly in cryogenic applications (including basic scientific research and magnetic resonance imaging), to purge and pressurize rockets, in chromatography, and to provide lift for weather (and party) balloons. Finally, CO_2 has many uses ranging from enrichment of greenhouse atmospheres in order to accelerate growth of vegetables, to enhanced oil recovery, and liquefied and pressurized CO_2 can also replace petrochemical-based solvents in dry cleaning.

As a result, industrial gases are used in sectors that account for more than half of the world's economic output and the value of their production has been growing faster than the growth rate of the global economy: in 2000 their global market was worth about $34 billion, a decade later it had nearly doubled as it exceeded $60 billion, and it is heading to about $80 billion by 2015 (Gale Group, 2012). Inclusion of industrial gases in a comprehensive appraisal of material flows is thus justified both because of the aggregate magnitude of annual use and because of the indispensability of those elements and compounds in virtually every important sector of the modern economy. I will take a closer look at the processes producing the three most important (and most voluminous) streams, those of oxygen, nitrogen, and hydrogen. The main reasons for the continuing expansion of these industries have been technical innovations that have cut costs and raised the efficiency of gas extraction, the shift of production to consumption sites, and greater economies of scale.

The most important pre-World War II innovations were Fränkl's regenerators (heat exchangers, patented in 1928) and expansion turbines replacing the piston engines used for refrigeration that were introduced during the 1930s. The first Linde–Fränkl oxygen plant began to operate in 1950, and diffusion of this process cut oxygen prices by an order of magnitude. Major progress in air separation took place during the mid 1980s when Linde introduced structured packings in rectification columns: compared to previously used tray sieves packed columns make it possible to enrich the downward-flowing liquid in oxygen, and the ascending stream to become increasingly richer in nitrogen with reduced energy consumption (The Linde Group, 2012).

Cryogenic extraction of the two dominant atmospheric constituents starts with the compression of ambient air to about 600 kPa (normal pressure is 101.3 kPa) and the removal of H_2O and CO_2 by molecular sieves. These are made of synthetic zeolites, aluminosilicates whose molecular-size pores and diverse chemical compositions make it possible to tailor them for highly specific uses; these materials are able to adsorb selectively H_2O, CO_2, and N_2 and produce a stream of 90–95% pure O_2 (Ruthven *et al.*, 1993). After removal of water and carbon dioxide, elements of cooled liquid air are separated by distillation, with N_2 boiling away first (at 77.4 K), followed by Ar (at 87.3 K) and O_2 (at 90.2 K). Cold gases then flow back to heat exchangers where they cool the incoming air while warming to near-ambient temperature and pressure, an exchange that reduces the plant's energy consumption and boosts its overall efficiency to about 35%.

A competing technique for gas separation that emerged during the 1950s is pressure swing adsorption (PSA), a process using a concomitant pressurization/adsorption and depressurization/desorption in two identical adsorber vessels filled with carbon molecular sieves and taking advantage of the fact that O_2 diffuses faster into their pore structure than N_2 does (The Linde Group, 2013). PSA popularity is due to its simplicity and low operating costs and the largest plants delivered by Linde have a capacity of up to 5000 Nm^3/h and yield N_2 and Ar with 97–99.0000% purity. Vacuum swing absorption (VSA) is another noncryogenic process that uses zeolites and produces oxygen with a lower energy cost than PSA (as it operates at near-normal pressure and temperature) and it has been widely adopted by plants delivering at more than 20–30 t/day. Synthetic zeolites are thus a perfect example of a relatively obscure but very important material; by 2010 their global output rose to 2.8 Mt (USGS, 2013).

When BASF began commercialization of Haber's ammonia synthesis process, it adopted a process of hydrogen production that was newly patented by Wilhelm Wild in 1912. This steam-reforming *Wasserstoffkontaktverfahren*, using catalysis to shift CO to CO_2, remained the standard process for large-scale H_2 generation until the late 1940s (Smil, 2001). It was a solution based on a resource constraint: German chemists knew that it would be best to reform pure methane, the compound with the highest H:C ratio, or naphtha, a mixture of light hydrocarbon liquids, but these feedstocks were not available in large quantities in interwar Germany, which was one of the world's leading coal producers.

Not surprisingly, the first steam reforming operation using CH_4 was built in hydrocarbon-rich America, in 1939, for an ammonia plant of Hercules Powder Company in California. After World War II, US production of hydrogen became entirely based on the steam reforming of methane, while in Europe and Asia coal and liquid hydrocarbons continued to dominate for several more decades. By the century's end, about half of the world's hydrogen output was based on methane. Reaction of CH_4 and steam at temperatures between 700 and 1100 °C and pressures of 0.3–2.5 MPa – $CH_4 + H_2O \rightarrow CO + 3H_2$ – is followed by shifting CO to CO_2 – $CO + H_2O \rightarrow CO_2 + H_2$ – and producing additional hydrogen, and the process is completed by removing CO_2 and unreacted CO. PSA, a standard choice since the 1960s, yields hydrogen that is 99.9999% pure, and Linde's plants with the largest capacities can produce more than $400\,000\,Nm^3$ of the gas every hour (Ruthven *et al.*, 1993; The Linde Group, 2013).

Liquid hydrocarbons (principally naphtha) are the feedstock for hydrogen production in crude oil refineries where the gas is needed for the catalytic conversion of heavier fractions to lighter fuels, and also in order to comply with ever stricter environmental regulation and to desulfurize the refined products. Steam reforming of coal is still done in China and India, and only small fractions of the global hydrogen output comes from highly energy-intensive electrolysis of water (needs $4.8\,kW\,h/m^3$ H_2) and from methanol cracking. Synthesis of ammonia remains the leading user of hydrogen, followed by refinery needs (above all in processing heavier oil and in desulfurization).

As in many other industries, the quest for economies of scale has progressed rapidly since World War II: while Linde's first plant, in 1902, produced 5 kg of oxygen per hour (120 kg/day), typical daily capacities of air separation plants rose from 50 to 100 t during the 1950s, to more than 200 t during the 1960s, and by the late 1970s Linde was operating plants with capacities of up to 2300 t/day of oxygen and 800 t/day of nitrogen. Another order-of-magnitude scaling was achieved by the 1990s: in 1997 each of the four nitrogen generation trains built by Linde was the largest air separation facility ever built (with total nitrogen flow of 40 000 t/day); in 2000 Linde built the world's largest five-train air separation plant in Mexico, with a nitrogen capacity of 63 000 t/day (equivalent to 17 500 t/day O_2), and in 2006 the company designed an eight-train project in Qatar with an oxygen capacity of 30 000 t/day (The Linde Group, 2013). This prorates to a unit capacity of 156 250 kg/h, more than 30 000 times the pioneering 1902 design.

Consumers requiring a steady supply of large volumes of industrial gases are best served by pipelines that carry highly pressurized gas. Major producers maintain their own pipelines, and as expected maps show high densities of O_2, N_2, and H_2 lines along the Gulf of Mexico to serve the region's exceptional concentration of refineries and petrochemical plants, and in the most industrialized regions of the USA, Germany, and Japan. At the same time, there has been a significant expansion of decentralized gas production. Leonard Parker Pool, of America's Air Products, pioneered this option during the 1940s: on-site custom-designed cryogenic units are much less expensive than distribution of gases in cylinders from central plants (Air Products, 2013).

As with most modern industries geared toward mass-scale industrial consumption, there is now a very high degree of concentration. In 2010 just five companies accounted for about two-thirds of the global output of industrial gases. The Linde Group (with head offices in Wiesbaden in Germany) became the global leader after purchasing the BOC Group in 2006; the former leader, Air Liquide (headquartered in Paris) claims second place, and both companies had very similar revenues in 2011. America has two large companies, Praxair (headquartered in Danbury, CT) and Air Products (head offices in Allentown, PA).

3.6   Fertilizers

As already described in the previous chapter, the real breakthrough in satisfying the rising demand for nitrogenous fertilizers came only in 1909, with Fritz Haber's demonstration of ammonia synthesis from its elements; this was followed by uncommonly rapid progress in turning the lab-bench reaction into industrial commercial synthesis. BASF, Germany's leading chemical company, had done so by 1913, under Carl Bosch's leadership, and I have argued that the resulting impact on global food production justifies the labeling of the Haber–Bosch synthesis as perhaps the most consequential technical innovation of the twentieth century (Smil, 2001). But its impact remained limited for decades, first due to the diversion of ammonia for the production of explosives during World War I, and then due to the Great Depression and World War II.

Post-1950 expansion was rapid, with global ammonia synthesis rising from less than 6 Mt in 1950, to about 120 Mt in 1989, before a stagnation in the early 1990s caused by the saturation of nitrogen use in many affluent countries and by the collapse of the USSR, at that time ammonia's largest producer. By the year 2000, synthesis had surpassed 130 Mt and it was 159 Mt in 2010 and 164 Mt in 2011 (USGS, 2013). In that year, the world consumed about 200 Mt of sulfuric acid, but because ammonia has a much lower molecular weight (17) than H_2SO_4 (98) NH_3, remains the world's most synthesized compound in terms of moles (nearly four times as much as H_2SO_4).

About two-thirds (65–57%) of all synthesized NH_3 has been recently used as fertilizer, with the total global usage more than tripling since 1970, from 33 to about 106 Mt N in 2010. Because ammonia is a gas under ambient pressure, it can be applied to crops only by using special equipment (hollow steel knives), a practice that has been limited to North America. The compound has been traditionally converted into a variety of fertilizers (nitrate, sulfate) but urea (containing 45% N) has emerged as the leading choice, especially in rice-growing Asia, now the world's largest consumer of nitrogenous fertilizers; ammonium nitrate (35% N) comes second.

The worldwide pattern of nitrogen fertilizer use has changed over the past two generations. Consumption reached plateaus in most affluent countries (in the USA it was 11.5 Mt in 1996, 10.5 Mt N in 2000, and 11.5 Mt N in 2010) and in some of it appreciably declined: from nearly 800 000 t N in 1979, to only about 450 000 t N in Japan in 2010 (FAO, 2013). Populous modernizing countries, led by China, emerged as the largest users. China surpassed the USA to become the world's largest consumer of reactive nitrogen in 1979, and a decade later it also became its largest producer: by the year 2000 it produced 22.1 Mt N, in 2010 35 Mt N. Large-scale availability of inexpensive nitrogen fertilizer made it possible to realize the yield potential of new crop varieties.

Hybrid corn was introduced first (during the 1930s in the USA), high-yielding, short-stalked wheat and rice cultivars were made available by the CIMMYT in Mexico and the IRRI in the Philippines during the 1960s (Smil, 2000). Compared to traditional harvests, the best national yields of these three most important grain crops have risen to about 10 t/ha for US corn (from 2 t/ha before World War II), 8–10 t/ha for European wheat (from about 2 t/ha during the 1930s), and 6 t/ha for East Asian rice (from around 2 t/ha). High-yielding US corn now receives, on average, about 160 kg N/ha, European winter wheat more than 200 kg N/ha, and China's rice gets 260 kg N/ha, which means that in double-cropping regions annual applications are about 500 kg N/ha. According to my calculations, in the year 2000 about 40% of nitrogen present in the world's food proteins came from fertilizers that originated from the Haber–Bosch synthesis of ammonia (Smil, 2001).

Efficient use of fertilizers depends not only on appropriate rates and modes of application but also on matching the crop needs for the three macronutrients (N, P, and K) and, if need be, on an adequate supply of micronutrients (above all boron, copper, iron, manganese, molybdenum, and zinc). The rising use of nitrogen had to be accompanied by a rising use of the other two essential macronutrients and, as with nitrogen, this rise was temporarily interrupted after 1989, mainly due to the demise of the Soviet bloc economies (where the overuse of nutrients was common) and owing to the higher efficiencies of fertilizer use in Western agriculture.

In mass terms, phosphorus is the second most important fertilizing material, used overwhelmingly in the form of beneficiated phosphate rock that is extracted mostly from surface mines. Mining of phosphates predates the synthesis of ammonia: global production reached 1 Mt in 1885, 10 Mt in 1928, and 100 Mt in 1974. In the year 2000 the output was 133 Mt and 10 years later it reached 181 Mt, with China (86 Mt), Morocco and the Western Sahara (26 Mt), and the USA (also about 26 Mt, mined mostly in Florida and North Carolina) dominating the world output (USGS, 2012). Nearly all (> 95%) of US phosphate rock is used to produce phosphoric and superphosphoric acid, the two intermediate feedstocks for the production of solid and liquid phosphate fertilizer and also feed additives. Treatment with H_2SO_4 produces single superphosphate (containing 8–9% P), phosphoric acid (H_3PO_4) is used to make triple superphosphate (with about 20% P), and H_3PO_4 and NH_3 are used to make ammonium phosphates (up to 23% P).

Agricultural phosphate consumption is usually expressed in terms of P_2O_5, but I prefer to make comparisons in terms of actual nutrient (multiply P_2O_5 by 0.436 to get P): worldwide agriculture consumed increasingly higher amounts of phosphates until 1989, when the annual rate peaked at 16.5 Mt, and after growth resumed (it was only 14.3 Mt in 2000) the total reached 20.3 Mt P in 2010. Potassium is applied to fields mostly as ground potash (KCl), obtained mostly by underground mining of sylvinite, a mixture of about a third KCl and two-thirds NaCl; Saskatchewan has the largest reserves of the rock and is the leading global producer. Worldwide extraction (expressed in terms of K_2O equivalent) rose to nearly 34 Mt by 2010, with Canada (nearly 10 Mt) and Russia (more than 6 Mt) being the largest producers and worldwide exporters. About 85% of all KC ends up as fertilizer, the rest is used by the chemical industry: crop applications have risen only modestly over the last four decades, from less than 18 Mt K in 1975 to about 22.5 Mt K in 2010 (FAO, 2013). The global nutrient (N:P:K) ratio in 2010 was thus 1:0.19:0.21 compared to 1:0.25:0.40 in 1975.

Increased crop yields were also made possible by the introduction of synthetic pesticides (mainly insecticides) and herbicides during the 1940s. Dichloro-diphenyl-trichloroethane (DDT) was the first important insecticide (Müller, 1948), but its excessive application led to resistance against the chemical and its bioaccumulation in body lipids caused a great deal of environmental damage. After some 6000 t of it were applied in the USA from the mid 1940s, its use was banned in 1972 (USEPA, 1972) but many new, more acceptable, insecticides are now in use (Hodgson, 1990; MacBean, 2013). Herbicide applications began in 1945 with 2,4-D and, while many new compounds followed, none became as important glyphosate synthesized in 1971 by John E. Franz at Monsanto Company (Franz, 1974).

This broad-spectrum herbicide (better known under its brand name Roundup^{TM}) inhibits the production of a key growth enzyme and destroys the roots of weed plants but it does not travel through soil to affect other plants and to contaminate water (Wesseler, 2005). Rapid adoption of transgenic varieties of the glyphosate-resistant crops introduced during the 1990s changed corn, soybean, and rapeseed farming in the USA and Canada, and later also in Brazil, Argentina, and China, but the diffusion of transgenic crops has met, so far, with persistent resistance in Europe as well as with obstacles in India (Ruse and Castle, 2002; Kole and Michler, 2010).

3.7   Materials in Electronics

If the classical designation of metal-specific eras, first proposed by Christian Thomsen in 1836, were to be extended to all materials, a strong case could be made to label the post-1954 world (after Texas Instruments released the first silicon transistor) – or perhaps the post-1971 world, after Intel released the first universal microprocessor – the Silicon Age. This was an unexpected development and, fortunately, one based on a superabundant material, the most important mineral element in the Earth's crust ubiquitously available in silica (SiO_2). Elaborate exploitation of this material is nothing new: silicon (as quartz sand) dominates the raw materials used to make glass (about 70% of the input mass), with the rest evenly divided between Na_2CO_3 (soda ash) or K_2CO_3 and $CaCO_3$ or (in heavy leaded glass) with Pb. In 1900, nobody could foresee that SiO_2, the commonest of all surface minerals, would acquire new importance as the quintessential material of a new electronic era. The element had been used for decades in metallurgy, above all in ferrosilicon alloys (containing 15–90% Si) for deoxidation in steelmaking. This market accounts for about 80% of the element's annual global production that surpassed 1 Mt during the late 1950s and 2 Mt by 1975; in the year 2000 the total reached 3.5 Mt and in 2011 it was 8 Mt (USGS, 2013). China has been by far the largest producer of both ferrosilicon (5.4 Mt in 2011) and silicon metal. The two main uses of silicon metal (polycrystalline silicon) have been in aluminum alloying and chemical syntheses, above all for the production of silicones, Si-O polymers with attached methyl or phenyl groups. Their commercial production began during the 1940s and they make excellent lubricants and resins as well as insulators and water repellents.

The raw material for producing silicon is abundant, but an energy-intensive high-temperature deoxidization with carbon – $SiO_2 + 2C \rightarrow Si + 2CO$ (using graphite electrodes in electric furnaces) – is required to yield element that is 99% pure. But even 99% purity is quite unacceptable for solar and electronic industries, and hence the metallurgical-grade Si has to undergo elaborate and costly processing that makes it many orders of magnitude purer in order to meet the specifications for producing semiconductors, solar cells, and optical fibers (Föll, 2000). Solar-grade polysilicon used to make photovoltaic (PV) cells has purity of 99.9999–99.999999% (6–8 N), while electronic-grade silicon used to make wafers (microchips) has purity levels of 9–11 N, that is up to 99.999999999% pure.

The first step to convert metallurgical-grade to electronic-grade Si is the catalytic production of trichlorosilane (Si + 3HCl → SiHCl$_3$ + H$_2$), developed by Siemens during the 1960s (McWhan, 2012). This is followed by trichlorosilane decomposition with hydrogen (SiHCl$_3$ + H$_2$ → H$_2$ + Si) and the concurrent addition of tiny amounts of impurities (dopants) whose presence will increase the element's conductivity. This is a demanding step given the high toxicity of common dopants (AsH$_3$ and PH$_3$), high combustibility of H$_2$ and Si HCl$_3$, and high corrosiveness of HCl vapors. After this exacting process the extremely pure polycrystalline silicon must be transformed into a crystal that can be cut into thin wafers (Duffar, 2010). A serendipitous discovery opened the way to this transformation when, in 1916, Jan Czochralski, a Polish metallurgist working in Berlin, dipped his pen into a small crucible of molten tin and lifted a thin thread of metal that turned out to be a single crystal.

When Czochralski used a small seed crystal to pull a molten metal from the melt, and combined the lifting with a slow rotation and gradual lowering of the temperature, he could produce crystals of tin, lead, and zinc with diameters a few millimeters across and lengths up to 1.5 cm (Czochralski, 1918). For three decades this remained just a useful way to study the growth of metallic crystals, but the invention of transistors made it the obvious candidate for pulling crystals of semiconducting materials. Germanium, rather than silicon, was the first semiconducting material and, in 1948, Gordon K. Teal and J.B. Little at Bell Labs deployed Czochralski's method to produce the first pure Ge crystals. By 1950, shortly after the appearance of first Si-based transistors, Teal and Ernest Buehler could grow pure silicon crystals and by 1956, with their improved crystal-pulling and crystal-doping methods, their largest crystal was eight times as heavy and had twice the diameter (2.5 cm) of their first products (Buehler and Teal, 1956).

And the growth still continues: a diameter of 7.5 cm and crystal mass of 12 kg were reached by 1973, 10-cm, 14-kg crystals followed by 1980, and by the year 2000 the largest crystals were 30 cm across and had a mass of 200 kg (Hahn, 2001; Zulehner, 2003). The next generation of crystals will have a diameter of 45 cm and a mass of 1000 kg, but the process is still at the developmental stage: Intel showed its first fully patterned wafer of that size in January 2013, with production runs planned sometime in 2015. In contrast to dark brown, powdery, amorphous Si, the crystals are black or black-grey and lustrous, with impurities being less than one part per billion. These perfect crystals are cut into layers about 1 mm thick (and deviating less than 1 μm from an ideal flat surface) and the polished wafers are then ready to undergo the final steps in the production of microprocessors (Intel, 2011).

Williams (2003) traced the global silicon production chain for the year 1998, and Takiguchi and Morita (2011) reconstructed worldwide material flows of silicon used in electronics during the years 1997 and 2009. In 2009 purification of metallurgical-grade Si by the Siemens process produced about 23 000 t of electronic-grade Si and that mass was converted to 16 100 t of single-crystal Si that yielded about 7500 t of wafers sent for microchip fabrication. By the year 2012, the mass of electronic-grade polycrystalline Si increased to about 28 000 t. Crowding of more transistors on a single wafer, a process subject to the much-publicized Moore's law, has been the key driver of advances in computing.

In 1965, when the number of transistors on a microchip had doubled to 64 from 32 in 1964, Gordon Moore predicted that this rate of doubling would continue, and forecast a chip with 65 000 transistors in 1975 (Moore, 1965). In 1975 he relaxed the pace of his already famous law to a doubling every two years (Moore, 1975). And that tempo has prevailed ever since as chip-making progressed with Intel's pioneering designs and with advances introduced by its competitors, selling chips of ever-increasing complexity. The world's first universal microprocessor, Intel 4004 released in November 1971, had 2250 metal-oxide semiconductor transistors and its operating speed was equivalent to that of the room-size ENIAC computer built in 1945 (Intel, 2013).

In 1979 came the company's 8088 microprocessor that had 29 000 transistors, soon the fabrication of microprocessors moved from large-scale integration (up to 100 000 transistors on a microchip) to very large-scale integration (up to 10 million transistors) and then to ultra large-scale integration (up to a billion transistors) by 1990. By 2000 Intel's Pentium 4 had 42 million transistors and by 2012 the count reached 5 billion in Xeon Phi Coprocessor (Intel, 2012). Mass deployment of these increasingly powerful microprocessors in conjunction with increasingly capacious memory devices has transformed every sector of modern economies thanks to unprecedented capacities for communication, control, storage, and retrieval of information.

Not surprisingly, the entire process of turning quartz into the substrate for microprocessors is one of the best examples of steeply adding value along a production sequence. Detailed analyses by Jackson (1996) and Williams (2003) showed that by the end of the twentieth century, pure quartz wholesaled at less than $0.02/kg, metallurgical-quality Si was available in bulk at $1.10/kg, trichlorosilane cost about $3/kg, polycrystalline Si was between $50 and 100/kg, monocrystalline Si was at least $500/kg, polished Si wafers were at least $1500, and epitaxial slices as much as $14 000/kg. By 2012 electronic-grade polycrystalline silicon sold for about $50/kg, so 28 000 t of the metal was worth about $1.4 billion. Wafer shipments for semiconductor applications rose from just over $3.5 Mm^2$ in the year 2000 to $5.9 Mm^2$ in 2010 and $5.7 Mm^2$ in 2012 (SEMI, 2013), and global accounts show the value of these shipments rising from less than $4 billion in 1977 (the first years for which worldwide data are available) to $50 billion by 1990, surpassing $200 billion in the year 2000 and reaching $292 billion in 2012 (SIA, 2013).

During the first decade of the twenty-first century, electronics ceased to be the major consumer of high-grade silicon as most of that material now ends up in PV cells. Using silicon to make PV cells is an application nearly as old as the making of transistors. Bell Labs had the first prototype in 1954 and in 1958 *Vanguard I* was the first satellite to receive PV electricity, just 0.1 W from about 100 cm^2 of cells, more than enough to power a 5 mW transmitter. By 1962 Telstar, the first commercial telecommunications satellite, had PV cells delivering 14 W, and in 1964 weather-monitoring *Nimbus* drew 470 W. Half a century later there are hundreds of PV-powered satellites used for weather and Earth monitoring, telecommunication, and spying; to give just one example, the latest Earth observation satellite, Landsat 8, launched in February 2013, has four arrays of high-efficiency PV cells delivering 3750 W (EMCORE, 2013).

Land-based PV applications began with less expensive (but still very costly) amorphous Si cells in 1976, and by 2012 the best conversion efficiencies of thin films rose to 20% in laboratory settings, while crystalline cells reached top performance around 25%, and the best commercially available models are rated at 19–22% (NREL, 2013; Solarplaza, 2013). For decades, PV cells were made with off-grade polycrystalline material that was not good enough for electronic applications, but as the heavily subsidized market for PV installation rose from less than 100 MW/year in 1995 to more than 10 GW/year in 2009, it was necessary to divert increasing amounts of purified polycrystalline metal into the solar cell industry. In 1997 the industry used only 800 t of such metal, by 2009 it required 69 100 t, three times as much as consumed by electronics, to produce about 44 500 t of solar cells, mostly by the casting of polycrystalline metal (Takiguchi and Morita, 2011).

4

How the Materials Flow

Materials move through societies and environments in ways that range from simple, fairly linear, and easy-to-trace flows through to complex and difficult-to-follow disappearances; and while some materials provide ephemeral services (are discarded after a single brief use) others are long-lived with life-spans measured not only in decades, but even in centuries and millennia. While some metals can be reused indefinitely (albeit with some mass losses) recycling of most materials often entails considerable loss of quality and functionality. There is nothing new about keeping detailed records of the material flows required for particular operations, factories, and companies: without them artisans could not fill their orders, and engineers and managers could not control production costs or strive to improve the productivity and profitability of their enterprises. But after setting aside the exceptional case of the world's earliest modernizer (in the UK, data on basic material flows were already both fairly comprehensive and fairly accurate by 1800), reconstruction of even a few basic material flows for the early decades of the nineteenth century entails a great deal of approximation and assumption for individual industries and faces even more difficulties when trying to assemble aggregates at a national level. Pre-1850 data on production, imports, and exports of individual raw materials are commonly questionable, unsystematic, or simply lacking at higher aggregation levels, be they functional for specific industries or spatial for entire regions and countries.

Between 1850 and 1870, the availability and reliability of European and American industrial statistics improved considerably, and before the century's end many rapidly modernizing countries had in place statistical services collecting a widening range of information that make it possible to reconstruct material flows of specific sectors (steel, paper, or textile industry) and input and output aggregates for the principal material requirements in at least half a dozen major European economies, the USA, Canada, and Japan. These nations also had good data on crop, wood, and fish harvests, and on extraction of fossil fuels. But nobody was interested in adding up all of these items and constructing aggregates of the material flows required for the growth of national

Making the Modern World: Materials and Dematerialization, First Edition. Vaclav Smil.
© 2014 John Wiley & Sons, Ltd. Published 2014 by John Wiley & Sons, Ltd.

economies. Statisticians, economists, engineers, and historians paid attention to sectoral material flows, particularly to those of dominant inputs.

This makes it possible to follow the evolution of fossil fuel extraction or the mining of phosphates and harvesting of wood during the twentieth century, and for some countries we can also reconstruct detailed material flows in some industries. The beginnings of a new approach to the study of material requirements in modern societies had their origins in concerns about increasing burdens of environmental pollution and the critique of economic thinking that tended to ignore such matters. Ayres *et al.* (1969, pp. 283–84), describing the reality in clear physical terms, noted that such omissions "may result in viewing the production and consumption processes in a manner that is somewhat at variance with the fundamental law of the conservation of mass," and pointed out the obvious consequences for the environment, namely that in the absence of trade and net stock accumulation "the amount of residuals inserted into the natural environment must be approximately equal to the weight of basic fuels, food, and raw materials entering the processing and production system, plus oxygen from the atmosphere."

But it took nearly two decades before this admonition was transformed into the first fairly comprehensive studies of material requirements on a national level, as it was only during the late 1990s that several research teams began to reconstruct direct material inputs (DMIs) as well as outflows, and total material requirements (TMRs) of the world's leading affluent economies. Fischer-Kowalski *et al.* (2011) provide a concise history of this material flow accounting, as well as a survey of the methodological foundations of these studies (system boundaries, compartments, stocks), major material flow indicators, and data reliability and uncertainties across various data sets. I will take a closer look at these material flow accounts, first in general analytical terms, and then I will review the results of global and national studies tracing the inputs, or inputs as well as outputs, in the world's largest economy, in European countries, and (inevitably, given the country's recent economic rise) in China.

There are other approaches to the investigation of material flows; one attempts to trace the life-cycles of individual commodities on a national, regional, or global level; another looks at the energy costs of commodities and products; and yet another traces the environmental impacts of their production, use, and abandonment (or recycling). Life-cycle assessments (or analyses, in either case the acronym is LCA) have been performed at different scales for many elements and compounds – for example, chlorine by Ayres (2008) and polyvinyl chloride (PVC) by the European Commission (EU, 2004) – and for products ranging from aluminum cans (Mitsubishi, 2005) to steel truck wheels (Alcoa, 2012). Not surprisingly, leading metals in general, and iron and steel in particular, have received particularly close attention. Because of their depth and complexity some of these analyses are far more revealing, and fundamentally far more informative, than the standard tracing of annual production, trade, and consumption flows.

Similarly, a comprehensive assessment of environmental impacts associated with the extraction, processing, use, and disposal of a specific commodity or manufactured product may be much more consequential than a simple tracing of its flows through a national or global economy. While there are many concerns specific to a particular material or group of commodities, there is now an important and truly universal consideration – the carbon burden of materials or, more accurately, greenhouse gas (GHG) emissions associated with their consumption. These emissions are dominated by releases of CO_2, with CH_4 and

N_2O as lesser contributors, and with other GHGs making a noticeable difference only in the case of a few materials.

4.1 Material Flow Accounts

Inquiries into TMRs can follow a hierarchy of scales – starting with accounts for a specific production process, progressing to those for an entire industrial sector, and culminating in national, continental, and global totals. Even a brief reflection about the challenges inherent in preparing these summaries and about the utility of such aggregate measures will reveal their dual qualities: they are useful and revealing, particularly in quantifying the extent of human acquisitiveness and its environmental consequences; but also inadequate and misleading, particularly due to their nondiscriminate preoccupation with quantity and inability to gauge many fundamental qualitative differences.

Four kinds of account appear to be most useful: detailed analyses on a national scale that can be compared within a group of economically similar countries or with states at different stages of modernization; longer-term historical perspectives that illustrate the impacts of technical innovation, efficiency improvements, managerial advances, and shifting market choices; combinations of these two approaches that make it possible to compare the differences in long-term national trajectories; and global accounts of TMRs as well as the resulting outputs into the environment that reveal the extent of human extractive and productive activities on a civilizational scale and that make us aware of the burdens that these processes put on the biosphere's finite (and still poorly understood) capacity to deal with the changes to the atmosphere, waters, and ecosystems.

I would argue that setting of the system's boundaries, rather than coping with the inherent complexity and heterogeneity of these accounts, is the greatest challenge in constructing national and global summaries of total material flows. This challenge is reflected by the different choices made by the pioneering studies of national material balances – they began during the late 1990s, and between 1997 and 2001 they included comparisons restricted to short-term accounts for a handful of affluent economies (Adriaanse *et al.*, 1997; Matthews *et al.*, 2000; Bringezu and Schütz, 2001), long-term assessments for the USA (Matos and Wagner, 1998), as well as the first deliberate annual effort by a statistical agency to provide data for DMI as well as outflows, and TMRs (sums of DMI and hidden flows, extractions that are not used in further processing) of European economies (European Commission, 2001).

There are problems with all of these indicators. Limiting the account to DMI will greatly underestimate the overall resource demand in all modern economies engaged in intensive international trade, and particularly in such major powers as the USA, Germany, or Japan that rely on imports for large shares of many materials. Correcting this by the inclusion of net imports of all raw materials is only a partial (and increasingly deficient) solution, because many metals and other minerals are not imported in the form of ores or concentrates or bulk shipments but are instead embodied in finished products. Identifying the specific material content of these products (even their limited inventory would run to many hundreds of individual machines, tools, components, and consumer items) presents a major challenge – but the adjustment should not end there, as many items imported from a particular country contain components made of materials in a number of other countries that, in turn, imported parts or raw materials from yet another country or, more likely, a set of countries.

A popular consumer item offers a perfect illustration of these, now truly global, accounting challenges. The tear-down of Apple's iPhone 5 shows that its major components are made by nearly 20 companies headquartered in the USA, Japan, South Korea, and the Netherlands, with manufacturing facilities in a dozen countries on three continents. For example, NXP Semiconductors, a Dutch company founded by Phillips and headquartered in Eindhoven that supplies the interface for display, manufactures its products in China, Germany, Malaysia, the Netherlands, the Philippines, Singapore, Thailand, and the UK and uses raw materials that originate in more than half a dozen countries.

I have already introduced the major material flow data series published by the Eurostat (2013) and by the USGS (2013). Starting in 2004, Eurostat expanded its coverage to all 27 member states as well as to Norway, Switzerland, Montenegro, Croatia, Macedonia, and Turkey, compiled according to detailed methodological guides for economy-wide material flow accounts (EW-MFA, or simply MFA). USGS releases periodic summaries of the country's material flows (Matos and Wagner, 1998; Matos, 2009; Kelly and Matos, 2013) and its annual *Minerals Yearbook* contains the latest production, trade, and consumption data for more than 80 minerals and materials for the USA and at the global level. The latest, on-line, addition to material flow databases was set up by the Sustainable Europe Research Institute in cooperation with the Wuppertal Institute for Climate, Environment and Energy: it begins in 1980 and it includes data classified in 12 categories for more than 200 countries; it also traces per capita consumption, material intensity, material productivity, reserves of nonrenewable materials, and forecasts of global resource extraction (SERI, 2013).

We also have five global MFA data sets aggregating the flows for between 173 and 203 countries and using different system boundaries: two of these studies are limited to domestic extraction (Schandl and Eisenmenger, 2006; Giljum *et al.*, 2008); two also include imports and exports for nearly 180 countries but in just a single year (Krausmann *et al.*, 2008; Steinberger *et al.*, 2010); and one presents global aggregates for the years 1900–2005 (Krausmann *et al.*, 2009). National MFA include studies by Weisz *et al.* (2006) for the EU-15 for 1970–2001; by Schandl *et al.* (2008) for Australia between 1970 and 2005 and Wood *et al.* (2009) between 1975 and 2005; by Singh *et al.* (2012) for India between 1961 and 2008; by Krausmann *et al.* (2011) for Japan between 1878 and 2005; by Schandl and Schulz (2002) for the UK between 1850 and 1997; and by Kovanda and Hak (2011) for Czechoslovakia between 1855 and 2007.

The SERI database and all of the previously cited global studies include biomass data derived from FAO statistics, mineral inputs (largely from USGS), and fossil energies (from the IEA series). I should also note that the global on-line database and most of the global and national studies of material flows have been produced by a small group of researchers from Austria and Germany and that most of them have been published in just two sources, in the *Journal of Industrial Ecology* and in *Ecological Economics*. Material flow accounts may deal with a universal physical basis of societies, but they have been, overwhelmingly, the product of just a handful of individuals, most of them associated with the Institute of Social Ecology in Vienna.

In conclusion, I question the utility of constructing these all-encompassing national or global flow accounts – those aggregates of oxygen, biomass, all fuels, all minerals, and all hidden and dissipative mobilizations where critical differences in qualities and no less important differences in environmental consequences are submerged in indiscriminating totals dominated by oxygen and hidden material flows – because I am not sure what other revealing conclusions to derive from these summations of disparate input and output categories besides the obvious confirmations of substantial differences in national aggregates and in the rates of long-term growth. Of course, the maximalist aggregates of the all-encompassing variety also have an undoubted heuristic and curiosity value and they do convey the truly massive scale of global mobilization of raw materials.

Half a dozen studies of global material extraction at the beginning of the twenty-first century, that include all harvested biomass, all fossil fuels, ores and industrial minerals, and all bulk construction materials (but exclude hidden flows, water, or oxygen), cluster fairly tightly around 50 Gt/year. This is hardly surprising given the fact that these studies derive the flows from the same sets of data: from the Food and Agriculture Organization's agricultural and forestry statistics for biomass, from USGS complemented by some national statistics for ores and industrial minerals, and from the International Energy Agency and the UN databases for fossil energy. The lowest value for the year 2000 was 48.8 Gt (Schandl and Eisenmenger, 2006), the highest was 58.7 Gt (Krausmann *et al.*, 2008); a year later Krausmann *et al.* (2009) – in their summation of flows for 1900, 1925, 1950, 1975, and 2005 – put the last total at 59.5 Gt. Other published values for the year 2000 ranged from 48.5 Gt (Steinberger *et al.*, 2010) to about 53 Gt (Behrens *et al.*, 2007).

Leaving the high outlier aside, the average of the five estimates for the year 2000 is almost exactly 50 Gt/year, with roughly 18 Gt coming from biomass, 10 Gt from fossil fuels, nearly 5 Gt from ores and other minerals, and more than 17 Gt from bulk construction materials. *Sensu stricto* global material flows in the year 2000 (restricted to direct inputs of raw materials and excluding all food, feed, energy, and hidden flows) were on the order of 25 Gt: about 22 Gt from minerals and 3 Gt from wood and other biomaterials. An estimate conforming to the USGS list should be increased by about 500 Mt in order to include all nonrenewable organics destined for nonenergy uses (the total dominated by naphtha and methane feedstocks and bitumens for paving). But given the uncertainties in estimating the mass of bulk construction minerals (above all for the extraction of sand and gravel) that account for at least two-thirds of the all material flows, the mass of 0.5 Gt is well within the minimal range of estimation error (±2 Gt), and the total of roughly 25 Gt thus remains my preferred aggregate of directly used global materials in the year 2000.

That total prorates to just over 4 t of materials per person (the global population was 6.08 billion in the year 2000), with at least 2.5 t (and perhaps as much as 3 t) accounted for by bulk construction materials and only about 0.8 attributed to all metals and nonmetallic minerals. These rates compare to nearly 1 t of food and feed crops (fresh weight), close to 0.5 t of wood (excluding fuelwood), and about 1.7 t of fossil fuels (roughly 0.8 t of coal, 0.6 t of crude oil, and 0.3 t of natural gas) extracted for every inhabitant of the world in the year 2000. Raw material inputs into the global economy, and hence the eventual worldwide output of finished products, have soared in absolute terms during the first decade of the twenty-first century. Growth of raw material inputs has been more subdued in affluent societies, partly because they have already put in place material-intensive infrastructures and partly because of the ongoing outsourcing of material-intensive (and often polluting) industries to foreign low-cost producers – but it has been still substantial in aggregate terms.

In the EU, the aggregate grew by 8% in just seven years between 2000 and 2007 (Eurostat, 2013). In the USA total material inputs (even after leaving all construction materials aside) grew by nearly 34% during the two decades between 1986 and 2006, and in per capita terms this growth translated to less than 0.5%/year. The country thus requires higher per capita flows of raw phytomass, metals, minerals, and fossil organics as it produces a greater mass of manufactured goods than at any time in its history: dematerialization can be demonstrated only when using a growing GDP as the denominator. This kind of dematerialization has been obviously beneficial – but it cannot prevent a further increase of total material inputs in any rapidly growing economy.

All of this is interesting, and if we had a reliable historical series we could also trace the rise of global material consumption for each of these major categories: all pre-1950 global totals are nothing but questionable estimates, and even the recent aggregates depend critically on what is included. For example, Krausmann *et al.* (2009) put the worldwide biomass extraction (crops, their residues, roughages, and wood) at 19.061 Gt in 2005, while in my detailed account of phytomass harvest (Smil, 2013) I showed that in the year 2000 the total for woody phytomass alone could be anywhere between 2 and 13.4 Gt depending on the boundaries chosen for the analysis.

Consequently, there can be no single accurate total, as the search for global totals will be always determined by assumptions, and even if everybody agrees on common boundaries the basic results will be largely predictable. Physical realities dictate that the mass of sand and gravel used to emplace and maintain modern concrete-based infrastructures must be substantially greater than the mass of metallic ores; and that the mass of iron, a metal of outstanding properties produced from abundant ores with a moderate energy intensity, must be orders of magnitude higher than the mass of titanium, an even more remarkable metal but one derived from relatively rare ores with a great energy expense. At the same time, it must be kept in mind that data for inexpensive, readily available bulk construction materials (particularly for sand and gravel) that are usually sold not far from their points of extraction are generally much less reliable than the statistics for metal ores and industrial nonmetallic minerals that are globally traded.

Economic realities dictate that the diffusion of successive changes of human organization – urbanization, industrialization, post-industrial societies characterized by mass consumption – promote, widen, and often also accelerate the use of materials. Although the resulting outcomes must be expected, some of the contrasts still amaze: for example, the world now consumes in one year nearly as much steel as it did during the first post-World War II decade, and (even more incredibly) more cement than it consumed during the first half of the twentieth century. At the same time, these admittedly fascinating but highly aggregate global-scale data series – that amalgamate quantities and ignore qualities – offer little guidance for future decision-making (beyond the obvious point that the recent high growth rates cannot continue for many decades).

Useful insights can be gained from two kinds of finer focus: through closer examination of material flows on the national level, and by putting more restrictive analytical boundaries on the set of examined materials and tracing the flows of individual commodities with some clear goals in mind. This can be done by detailing their uses, dispersal, and persistence in a society, and by attempting life-cycle analyses of those materials that circulate on a human timescale, that is by quantifying their direct and indirect requirements for energy or by identifying and assessing the environmental impacts of their production and use.

4.2   America's Material Flows

The world's largest economy is obviously the leading candidate for a closer examination of national material flows and, fortunately, the task of assembling the primary data is made easy thanks to the ready availability and high accuracy of the country's production, trade, and consumption statistics. But even this simple additive chore is unnecessary thanks to the USGS researchers who have collated the primary information from more than 100 individual material streams into clearly defined categories and, moreover, have done so in a fully comparable manner, originally for the period from 1900 to 1995 (Matos and Wagner, 1998), with subsequent extension to 2006 (Matos, 2009), and with the latest disaggregated statistics available in the regularly updated historical series (Kelly and Matos, 2013).

As already explained, the USGS series contains data for metals and minerals (subdivided into primary and recycled metals, and industrial and construction materials) and nonrenewable organics (asphalt, waxes, oils and lubricants, and fossil fuels used as feedstocks) that were used to make products in the United States: it leaves out food and the bulk of the fossil fuels that are used as a source of energy (as large shares of these materials are consumed directly) but it includes all phytomass that is used in manufacturing (ranging from lumber to new and recycled paper, and from cotton to wool). A centennial perspective traces material demands from the peak of the first phase of the Industrial Revolution (energized by coal and steam engines and characterized by growing but still limited mass consumption) through the rise and maturation of the second phase (increasingly energized by refined oil products and natural gas, with steam turbogenerators, internal combustion engines, and gas turbines as the dominant prime movers and resulting in the unprecedented extent of mass consumption), to the post-1950 rise of service economy.

This perspective makes it possible to make several important generalizations that are undoubtedly valid for other major affluent economies: a rise of absolute consumption in every material category; a substantial decline in the relative importance of biomaterials; the dominance of the total mass by bulk construction materials; an expanded demand for metals; the rising importance of recycling; the emergence and a huge increase in consumption of nonrenewable organics; and material-specific shifts (fluctuating stagnation, steady rise, or long-term decline) in per capita consumption. During the twentieth century, natural growth potentiated by immigration increased the US population nearly 4-fold (3.7), and the country's GDP (expressed in constant monies) was 26.5 times higher in 2000 than in 1900: not surprisingly, the combination of these two key factors drove absolute consumption increases in all material categories, with the multipliers ranging from 1.7 for materials produced by agriculture to more than 90 for nonrenewable organics (and 8 for primary metals, 34 for industrial minerals, and 47 for construction materials).

The importance of renewable materials (wood, fibers, leather) fell from about 46% of the total mass (when bulk construction materials are included) or 74% (with stone, sand,

and gravel excluded) in 1900 to just 5% (or 22%) for analogical rates in the year 2000, a trend that was expected given the increasing reliance on light metals and plastics. Aggregate wood demand rose less than 1.4-fold during the twentieth century, but consumption of primary paper and paperboard multiplied about 19 times and was supplemented by rising quantities of recycled paper: when the data collection in the latter category began in 1960, recycled paper accounted for about 24% of all paper and paperboard use, but by the year 2000 its share was up to 46% even as large quantities of waste paper are exported (in 2000 this amounted to about 22% of all domestic collections), primarily to China (FAO, 2013).

The fact that bulk minerals used in construction (crushed stone, sand, and gravel) have increasingly dominated America's annual flows during the twentieth century – in 1900 they accounted for 38% all materials, by 2006 their share reached 77% – is not surprising given the enormous expansion of material-intensive transportation infrastructures after World War II. Construction of the Interstate system began in 1956 and required the building of many new bridges (USDOT, 2012), while the introduction of commercial jetliners led to the rapid expansion of airports, a process recently repeated in China. Large demands for bulk construction materials also came from the building of new container ports, stream regulation (above all in the Mississippi basin), electricity generation (hydroelectric dams, nuclear power plants), new factories, commercial real estate (warehouses, shopping centers), and housing.

The mass of construction materials used in the USA rose about 7-fold between 1900 and 1940 and then doubled between 1945 and 1951, doubled again by 1959 to 1.1 Gt, but the next doubling, to 2.26 Gt, took until 1997. Use of these materials shows notable decline during economic recessions: during the Great Depression the peak-to-bottom decline between 1929 and 1933 was 50%, between 1979 and 1982 the total demand fell by a third, and during the latest recession US consumption of sand and gravel fell by 44% between 2007 and 2010 while demand for crushed stone declined by 29% during the same period (Matos, 2009).

End-use data indicate that the largest identifiable category of sand and gravel consumption (about a fifth of the total surpassing 1 Gt/year) is as aggregate added to cement in the production of concrete, followed by road base and coverings, fill, and as aggregate added to asphaltic and bituminous mixtures; but unspecified uses make up the largest category, accounting for about a quarter of the total. Differently-sized aggregates used in the production of concrete are also the leading final uses for crushed stone, and railroad ballast is another indispensable application. With a ballast minimum depth of 15 cm and up to 50 cm for high-speed lines, and an overall width of roughly 4.5 m, this amounts commonly to more than $1000\,m^3/km$ or (with density of $2.6\,t/m^3$) to around $3000\,t$ of crushed stone per kilometer. Dimension stone (rough or dressed) used in construction (for facing, curbing, flagging, and monuments) is a smaller category and some four-fifths of the quarried mass actually never leaves the site (constituting a rare instance of a hidden flow quantified by the USGS data).

In comparison to construction sand, the total use of industrial sand is minuscule but qualitatively very important. Annual consumption has recently fluctuated around 25 Mt/year: about 40% of this total is pure silica used in glassmaking, and a fifth goes to foundries to make moldings and refractories as well as silicon carbide for flux and metal smelting. Smaller but functionally irreplaceable uses include abrasives used in blasting and sanding, sands for water filtration, and sands for creating artificial beaches and sporting areas. A new, and rapidly rising, market is for the special kinds of sands used in hydraulic fracturing of gas- and oil-bearing shales, well-packing, and cementing. Other materials aggregated by the USGS into the heterogeneous group of industrial minerals include elements (carbon in the form of graphite and diamonds, boron, bromine, hafnium, helium, lithium, nitrogen, sulfur, strontium, zirconium) as well as both abundant (phosphate, potash, salt) and relatively rare (gemstones, industrial garnet, quartz crystal) compounds.

In total mass terms this is the second largest material group, with recent domestic use amounting to roughly 2.5 times the mass of all (primary and recycled) metals. Its total rose only about 10 Mt in 1900, to 150 Mt in 1950, it surpassed the 300 Mt mark in 1972 and then, after two decades of stagnation, it rose to 370 Mt in 2000, and 390 Mt in 2010. Its largest constituents include salt (about 55 Mt in 2010), phosphate rock (about 30 Mt), nitrogen (about 14 Mt), and sulfur (about 11 Mt). America's salt consumption is remarkably high (20% of the world total in 2010); the two dominant uses (each about 18 Mt in 2010) are production of alkaline compounds and chlorine, and for road deicing; amounts an order of magnitude smaller (both about 1.8 Mt) are used in food production and in animal feed, and more than 1 Mt/year is used in water treatment (in water-softening to remove mineral ions).

The rising consumption of metals and the changing shares of individual elements and alloys in this broad material category have been other, expected, developments. Consumption of 10.3 Mt of primary metals in 1900 doubled in just six years and then doubled again to 42 Mt in 1929, before it was cut drastically by the Great Depression to just 11.3 Mt in 1932. The pre-depression record was surpassed in 1940 and the annual rate remained around 59 Mt until 1944; after years of decline and fluctuation it moved above 60 Mt only by 1963. By that time, recycled metals provided almost a third of the total supply of roughly 87 Mt, and by the year 2000, with total metal consumption at nearly 144 Mt, they supplied 44%.

But while metal recycling has shown a generally upward trend (albeit with some notable recession-induced downturns), domestic consumption of primary metals reached a fluctuating plateau in the late 1960s and the early 1970s, followed by a general decline for the next 30 years that is ascribable to the first wave of deindustrialization of America, that is to the country's loss of global dominance in iron and steelmaking and the declining output of color metals. Steel production peaked in 1979 at 91.9 Mt, a decade later it was 52% lower, and by the century's end it was, at 47.9 Mt, barely above that level. The peak years of domestic output were 1969 for zinc, 1972 for copper and nickel, and 1973 for lead (only copper production was eventually surpassed, for eight years starting in 1992, before it fell back again). As a result, by 1979 domestic metal consumption was nearly 60% lower and the 1968 peak of 78.2 Mt was not surpassed until 1998.

As for the individual metal shares within this diverse category, steel retained its overwhelming dominance among metals during the first half of the twentieth century (92% of the total in both 1900 and 1950) and its share declined to about 83% by the year 2000. The second most important metal in mass terms is aluminum, whose domestic consumption rose from less than 900 000 t in 1950 to more than 7 Mt by 1999: this was owing to aircraft production and the widespread adoption of aluminum and its alloys in the automotive industry and in many consumer products (from cookware to computers). Afterwards the domestic use fell back to about 3.5 Mt by 2010, but that total hides a complex combination of production methods and trade. In 2010, output of the primary metal was just 1.7 Mt but secondary production was 2.8 Mt (1.25 Mt from old and 1.55 Mt from new scrap), imports were 3.6 Mt but exports were not far behind (3 Mt), and metal stocks reached 3.2 Mt resulting in an apparent domestic consumption of 3.46 Mt. USGS (2005) also estimated cumulative mass of aluminum stocks in structures and products: in 2002 it reached 142 Mt, compared to 117 Mt for copper and 4.13 Gt for steel.

Copper has seen similarly large shifts, with 2010 domestic output nearly halved compared to the late 1990s, with apparent consumption of just 1.74 Mt compared to 3.1 Mt in 1999. The only other metals whose recent annual consumption has surpassed 1 Mt are zinc (1.3 Mt in 2000) and lead (1.4 Mt in 2000). Manganese, whose dominant use is to alloy steel, was in the same group when America's steelmaking dominated global output (it peaked at 1.35 Mt in 1974). As the American production of specialty steel declined, so did the consumption of manganese: by 2010 it fell to less than 800 000 Mt, all of it imported (its small domestic production stopped in 1990). In contrast, molybdenum consumption, whose annual rate had fluctuated mostly between 15 000 and 25 000 t for decades prior to the year 2000, has risen to record levels since 2005 (46 400 t in 2010), the only metal to have done so. The metal is another important ingredient in steel alloys (used for hardening and heat resistance; catalysts are another important Mo market) and the USA is its second largest producer (after China) and a significant net exporter of the metal.

Before moving on to the last material category monitored by the USGS, I must reiterate that actual domestic US consumption of virtually all metals is, often significantly, higher than shown by the USGS balances because substantial amounts of various metals reach the country not only as imported raw materials (all of those are included in the totals of domestic consumption) but also embedded in products, and that presence is excluded from nationwide aggregates of apparent domestic consumption. Major components of these unaccounted flows include not only such leading metals as steel and aluminum in cars, airplanes, machinery, and appliances, but also such toxic heavy metals as lead in automotive lead-acid ($PbSO_4$-H_2SO_4) batteries and cadmium in rechargeable Ni-Cd batteries.

In 1900, the total consumption of nonrenewable organics (mostly paving materials and lubricating oils) was less than 2 Mt, but subsequent extension of paved highways, mass ownership of cars, the rise of the trucking industry, and, above all, rapid expansion of crude oil- and natural gas-based synthetic materials made this the fastest growing material category in the USA: by 1950 the flow surpassed 30 Mt and by 1999 it had reached 150 Mt, with nearly two thirds being hydrocarbon feedstocks (naphtha and natural gas) used to make ammonia, the starting compound for all synthetic nitrogen fertilizers. The second largest input by mass is asphalt and road oil; consumption of these paving and surfacing materials rose from less than 10 000 t in 1900 (when few paved roads existed outside cities) to 100 times that mass in less than two decades, it reached more than 10 Mt by 1950 and, until the 2008 recession, it was on the order of 30 Mt/year. The other two major classes of organics are lubricants and waxes.

I will address the levels of America's material intensities – that is material flows per unit of economic product – and their long-term trends in the next chapter dealing with apparent dematerialization of modern economies, but before leaving this section I will sketch some notable per capita consumption levels. In aggregate terms, the USGS accounts translate to a domestic consumption of about 1.9 t/capita in 1900, 5.6 t in 1950, and 12 t/capita in the year 2000; after leaving out bulk construction materials these rates are reduced, respectively, to 1.2, 2.3, and 3 t/capita, which means that the use of construction materials rose from about 0.7 t/capita in 1900 to 3.3 t in 1950, and 9 t in the year 2000.

Wood is the only material category showing a century-long decline of per capita consumption, from about 800 kg in 1900 to about 400 kg by 1950 and about 300 kg/capita in 2000. Materials produced by agriculture rose slightly from 40 to 47 kg/capita during the first half of the twentieth century, but afterwards they declined to just 18 kg (but cotton and wool in imported finished apparel, now the dominant source of clothes worn in the USA, is not included in these totals). Paper consumption rose from 35 to about 175 kg/capita between 1900 and 1950 and afterwards leveled off (it stood at 179 kg/capita in 2000). Consumption of all metals has shown a similar pattern, rising from 135 kg in 1900 to 515 kg/capita in 1950, but by the year 2000 were essentially the same at 510 kg/capita (once more a somewhat misleading rate given the country's large post-1970 net imports of cars, airplanes, and machinery).

As far as the three leading metals are concerned, per capita consumption of steel kept on rising, with some fluctuation, from 125 kg in 1900 until 1974 when it just surpassed 500 kg, but the subsequent decline brought it to 425 kg in the year 2000 and to only 260 kg/capita in 2010. Copper reached a domestic consumption peak in 1972 at 10.5 kg/capita (compared to 2/2 kg in 1900), after a temporary decline it rose to 11 kg in 2000 but then fell to less than 6 kg by 2010. Aluminum, too, has experienced a recent drastic fall, from a peak of more than 26 kg/capita in 1999 to less than 11 kg in 2010. Are these declines just the latest examples of temporary downturns accompanying economic recessions, or are they characteristic of a new phase of material consumption in affluent, material-rich economies?

4.3 European Balances

Although several European countries have excellent historical statistics, none of their national institutions has produced a study of long-term material flows comparable to the USGS series. Instead, we have four distinct kinds of material accounts: a few long-term reconstructions done by groups of researchers for a few economies, for the UK (Schandl and Schulz, 2002) and the Czech Republic (Kovanda and Hak, 2011); a set of EW-MFA for the EU-15 between 1990 and 2004 (Weisz *et al.*, 2006); Eurostat's series for more than 30 countries that (as already noted) begins only in the year 2000; and a new study of raw material consumption (RMC) expressed in terms of raw material equivalent (RME) for the EU-27 for a single year, 2005 (Schoer *et al.*, 2012).

As different as these studies are, they share most of the underlying statistical sources as well as a maximalist approach to material accounting: they include all food and all fuels besides the materials used in production and services. As in all modern economies, bulk construction materials are the single largest category of Europe's material flows: at 4.6 t/capita for EU-27 they are an order of magnitude greater than the consumption of metal ores and about 70% higher than the flow of all nonmetallic minerals. EU countries are highly dependent on imports, particularly for metal ores, concentrates, and semi-finished products: in 2009 the union imported 58% of the final consumption of these materials (compared to just 4% for nonmetallic minerals), with national dependencies ranging from just 4% for Sweden to 59% for the UK.

In comparison with the USA, the EU-27 has similar metal consumption (0.4 vs. 0.5 t/capita) but a much lower demand for construction minerals (4.6 vs. nearly 10 t/capita), a difference that is due mostly to the continent's much higher population density and more compact transportation infrastructure. Some European accounts have also compared the nations based on the spatial intensity of their material use (domestic material consumption, DMC per square kilometer) but this is a dubious measure because the extraction of both construction materials (dominant category by aggregate mass) and metallic ores are highly localized affairs, especially compared to much more extensive crop cultivation or forestry harvests whose density is a valid indicator of the prevailing intensity of harvests. As expected, large countries with relatively small populations will have (even if they have considerable mining sectors) a low rate of DMC per unit of area, and the opposite will be true for small densely-populated nations. In Europe, this difference spans an order of magnitude: in 2000 the Swedish rate was about 200 t/km^2 compared to about 2600 t/km^2 for Belgium and Luxembourg.

The EU's material flows include all food, and this inclusivity is carried to excessive lengths by counting wild berries and mushrooms and all hunted animals. Indeed, the EU's compilation guide (Schoer *et al.*, 2012) contains a list of scores of mammals and birds with recommended live weights to be used in aggregating their consumed biomass (but I have to wonder how many marmots, brown bears, or pink-footed geese are actually killed for food every year to make any difference to the mass of available food, especially given the well-known impossibility of accounting for millions of tons of staple grain harvests with an uncertainty lower than ±3%). Even more importantly, in contrast to these irrelevant game hunting details, the guidelines contain single conversion factors for translating gross ores into metal content or ore concentrate, assuming (quite unrealistically) that all iron ores have 43.32% of Fe and all copper ores have 1.09% of Cu.

In any case, given the brevity of the Eurostat series, and the fact that all post-2007 material flows have been affected by the worst post-World War II economic recession, its value lies, so far, in allowing us to compare national levels of material use based on uniform accounting methods, and in revealing some notable national peculiarities (Eurostat, 2013). These include the expected high per capita consumption of forest phytomass in Finland (5 t/year) and Sweden (4.2 t/year), more than twice the EU-27 mean of 1.9 t/year, and high consumption of metal ores in Sweden (2.9 t/capita compared to the EU mean of 0.4 t) and nonmetallic minerals in Ireland (due to still considerable extraction of peat).

There are also some unexpected outliers, such as an exceptionally high consumption of sand and gravel in Cyprus (nearly 19 t/year) and Finland (almost 16 t/year compared to the EU-27 mean of 4.6 t/capita) and high consumption of metal ores in Bulgaria (3.7 t/capita). Per capita consumption of construction materials tends to be higher in less populous countries, mostly because of higher demands for building more extensive transportation infrastructure. In Europe this difference is obvious when comparing Finland (sand and gravel consumption of nearly 6 t/capita in 2009) with the Netherlands (1.8 t/capita), or Sweden (9 t/capita) with the UK (1.9 t/capita).

I feel that the latest attempt at disentangling material flows of a continent heavily dependent on imports (Schoer *et al.*, 2012) tries to do too much, and that it can do so only by resorting to many debatable substitutions and assumptions. It actually does not result in any stunningly new insights except for presenting the totals of material equivalents adjusted for gross capital formation, although those calculations may be the least reliable part of the study. The easiest way to outline what the authors attempted in order to measure economy-wide RMC of the EU in terms of RMEs used in the entire production chain of consumed goods, is to quote their explanation:

> We first developed a highly disaggregated and environmentally extended mixed unit input output table and then applied life cycle inventory data for imported products without appropriate representation of production within the domestic economy. Lastly, we treated capital formation as intermediate consumption.

Aggregate RME is essentially the material footprint of the EU and it is, undoubtedly, a more comprehensive and hence a more realistic approach than Eurostat's values of DMC. DMC accounts for internal raw material extraction, but it measures all imports merely in terms of their product weight while RMC avoids this asymmetry by quantifying imports and exports in terms of RME. But readers familiar with the advantages and weaknesses of manipulating even fairly disaggregated input–output matrices, with aggregations and simplifications inevitable in the construction life-cycle and with many pitfalls in converting monetary values into mass flows – moreover, in this case based on unpublished input–output tables for Germany whose economic structure was deemed an excellent model of the entire EU and inventories – would see right away that this concatenation of arguable approaches is far from being the most reliable way of quantifying the impacts of material consumption.

A no smaller challenge was presented by the decision to quantify the RME of consumed capital goods (the paper and its supplementary information describes how these matters were resolved). And while metal recycling has much lower material demands and environmental consequences than does primary metal production, I find the assumption to make the RME of all secondary metals equal to zero quite unrealistic. I have no idea to what extent this overly ambitious effort has been compromised by the many assumptions and simplifications needed in order to arrive at, to cite just one key example, world averages applicable to the import structure of ores and metals, whose extraction has so many unique, site-specific attributes. And when checking the expanded hybrid input–output table I fail to imagine how 217 000 t of straw, 452 000 t of fodder crops for grazing animals, and 7000 t of fish catch are associated with the production of radios and TVs.

In conclusion we are told what has been always known, that the EU is nearly self-sufficient in biomass, sand, gravel, and many other nonmetallic minerals, and extremely dependent on direct and indirect imports of fossil fuels and metal ores; and the findings of the highest RME associated with the imports of crude oil and the largest export-generated RME due to machinery and motor vehicles (each one nearly 140 Mt) are as expected. But an interesting result is that, after excluding fuels, the highest RME associated with imports is for gold and gold products, with the total (about 300 Mt) far higher than the aggregate attributable to iron ores (less than 200 Mt), iron, steel and ferro-alloys (about 130 Mt), and copper products (roughly 100 Mt in 2005).

The most interesting results come from comparing grand aggregates without and with internalized capital formation: the study's key conclusion is that a substantial part of material extraction attributable to consumption is missed without including the consumed capital in the analysis (as is the norm with input–output studies). Raw materials embedded in the EU's final consumption in 2005 added up to about 4.7 Gt, with nearly 60% embodied in goods, when the capital formation is excluded. With the internalized capital formation (gross fixed capital formation is treated as intermediate consumption) the grand total of RME is 8.26 Gt, and when classified by final consumption expenditures this total is dominated by real estate services (1.58 Gt or nearly 20%; without internalized capital formation it is just 246 Mt), an outcome that is directly related to construction activities and to trade in construction products.

The two next largest categories are trade and public administration and defense (each about 700 Mt), followed by three roughly equal final consumption sectors, hotels and restaurants, health and social work, and meat and meat products (all roughly at 400 Mt). When division is made into large sectoral segments, then services dominate (4.95 Gt or 60% of the total), followed by manufactured products (2.4 Gt or 29% of the total), followed by roughly equal shares (about 360 Mt or a bit over 4% of the total) for utilities (electrical energy, gas, steam, and water) and agricultural and forestry products, with mining and quarrying placed last. While interesting, the result is hardly surprising as it corresponds fairly closely to the continent's shares of gross economic product.

European data also make it possible to contrast national differences in resource productivities (expressed in purchasing power standard, PPS, per kilogram of all materials used): the latest available data (for 2009) show EU-27 means of 1.6 (17% gain since the year 2000) compared to 3.28 in the Netherlands, 2.53 in the UK, 2.06 in France, 1.80 in Germany, 1.72 in Spain, and 0.87 in Poland (Moll *et al.*, 2012). But this comparison also reveals the limits of the aggregate approach where all materials are included without any regard to their fundamental qualitative differences. Does Germany really have a resource productivity barely higher than Spain and lower than France or the UK? Well, it does not once the account is properly adjusted. After leaving out fossil fuels and sand and gravel the average material intensity for the EU-27 was roughly 280 g/PPS compared to 165 g in Germany, 285 g in France, and 190 g in the UK, with Spain at 385 g and the Netherlands being the least material-intensive major economy with about 130 g.

4.4 Materials in China's Modernization

China's modernization began with Deng Xiaoping's reversal of the longstanding autarkic Maoist policies that were largely responsible for the greatest famine in history (1959–61) and for three decades of mass poverty: reform gathered pace between 1980 and 1985, then it appeared that its advance would be affected by the mass protest and Tian'anmen killings in 1989. But while those events set back the cause of human rights, economic growth accelerated during the 1990s and was kept at a high level during the first decade of the twenty-first century (actual rates were lower than the officially exaggerated statistics, but still as high or higher than the fastest pace achieved during the peak-growth spells of Japan or South Korea). Looking back, it is now clear that China's three post-1980 decades added up not only to the fastest, but also to the largest, economic advance in history.

According to official statistics, between 1980 and 2010 China's annual rate of economic growth was only below 5% three times (1981, 1989, and 1990) and was above 10% 16 times, while the average for the three decades was 9.6% (IMF, 2013). This implies a doubling every 7.3 years resulting in a 2010 GDP (in constant prices) 17.8 times higher than in 1980. In per capita terms, the multiple was still roughly 13-fold (NBSC, 2013; IMF, 2013). Using official exchange rates these sums convert, respectively, to about $200 billion and $5.9 trillion (in current terms) and purchasing power parity (PPP) conversion boosts them to about $250 billion and $10.1 trillion. In 1980, China's economy was (in PPP terms) only about half the size of Italy's, by 2010 it was the world's second largest, more than twice as large as that of Japan (IMF, 2013).

This unprecedented achievement is easier to comprehend when seen primarily as a combination of the following factors: the tremendous pent-up desire of the world's largest population, kept for decades in Maoist economic misery, to improve their quality of life; a huge population increase together with delayed urbanization that released hundreds of millions of young productive workers from the countryside to cities, powering the expansion and new manufacturing capacities that made the country the world's largest exporter; and massive direct foreign investment amounting annually to more than $50 billion since the mid-1980s and accompanied by an unprecedented transfer of modern extraction, industrial, and transportation techniques, a perfect case of advantages enjoyed by a late-starter to modernization.

Given the relatively low starting point of post-Maoist expansion (in 1980 China's per capita GDP, at just $250, averaged less than that of Pakistan) it must be expected that all material flows would multiply in order to enable four successive doublings of the country's economy in three decades, and that China's domestic endowment would not be able to satisfy most of these expansions and would have to resort to massive imports. Not surprisingly, consumption of some materials has grown at an even faster rate than has the expansion of GDP, and the emergence of China as the world's leading manufacturer made the country the world's largest importer of commodities, ranging from iron ore to potash fertilizer.

China's huge post-1980 construction boom – by any measure the world's largest and fastest expansion of housing, commercial buildings, industries, and transportation infrastructure – has engendered unprecedented demand for all construction materials. A rough estimate is that the domestic extraction of construction minerals increased 25-fold between 1980 and 2010, and that it nearly tripled during the first decade of the twenty-first century (SERI, 2013). This overall output multiple was almost exactly matched by the output of two major manufactured construction materials, cement needed to make concrete, and plate glass.

The pace of China's frenzied concretization and its overall scale has been stunning. In 1980 the country produced just short of 80 Mt of cement, a decade later it had more than doubled the total to about 210 Mt, by the year 2000 it rose to 595 Mt and by 2010 that total had tripled and reached 1.88 Gt (nearly 24 times the 1980 total and 57% of the global production for less than 20% of the world's population), and it rose further to 2 Gt in 2011 (NBSC, 2013). Perhaps no other comparison illustrates the scale of this concretization better than this one: consumption of cement in the USA totaled about 4.56 Gt during the entire twentieth century – while China emplaced more cement (4.9 Gt) in new construction in just three years between 2008 and 2010, and in the three years between 2009 and 2011 it used even more, 5.5 Gt (NBSC, 2013).

Inevitably, such a pace of construction guarantees that a substantial share of newly poured concrete will be of substandard quality, a conclusion confirmed by the obvious dilapidation of China's concrete structures built during the late 1980s and the early 1990s, the first period of China's construction boom. The quality of concrete used to construct many of China's new dams (by 2010 the total stood at more than 87 000 structures of all sizes including the world's largest dam, Sanxia) is of particular concern, even more so as thousands of them are located in areas of repeated, vigorous seismic activity. Even if the initial quality of concrete posed no problems, the US experience of extensive concrete-based infrastructure offers a sobering view of what lies ahead for China.

Poor quality and enormous overcapacity have also been problems in the plate glass industry, whose output increased 25 times between 1980 and 2010, from about 25 to 630 million weight cases, or from just 1.25 to 31.5 Mt (NBSC, 2013). China thus produced 60% of the world's flat glass demand of about 55 Mt in 2010 when its plate-making capacity rose by 6.5–44.5 Mt alone; that pushed the industry's average capacity factor to just 70% and resulted in new approval rules governing any additional plant construction (China Daily, 2011). Moreover, a major share of that excess capacity has been in energy-intensive and highly polluting enterprises; their output of roughly 20 Mt, or nearly two-thirds of the total production, was low-quality float glass (Pilkington, 2010).

Expansion of China's steelmaking has matched almost exactly the pace of GDP growth: raw steel output rose 17.2 times between 1980 and 2010, from 37.1 Mt in 1980 to 637.4 Mt in 2010, when it accounted for nearly 45% of the global output (WSA, 2013). But as the extraction of iron ores increased about 14 times (from 75 000 t to 1.07 Gt) an increasing share of this output has come from imported materials. In 2010, China imported 618 Mt of iron ore, more than a third of the total input into its blast furnaces, and it has been by far the largest iron ore importer (nearly 60% of the global total and close to 70% of the domestic demand) with Australia and Brazil being the major suppliers. And while the country has been the near-monopolistic exporter of rare earths and a major exporter of molybdenum and magnesium (also of graphite), it has been also the world's largest importer of bauxite (44 Mt in 2010) and, at nearly 1.2 Mt in 2010, of copper ores and concentrates.

The only major construction material that has seen only a modest production increase is industrial roundwood, whose output rose by less than 30% in three decades, from less than 80 to about 102 Mm^3 (FAO, 2013). China's extensive pre-Communist deforestation, pre-1980 overexploitation, mismanagement of remaining natural forests, and mass-scale reforestation campaigns producing spindly growth of a few widely planted species (pines, eucalyptus) explain the continuation of an inadequate domestic timber supply that necessitated more than quadrupling of imports between 1980 and 2010, from just over 8 to more than 35 Mm^3, with Canada being the largest exporter and with a rising share of imports coming from Africa and contributing to the destruction of the continent's tropical rainforests (FAO, 2013; Smil, 2013).

But the material category that has seen the greatest production increase has been the synthesis of plastics, with a nearly 70-fold rise between 1980 and 2010. Of course, that large multiple is due to a rapid development from a very low base (less than 900 000 t in 1980) but the absolute output of 62 Mt in 2010 was larger than the production of about 57 Mt in EU27 (Europe Plastics, 2011). Finally, the need to secure more food and better nutrition for a still-growing population has led to substantial gains in the production (and imports) of fertilizers. New Haber–Bosch plants were added to raise the output of nitrogenous fertilizers from 10.3 Mt N in 1980 to 45.2 Mt in 2010, but in 2009 the record output was 48.6 Mt N, a nearly 5-fold increase in three decades, while production of phosphate fertilizers posted a roughly 8-fold increase to 19 Mt. Disparity between N and P growth rates is explained by China's attempt to move away from excessive nitrogen uses toward more balanced fertilization with N:P:K ratios improving the efficiency of applications. As a result, China has been buying record amounts of potash from Canada.

China has also become a prominent importer of materials for recycling, and the USA has been their greatest supplier. In 2010, Chinese imports of waste paper were nearly 25 Mt/year, with the USA as the leading exporter (Magnaghi, 2011). Similarly, in 2010 China bought almost 6 Mt of scrap steel – becoming the world's third largest importer of the material after Turkey and South Korea (WSA, 2013) – with the USA again as the leading supplier. This trade is certainly one of the most remarkable indicators of changing national fortunes, as the world's largest affluent economy has become the primary supplier of waste materials to the second largest economy experiencing a rapid rate of growth. In 2011 the USA exported more than $11 billion of waste and scrap (materials belonging to the 910 category of the North American Industry Classification System) to China. This was less than the exports of transportation equipment or agricultural products – but more than the exports of all nonelectric machinery and more than five times as much as the shipments of all electrical equipment and appliances (Smil, 2013). China is also the world's largest importer of plastic and electronic waste.

Similar national reviews of major material flows could be presented for other leading economies, a task that would be particularly easy in Japan's case: the country has excellent statistics and there are also many recently published studies on Japan's flows of specific materials. Most notably, (Krausmann *et al.*, 2011) used Japan's historical statistics to follow the country's material use between 1878 and 2005; copper's stocks and flows were analyzed by Terakado *et al.* (2009) and Daigo *et al.* (2009); Daigo *et al.* (2010) traced the flow of chromium and nickel in stainless steel; and Kuriki *et al.* (2010) looked at the potential of recycling platinum group metals.

Major findings of these studies are easy to summarize. The country's total mass of consumed materials increased 40-fold in the 127 years between 1878 and 2005, with most of this expansion coming only during the 1950s and 1960s and supplied overwhelmingly by imports (a dependence made even more pronounced by the inclusion of fossil fuels). This great post-World War II surge in material requirements peaked at the time of the first round of OPEC-driven oil price increases (1973–74). After a temporary decline in fuel imports total primary energy supply (TPES) resumed its growth (albeit at a slower rate) but the consumption of metallic ores and nonmetallic minerals began to stagnate and after 1990 to decline. Japan remains an affluent society based on mass consumption of mostly imported materials, but the post-1990 combination of population aging, economic setbacks, and deindustrialization have brought reduced imports and stagnating or declining output and consumption rates.

A more profitable task than to offer such brief summaries for other leading industrialized nations (most of them sharing heavy dependence on imports of essential materials) is to turn to some fundamental attributes of material flows, to their energy costs and to their environmental impacts. I will do this in two separate sections, the first quantifying fuels and electricity requirements for many raw and processed materials, the second introducing results of revealing life-cycle assessments, a relatively new analytical tool whose aim is to present costs other than capital and operating expenditures.

4.5   Energy Cost of Materials

Before any materials can start flowing through economies, energies must flow to power their extraction from natural deposits or their production by industrial processes ranging from simple mechanical procedures to complex chemical reactions. These energies belong to two distinct streams: direct flows of fuels and electricity used to energize the production processes (producing mechanical energy, heat or pressure and lighting, and electronically controlling a process) and indirect flows (embedded energies) needed to produce the requisite materials, machines, equipment, and infrastructures. In modern production systems the first category of flows is usually dominant.

For example, the energy needed to smelt a ton of iron from its ore in a blast furnace (as coke and supplementary coal, gas, or oil) will be vastly greater than energy embedded in the furnace's steel, lining, and charging apparatus and prorated per unit of output. Modern blast furnaces can operate without relining for two decades, and during that time can produce tens of millions of tons of hot metal. Similarly, the energy needed to create the combination of high temperature and pressure that is required by many chemical syntheses will be far greater that any prorated energy embedded in the initial construction of reaction vessels, pipes, boilers, compressors, and computerized controls.

This explains why the second category of flows is almost always neglected – and until two generations ago most industries had little reason to examine even the direct energy flows of their energy use: for decades real fuel and electricity prices were low, even falling, and hence only the sectors with inherently high electricity and fuel inputs (such as electrochemical processes, high pressure, and high temperature chemical syntheses) have tried to reduce their overall energy cost in order to improve their earnings. This common neglect of energy uses ended abruptly when OPEC's first round of oil price increases (1973–74) quintupled the world oil price in a matter of months, and this new concern only increased when the second round of OPEC's price manipulation (driven by the overthrow of Iranian monarchy) pushed the price of an average barrel of the Middle Eastern oil from about \$12 in 1977 to nearly \$36 in 1980 (BP, 2013).

The new discipline of energy analysis aimed to trace and quantify energy used not only to extract and process natural commodities and to produce manufactured items, but also to grow food and to provide services (IFIAS, 1974; Chapman *et al.*, 1974; Verbraeck, 1976; Thomas, 1979). The chosen label was misleadingly broad, as energy analysis should stand for a much wider range of inquiries. Embedded (or incorporated) energy is a descriptively more correct term, but I have always preferred to use a simple label of energy costs. I was an early practitioner of this method – among other efforts being the principal author of the first comprehensive energy analysis of US corn, the country's most important crop (Smil *et al.*, 1983) – and hence I have a good appreciation of associated problems and weaknesses.

Most appraisals of energy costs have followed one of two distinct approaches: either a quantification based on input–output tables of economic activities, or a process analysis that traces all important energy flows needed to produce a specific commodity or manufactured item. In the first instance, relevant prices are used to convert values of energy flows in a matrix of economic inputs and outputs (for major industrial sectors or, where available, disaggregated to the level of product groups or individual major products) to energy equivalents in order to assemble direct and indirect energy requirements. In contrast to this aggregate approach, process analysis can focus on a particular product in specific circumstances as it identifies all direct energy inputs and as many relevant indirect needs as possible, a process that is in itself quite valuable as a management tool.

As with all appraisals that deal with complex inputs and encompass sequential processes, the setting of analytical boundaries will affect the outcome of process analysis. In most cases, the truncation error inherent in counting only direct energy inputs (purchased fuels and electricity) will be small, but in some instances it could be surprisingly large. For example, Lenzen and Dey (2000) found the energy costs of Australian steel to be 19 GJ/t with process analysis, but 40.1 GJ/t with input–output analysis. Similarly, input–output analysis by Lenzen and Treloar (2002) put the total embodied energy in a four-story Swedish apartment building at roughly twice that of a process analysis by Börjesson and Gustavsson (2000).

Another complication is introduced due to increasing shares of globally traded commodities and products: in some cases the additional energies required for import of raw materials and export of finished products will be a negligible share of the process energy cost, in other cases their omission will cause a serious undercount. For example, two identical looking steel beams used at two construction sites in New York may have two very different histories: the first being a domestic product made by the scrap-EAF-continuous (electric arc furnace) casting route in an integrated operation in Pennsylvania, the other coming from China where Australian iron ore and coke made from Indonesian coal were smelted in a blast furnace in one province and the beams were made from ingots in another one before loaded for a trans-Pacific shipment and then transported by railroads across the continent.

Approximate energy costs of long-distance transportation can be easily calculated by assuming the following averages (all in ton-kilometers for easy comparability, ranked from the highest rates to the lowest): air transport 30 MJ, diesel-powered trucks (depending on their size) mostly between 1 and 2.5 MJ, diesel-powered trains 600–900 kJ, electricity-powered trains 200–400 kJ, smaller cargo ships 100–150 kJ, and large tankers and bulk cargo carriers just 50 kJ/tkm (Smil, 2010). Obviously, energy-intensive air shipments will be restricted to high value-added products, while bringing iron ore by a bulk carrier from a mine 3000 km from a Chinese blast furnace would entail energy expenditure equal to less than 10% of overall requirements for steel production – while the energy cost of shipping construction stone from Europe or Asia to the USA may be equal to 25–50% of the energy used to cut and polish it.

These realities should be kept in mind when examining and comparing the values reviewed in this section. Energy costs – presented here in a uniform way as gigajoules per ton (GJ/t) of raw material or product – offer a revealing insight into the requirements and impacts of diverse materials and they are useful comparative guides. But, as with any analytical tool, they alone cannot be used to guide our choices and preferences of material use without concurrent considerations of affordability, quality, durability, or esthetic preference; if the latter were to be ignored concrete, a material of low energy intensity, would rule the modern world even more than it actually does. The presentation of energy costs here follows the same material order as the preceding chapter, starting with biomaterials and concluding with silicon.

The energy cost of market-ready **lumber** (timber) is low, comparable to the energy cost of many bulk mineral and basic construction materials produced by their processing. Tree felling, removal of boles from the forest, their squaring and air drying will add up to no more than about 500 MJ/t, and even with relatively energy-intensive kiln-drying (this operation may account for 80–90% of all thermal energy) the total could be as low as 1.5 and more than 3.5 GJ/t (including cutting and planing) for such common dimensional construction cuts as 2×4 studs used for framing North American houses. These cuts are rectangular prisms whose actual dimensions are smaller (1.5×3.5 in., or 38×89 mm) and whose standard length for house construction is 8 or 9 ft. The low energy cost of wood is also illustrated by the fact that, in Canada, the energy cost of wood products represents less than 5% of the cost of the goods sold (Meil *et al.*, 2009). Energy costs on the order of 1–3 GJ/t are, of course, only small fractions of wood's energy content that ranges from 15 to 17 GJ/t for air-dry material.

Obviously, the energy cost of **wood products** rises with the degree of processing (FAO, 1990). Particle board (with a density between 0.66 and 0.70 g/cm^3) may need as little as 3 GJ/t and no more than 7 GJ/t, with some 60% of all energy needed for particle drying and 20% for hot pressing. Energy needs for the production of **plywood** vary by a factor of 2 (depending on the wood and specific manufacturing procedure), from 7 to 15 GJ/t, with, once again, some 60% of all energy required for drying and 10% for pressing. The increasingly popular **glulam** used in commercial construction requires between 5 and 12 GJ/t.

Paper is a relatively energy-intensive product. Chemical pulping entails pressurized boiling of ground wood in acid solutions followed by even more energy-intensive dewatering of freshly laid fibers to make new paper layers; it consumes about 2.2 t of wood per ton of paper, but it produces about 22 GJ of black liquor per ton of pulp, and burning that by-product energy can make a mill a net energy producer. In contrast, mechanical pulping needs less than 1.1 t of wood per ton of pulp and it is a large consumer of electricity. Integration of the two processes is the best way to maximize energy efficiency. Primary energy requirements for the most efficient pulping are about 11 GJ/t of air-dried kraft **pulp** (including about 650 kWh of electricity) but as much as 23 GJ/t for the thermo-mechanical process, while the most efficient repulping of waste paper consumes only 4 GJ/t (Worrell *et al.*, 2008).

The energy cost of papermaking varies with the final product and, given the size and production scale of modern papermaking machines (typically 150 m long, running speeds up to 1800 m/min., and annual output of 300 000 t of paper), is not amenable to drastic changes (Austin, 2010). Unbleached packaging **paper** made from thermomechanical pulp is the least energy-expensive kind (as little as 23 GJ/t); fine bleached uncoated paper made from kraft pulp consumes at least 27 GJ/t and commonly just over 30 GJ/t (Worrell *et al.*, 2008). Most people find it surprising that this is as much as a high-quality steel. In contrast, the best printing paper, the coated stock used for arts books, may be actually less energy intensive (about 25 GJ/t) because the production and application of the filler (mostly chalk) is cheaper than making fiber from wood. Recycled and de-inked newsprint or tissue can be made with less than 18 GJ/t, but the material is often down-cycled into lower quality packaging materials.

Construction aggregates whose production requires only extraction and some physical treatment (sorting, sizing, crushing, milling, drying) have generally very low to low energy costs, and higher fuel and electricity use comes only with the pyro-processing required to make bricks, tiles, glass, and, above all, cement. The energy cost of natural **stone** products is low, usually just around 500 MJ/t for quarried blocks, somewhat less for crushed stone, but twice as much for roughly cut or split stones, and three to four times as much (up to 2 GJ/t) for accurately cut and polished ornamental stones. The energy costs of **sand** extraction and processing can easily vary by a factor of 2, but even the higher costs leave them in the category of the least energy-intensive materials when compared in mass terms.

The simplest mining and preparation sequence to produce fairly clean sand and uniformly-sized sand may require no more than 100 MJ/t, and even more costly gravel sorting (or crushing as needed) should have an energy cost well below 500 MJ/t. The highest energy input is required for the preparation of the industrial sand that is used in glassmaking, ceramics and refractory materials, metal smelting and casting, paints, and now also increasingly in hydraulic fractioning of gas- and oil-bearing shales: its moisture must be reduced (in heavy-duty rotary or fluidized-bed dryers) to less than 0.5%, and this may consume close to 1 GJ/t. **Bricks** fired in inefficient rural furnaces in Asia may require as much as 2 GJ/t, just 1.1–1.2 GJ/t is typical for Chinese enterprises (Global Environmental Facility, 2012; Li, 2012) while US production of high-quality bricks requires 2.3 GJ/t (USEPA, 2003).

Cement production is fairly energy-intensive because of high temperatures required for the thermo-chemical processing of the mineral charge. Limestone supplies Ca and other oxides, and clay, shale, or waste material provide silicon, aluminum, and iron; in order to produce a ton of cement about 1.8 t of raw minerals are ground and their mixture is heated to at least 1450 °C. This sintering process combines the constituent molecules, and the resulting clinker is ground again with the addition of other materials to produce 1 t of clinker that is then ground to produce fine Portland cement. Fly ash (captured in coal-fired power plants) or blast furnace slag can be used to lower the amount of clinker. Additional energy is to needed to rotate large kilns. These inclined (3.5–4°) metal cylinders are commonly around 100 m, up to 230 m, long, with diameters of 6–8 m, and they turn typically at 1–3 rpm, with the charged raw material moving down the tube against the rising hot gases (Peray, 1986; FLSmidth, 2011).

Disaggregation of all energy inputs shows that the extraction of minerals (limestone, clay, shale) and their delivery to cement kilns is a minimal burden. Kiln feed preparation is electricity-intensive as crushing and grinding of the charge consumes about 25–35 kWh/t and the grinding and transportation of the finished product (clinker) claims at least 32–37 kWh/t (Worrell and Galitsky, 2008). This leaves the bulk of energy consumption for the pyro-processing, a sequence of water evaporation, decomposition of clays to yield SiO_2, decomposition of limestone or dolomite (calcination) that releases $CaCO_3$, formation of belite (Ca_2SiO_4, making up about 15% of clinker by mass), and finally sintering, production of alite ($Ca_3O \cdot SiO_4$, that makes up some 65% of the clinker mass) (Winter, 2012).

Total energy use in cement production varies with the principal fuel used, the origin of the electric supply, and the method of production. Average specific energy consumption in the cement industry has declined as a more efficient dry process replaced the old wet method. The highest electricity consumption in the dry process is for the grinding of raw materials and clinker and for the kiln and the cooler, in aggregate more than 80% of the total that averages mostly between 90 and 120 kWh/t of cement (Madlool *et al.*, 2011). Heating of dry kilns (mostly with coal, petroleum coke, and waste materials in the USA, and with coal in China) consumes between mostly 3 and 4 GJ/t; the range is 3.0–3.5 GJ/t for kilns with four or five stages of preheating, while a six-stage process could work with as little as 2.9–3 GJ/t (IEA, 2007; Worrell and Galitsky, 2008).

World best practice can now produce Portland cement with total primary energy inputs of 3.3–3.5 GJ/t, while the rates for fly-ash cement and blast furnace slag cement can be as low as, respectively, 2.4 and 2.1 GJ/t (Worrell *et al.*, 2008). In contrast, many plants in low-income countries still need around 4.5 GJ/t for Portland cement. As expected, energy consumption for cement production has the greatest impact in China. Despite gradual improvements in specific efficiency the average nationwide rate was still close to 5 GJ/t in 2000 and just over 4 GJ/t in 2007 when the industry claimed 7% of the nation's total energy use (Sui, 2010); introduction of more efficient dry kilns reduced the nationwide mean to about 3.3 GJ/t in 2010, and many efforts are underway to limit the industry's emissions.

Energy requirements for **glass** production range mostly between 4 and 10 GJ/t, with about 7 GJ/t being a typical value and with little difference between container and flat glass (IEA, 2007). The theoretical minimum for glassmaking (chemical reactions among the constituent minerals and the melting) is about 2.4 for borosilicate and crystal glass and 2.8 GJ/t for common soda-lime glass. As expected, the energy cost of ceramic products rises with the degree of pyro-processing and the quality of items: unglazed tiles need only 6 GJ/t, glazed tiles up to 10 GJ/t, fine ceramics as much as 70 GJ/t, while sanitary stoneware rates at about 30 GJ/t.

The usual approach for quantifying the energy costs of the **iron and steel** industry is to include the energy costs of coke, pelletizing and sintering of ore, iron and steel making, cold and hot rolling, and galvanizing and coating; this leaves out the energy costs of coal and ore mining and transportation, of such energy-intensive inputs as electrodes and refractories, as well as the embodied energy cost of scrap metal. Analyses performed (more or less) within these boundaries show that average energy consumption in the global steel industry was about 60 GJ/t in 1950, just over 30 GJ/t by 1975, and about 20 GJ/t by the year 2000 (Yellishetty *et al.*, 2010).

Reduced charging of coke to blast furnaces has been the main reason why the lowest energy requirements in iron smelting declined from about 30 GJ/t in 1950 to as little as 12.5 GJ/t by the late 1990s, the rate only twice as high as the thermodynamic minimum of 6.6 GJ/t needed to reduce iron from hematite (de Beer *et al.*, 1998). Typical ranges are 10–13 GJ/t for blast furnace operation, 2–3 GJ/t for sintering, 0.75–2 GJ/t for coking, and 1.5–3 GJ/t for rolling (IEA, 2007). A review of best industry practices for the entire iron–steel sequence ended up with 16.3–18.2 GJ/t for the blast furnace-BOF-continuous (basic oxygen furnace) casting route, 18.6 GJ/t for direct iron reduction followed by EAF steelmaking and thin-slab casting, and 6 GJ/t for melting scrap metal in EAF and thin-slab casting (Worrell *et al.*, 2008).

The overall national energy intensity of steelmaking will be thus considerably lower in countries relying more on the most efficient route. The IEA (2007) has estimated global energy intensities of crude steel by country, and the resulting stepwise progression shows averages ranging from less than 10 GJ/t (steelmaking based only on EAF) to almost 40 GJ/t, with the modal rate of 26 GJ/t and 90% of the output in the still rather wide range of 14–30 GJ/t. A comparative analysis of the energy costs of the iron and steel industry in the USA and China illustrates this reality: it shows that the aggregate input in 2006 was, respectively, 14.9 GJ/t and 23.11 GJ/t of crude steel (Hasanbeigi *et al.*, 2012).

Generally superior efficiencies were only a minor reason for this difference, which can be explained largely by a much higher share of more efficient electric arc furnaces in US steelmaking: in 2006 they produced 57% of US steel compared to only about 11% in China, where primary smelting in blast furnaces still dominates. As a result, electricity's share is 20% of the total primary energy used in the US steelmaking, but only 10% in China's industry. Taking 25 GJ/t as a mean would suggest that in the year 2010 the global iron and steel industry needed roughly 36 EJ of energy, or about 6% of the worldwide consumption of primary commercial energy. For comparison, Allwood and Cullen (2012) put the global energy use in steelmaking at 38 EJ.

Aluminum production is much more energy intensive than making steel. Fuel and electricity consumption in the Bayer process, between 10 and 13 GJ/t of alumina, is a small share of the overall cost that is dominated by the electrolysis that is done, preferably with the cheapest kind of electricity produced in large hydro stations (it supplies about 60% of the industry's needs worldwide). A version using electrodes that are baked *in situ* is slightly more electricity-intensive than the dominant method using pre-baked anodes made from tar pitch or coke in oil. At least 400 kg of anode are required to produce a ton of the metal and anode production needs about 2.5 GJ/t of fuel and 140 kWh/t of electricity.

The efficiency of electrolysis has been steadily improving: between 1900 and 2000 the size of the average cell used in the Hall–Héroult process doubled every 18 years, while electricity use declined from about 50 MWh/t in 1900 to 25 MWh/t by 1950, and to less than 13 MWh/t by the year 2000 (Beck, 2001). Recent requirements have been between 13 and 18 MWh/t of metal, with a weighted global mean of 15.3 MWh/t in 2004 and with the lowest cost (14.3 MWh/t) in Africa (IEA, 2007). Compared to the theoretical minimum of 6.3 MWh/t, this means that the process is now, at best, about 45% efficient.

The lowest requirement (13 MWh) supplied by hydroelectricity is equivalent to nearly 50 GJ/t, the highest would require nearly 200 GJ/t of primary energy for fossil-fueled electricity generation. Additional energy is needed for casting and rolling of ingots, and other smelter activities add 7–8 GJ/t. The IEA put the weighted energy cost of the entire sequence at 175 GJ/t in 2004, and a review of best industry practices came up with a nearly identical rate of 174 GJ/t of metal (Worrell *et al.*, 2008). That is nearly twice the energy intensity of copper and almost 10 times as much as the least energy-intensive production of steel using the blast furnace-BOF-continuous casting route. Global 2010 production of 40.8 Mt of Al would have thus required about 7.1 EJ, less than 1.5% of the world's total primary commercial energy supply. Metal's high electricity requirements steer the location of primary aluminum production to countries with abundant hydro resources, and the four such largest producers (China, Russia, Canada, and the

USA) account for half of the world output. Secondary aluminum requires only 7.6 GJ/t (for remelting only).

As already noted, titanium has the highest energy cost among the other relatively commonly used metals (400 GJ/t), followed by nickel at about 160 GJ/t and copper (global average of 93 GJ/t), while chromium, manganese, tin, and zinc have a very similar energy cost of about 50 GJ/t (IEA, 2007). Not surprisingly, very low metal concentrations of even the best exploited deposits raise the energy intensities of silver and gold orders of magnitude above common metals: the average for silver is about 2.9 TJ/t (30 times higher than for copper) and for gold it is 53 TJ/t, roughly 300 times the energy cost of aluminum.

Plastics are, without exception, energy-intensive materials, but the published rates have been surprisingly divergent. During the mid 1970s the first comparative studies of energy used in material production showed final process energy requirements of polyethylene (PE) to be as low as 50 GJ/t in the Netherlands and as high as 116 GJ/t in the USA (Berry *et al.*, 1975). For polypropylene (PP) the range was much narrower, between 111 and 125 GJ/t, and for PVC the Dutch, British, and US rates were very close at, respectively, 69.8, 66.0, and 60.8–70.8 GJ/t. The most detailed industry-wide analysis of energy needs in the US chemical industry found that, in 1997, PE production averaged 77.4 GJ/t (with feedstock energy being 20.1 GJ/t), that PP embodied, on average, 54.7 GJ/t, and that PVC's total average energy cost was 44.7 GJ/t, with feedstock accounting for 17 GJ/t (Energetics, 2000).

A World Bank review of PE energy costs found ranges of 87.4–107.8 GJ/t for high density polyethylene (HDPE) and 74.4–116.3 for low-density polyethylene (LDPE), with the processing energy being as low as 25–28 GJ/t and as high as 45 GJ/t (Vlachopoulos, 2009). Berge (2009), in his survey of construction materials, cited European rates of 110–115 GJ/t for PE and PP and 80–90 GJ/t for PVC. And in their review of best current practices, Worrell *et al.* (2008) found that the energy costs of ethylene made from naphtha or ethane were very similar, respectively 14–22 GJ/t and 12.5–21 GJ/t. The heavy dependence of modern production of plastics on hydrocarbon feedstocks has not been, so far, a major burden, as the industry still claims less than 5% of the world's natural gas and crude oil output.

Rising demand for plastic materials and higher costs, particularly of crude oil, will change this, and in the long run plant-based bioplastics appear to be the only practical answer both to the eventually less abundant petrochemical feedstocks and to the presence of nonbiodegradable materials in the environment. But the higher demand for composite materials increasingly used in the aerospace and automotive industries will result in higher energy costs even after adjustments are made for the lower density and superior strength of these new materials. For example, the lignin carbon fiber used in reducing the weight of passenger cars and other vehicles costs 670 GJ/t and carbon-fiber reinforced polymer (polyacrylonitrite) fiber requires just over 700 GJ/t (Das, 2011), more than three times that of aluminum.

Because of the high pressures and high temperatures required for the catalytic synthesis of **ammonia** from its elements, the Haber–Bosch process used to be among the most energy-intensive sectors in chemical industry: the first commercial Haber–Bosch plants (using coke as their feedstock) required more than 100 GJ/t of NH_3, and pre-World War II coal-based plants still needed about 85 GJ/t. Post-1950, the shift to natural gas and low-pressure reforming using reciprocating compressors lowered the rate to 50–55 GJ/t. In modern plants, nitrogen is obtained by fractional separation from air, and hydrogen comes from the steam reforming natural gas; natural gas is also used to drive compressors, acting as both feedstock and energy source, and the specific energy cost of the best practices has been reduced to rates remarkably close to the stoichiometric minimum of 20.9 GJ/t (Smil, 2001; Worrell *et al.*, 2008).

The most important innovation came after 1963 with the introduction of centrifugal compressors: while electricity-powered reciprocating compressors consumed 520–700 kWh/t NH_3, steam turbine-driven centrifugal compressors need as little as 20–35 kWh/t NH_3, a 95% reduction. High-pressure reforming (with pressures above 3 MPa) was another innovation of the 1960s, and by the 1970s this combination lowered the energy use to just 35 GJ/t NH_3. The combination of new plant designs, higher efficiency of individual plant processes, and better catalysts cut the best performances to as little as 27 GJ/t NH_3 in the year 2000. When Worrell *et al.* (2008) reviewed the best commercial practices, they rated natural gas-based synthesis at 28 GJ/t (roughly a third higher than the stoichiometric minimum), and coal-based process at 34.8 GJ/t.

Naturally, typical performances are higher, around 30 GJ/t NH_3 for gas-based plants, 36 GJ/t for heavy fuel oil feedstock, and more than 45 GJ/t NH_3 for coal-based synthesis (Rafiqul *et al.*, 2005). The IEA (2007) used regional means ranging from 48.4 GJ/t in China to 35 GJ/t in Western Europe, resulting in a global weighted mean of 41.6 GJ/t for the year 2005. The best weighted worldwide rate for ammonia synthesis in the year 2000 would be about 45 GJ/t NH_3, for those working in 2010 about 40 GJ/t. Conversion of NH_3 to $CO(NH_2)_2$ (urea) requires 10–12 GJ/t N and the prilling (production of small uniform spheres from molten solids commonly used to make granular fertilizers), bagging, and distribution of the compound raise the total energy cost of this solid fertilizer is typically 55–58 GJ/t N.

Surface-mining of **phosphates** takes only 4–5 GJ/t, but treating the insoluble rocks with sulfuric and nitric acids in order to produce water-soluble phosphorus compounds is much more energy intensive. Overall energy costs range from 18 to 20 GJ/t for superphosphates (single superphosphate with just 8.8% P, triple superphosphate with 20% P) to 28–33 GJ/t for diammonium phosphate containing 20% of soluble P (Smil, 2008). The energy cost of **potash** (sylvinite) extraction is low: in Saskatchewan, conventional underground mining followed by milling needs only 1–1.5 GJ/t, and surface mining and milling averages only about 300 MJ/t (NRC, 2009).

Because electricity accounts for most of energy used to produce **electronic-grade silicon**, the energy intensity of the production chain is usually expressed in kilowatt-hours rather than in megajoules per kilogram and the conversion to primary energy will depend on the source of electricity. Crystals of electronic-grade silicon are 1 billion times purer than the metallurgical grade of the element that is used for deoxidation and alloying (Föll, 2000). Making metallurgical-grade Si from quartz requires 12–13 kWh/kg, the Siemens process consumes about 250 (200–300) kWh/kg, the Czochralski process has roughly the same energy need, and processing wafers from ingots adds 240 kWh/kg for a total of about 750 kWh. But the actual rate is much higher because of significant material losses along the processing chain.

Conversion from quartz to metallurgical Si has a 90% yield, but the Si yield for the Siemens process is less than 50%. Monocrystalline Si produced by the Czochralski process from polycrystalline Si (whose manufacture uses at least 400 MJ/kg) consumes on the order of 1 GJ/kg of electricity for the total of nearly 1.5 GJ/kg (Kato *et al.*, 1997). Conversion of Czochralski crystals to wafers is fairly wasteful, with Si lost from the ingot's top and tail, due to kerf and pot scrap and test wafer cuts. As a result, Williams *et al.* (2002) found that 9.4 kg of raw Si were needed to make a kg of final microchip wafer.

This means that the entire production chain – starting with Si made from quartz and carbon through trichlorosilane, polysilicon, single crystal ingot, Si wafers, and actual fabrication and assembly of a microchip – consumes about 41 MJ for a 2-g chip. This implies a total electricity cost for shipped wafers of at least 2100 kWh/kg. Even if using only hydroelectricity this would prorate to about 7.6 GJ/kg, and energizing the entire process by electricity generated from fossil fuels would push the total primary energy to more than 20 GJ/kg for finished Si wafers, 2 orders of magnitude more than aluminum made from bauxite, and 3 orders of magnitude more than steel made from iron ore. Schmidt *et al.* (2012) offer a detailed assessment of life-cycle energy costs for both solar- and electronic-grade silicon wafers but express their results per hour of production and per square meter of wafers.

The typical rates presented in this section can be used (after rounding, to avoid impressions of unwarranted accuracy) to assess the global energy needs of major material sectors and to calculate their fractions of TPES (whose total was just over 500 EJ) in 2010 (BP, 2013). Not surprisingly, steel's relatively high energy intensity (25 GJ/t) and its massive output (1.43 Gt in 2010, 1.5 Gt in 2011) make it the material with the highest total energy demand that dominates the total of about 50 EJ (or 10% of TPES) required to produce all metals in 2010. Plastics are next (assuming 80 GJ/t and an output of 265 Mt in 2010) with roughly 20 EJ (4% of TPES), well ahead of construction materials (cement, bricks, glass) with about 15 EJ, or 3% of TPES. Paper production required about 10 EJ and fertilizers added less than 8 EJ for the grand total of just over 100 EJ or 20% of the world's TPES in 2010. For comparison, the IEA (2007) estimated energy input for the entire global industrial sector at almost 88 EJ for the year 2005.

Paper (and paperboard) and aluminum each end up with very similar totals of close to 10 EJ (2% of TPES) as a result of aluminum's much higher energy intensity (175 vs. 25 GJ/t) but much lower total output (53 vs. 400 Mt in 2010). Perhaps the most interesting result concerns the energy cost of inorganic fertilizers: given their truly existential importance it is reassuring to realize that the energy needed to produce them adds up to a surprisingly small share of global supply. Assuming averages of 55, 20, and 10 GJ/t for, respectively, N, P, and K (all including the cost of final formulation, packaging, and distribution) would result in a total demand of a bit more than 5 EJ in the year 2010 (with nitrogenous fertilizers accounting for about 90% of the total) – or only about 1% of the TPES.

This must be one of the most rewarding energy investments in history because, without the application of these compounds, we could not support the existing population of 7 billion people; I hasten to add that a still high level of global malnutrition is due to unequal access to food, not to inadequate food supply. Another case of a highly rewarding energy pay-off is the cost of silicon wafers. As explained, their embodied energy is orders of magnitude higher than for any of the commonly used materials (on the order of 20 TJ/t compared to 20 GJ/t for steel), but the steadily increasing crowding of transistors has limited the annual mass of wafers needed to produce all of the world's microchips to only about 7500 t in 2009 and to an aggregate energy expenditure of just 150 PJ, or about 0.03% of TPES.

These calculations also make it clear that modern civilization can afford all this steel and fertilizers and microchips because scientific discoveries and technical advances have greatly reduced their energy intensities. For example, had the steelmaking intensity remained at the 1900 level, the global production of 1.5 Gt of the metal in the year 2010 would have claimed almost 20% of the world's TPES – rather than just over 6%. Another perspective is even more impressive: in 2010 the aggregate energy cost of producing steel, aluminum, paper, plastics, and fertilizers was about the same (90 EJ) as would have been the cost of 1.5 Gt of steel requiring average energy inputs in the year 1900. Even when using liberal rates for average energy intensities of all biomaterials other than paper, construction materials other than cement, and metals other than steel and aluminum we end up with a grand total of no more than 120 EJ, or less than 25% of the world's TPES: we create the modern world's material wealth with no more than a quarter of all energy we use.

4.6 Life-Cycle Assessments

Quantifications of energies embedded in materials are useful and revealing companions to standard analyses of material flows using mass or value. But a brief reflection shows that even the combination of mass, value, and energy does not provide a sufficient basis for informed and balanced assessment because it leaves out the broad category of environmental impacts. Moreover, such assessments should not be limited to costs and impacts at the production stage, because many materials are parts of products and structures that will serve for years and decades and the full consequences of their use can be appreciated only by appraising their life-long trajectories. As the OECD put it, "all the phases of organized matter and energy that are in some way related to the making and use of a product can also be linked to an impact on the environment" (OECD, 1995).

The discipline that is doing precisely this – quantifying objective indicators of the environmental impacts of materials during their entire life-span – is life-cycle assessment (or life-cycle analysis, with the identical acronym, LCA), now a widely accepted procedure that has been codified as part of international environmental management standards that are routinely used to evaluate the various burdens imposed by production, use and disposal (or reuse) of products, and by the performance of services (ISO, 2006a,b). LCA is now a mature analytical discipline that has its own periodical, *International Journal of Life Cycle Assessment*, published since 1995, and an increasing number of how-to books (Horn *et al.*, 2009; Jolliet *et al.*, 2013), as well as extensive life-cycle inventory databases maintained by the US National Renewable Energy Laboratory (NREL, 2009; 2013a), by the European Commission (EC, 2009), and by the Swiss Centre for Life Cycle Inventories (Ecoinvent, 2013).

Varieties of LCA include complete cradle-to-grave sequences (from raw materials to final disposal) and partial cradle-to-gate assessments (that is for finished products before their distribution for sale). Inventory databases contain information on input and output flows for many materials, but a systematic look at the recently published LCAs (done in perhaps the most representative way by using appropriate searches of publications on the Web of Science) quickly reveals that the choice of assessed items (products or processes) has been guided neither by their importance to modern civilization (in that case studies of steel, concrete, and various plastics would dominate) nor by the quest to quantify the potentially most worrisome cases (in that case there would be many assessments of heavy metals). Instead, recently published life-cycle analyses offer a diverse choice: they are now available for a number of common crops; for foodstuffs including bread, milk, and tomatoes; for renewable energy production systems (biomass, wind, plant oils) and from tiny microchips and granite sidewalks to entire residential and office buildings.

Obviously, some of these findings will be fairly durable while other LCAs will require relatively frequent reassessments: making steel via the blast furnace-BOF-continuous casting route is not subject to the same rate of change as is the production of consumer electronics. LCA offers many benefits but it also has a number of inherent problems and it is often performed in a truncated or superficial manner. Certainly its most obvious contribution is to provide a revealing reminder of post-purchase (or post-installation) costs of a product, as well as the fact that in many instances costs associated with the operation and use of products and infrastructures are far higher than the costs and impacts of its creation; that, obviously, is key to devising rational response strategies. These realities are well illustrated by examples ranging from LCA of buildings and cars to production and washing of garments.

Given the differences in house sizes and in energy intensity of lumber, steel, or concrete (the three principal structural materials in modern housing) it is not surprising that the life-time energy cost (construction, maintenance, operation, demolition) of residential space ranges roughly 3-fold ($3-9 \, GJ/m^2$) and that the total embedded energy in single family houses can be as low as 200 GJ for a small wooden structure, average around 500 GJ for a three-bedroom, wood-framed bungalow, and surpasses 2 TJ for large houses or two-story buildings (Canadian Wood Council, 2004; Smil, 2008). The quest for energy savings in buildings (regulation of temperature, air flow, and moisture) has led to more rigorous construction codes and to the deployment of a wider range of insulating materials. Fiberglass, foamed plastics, and loose fill (made of plant fibers and cellulose) are commonly used to insulate walls and ceilings, plastic sheets and foils are applied as moisture barriers, and aerated concrete and gypsum boards are used to reduce noise.

As must be expected, LCAs of housing in cold climates show embodied energies as a small fraction of life-time total. The LCA of three types of residential buildings (concrete, steel, and wood framework) in Beijing (hot summers but often very cold winters) showed that embodied energy was no more than 13–27% of life-long energy consumption (Gong et al., 2011). And even for a larger ($200 \, m^2$) Canadian house that cost 1.5 TJ to build, heating and lighting (averaging about $25 \, W/m^2$) will claim about 9.5 TJ in 60 years, reducing the construction share to just 14% of the overall cost. By coincidence, that share is nearly identical to the construction share of a medium-sized American car: it takes about 100 GJ to produce and (at about 8 l/100 km and 20 000 km/year) it will need about 550 GJ of fuel and oil in 10 years, while the initial construction will claim only about 15% of the overall cost, and even less once repairs and garaging are included (Smil, 2008). Embodied energies make up even lower shares in the life-cycles of machines that are in nearly constant operation: only 6–7% for jetliners, freight trains, and cargo ships (Allwood and Cullen, 2012).

But the shares of initial and operating costs are not always clear. While several LCAs show the energy costs of operating a personal computer or a laptop as substantially higher than those of manufacturing it – the split as low as 74:26 and as high as 93:7 (Andrae and Andersen, 2010) – Williams (2004) ended up with the reverse ratio. According to his analysis, the energy used to make a desktop with a Pentium III processor, 30 GB hard drive, and 42.5 cm monitor added up to 6.4 GJ, while during its relative life of three years the desktop would consume about 420 kWh of electricity or roughly 1.5 GJ of primary energy, yielding a manufacturing:usage energy split of 81:19. For a Swiss desktop computer, the split was much closer at 46:54 (Ecoinvent, 2013).

LCAs also make it clear that over their life-time many infrastructures will cost nearly as much, or more, to maintain than their initial construction. A Canadian LCA for a high-volume two-lane concrete highway shows initial construction costs of 6.7 TJ/km and a rehabilitation cost of 4.1 TJ/km or 38% of the total 50-year cost of 10.8 TJ/km, and the burdens are actually reversed for a roadway made of flexible asphalt concrete that needs 15 TJ/km to build and 16% more (17.4 TJ/km) to rehabilitate over a 50-year life-cycle (Cement Association of Canada, 2006).

The LCA of repeatedly washed garments is yet another excellent illustration of the boundary problem noted at the beginning of the energy costs section, as products made from different materials have different durabilities and maintenance requirements and a complete account of these realities may shift the overall advantage from a material that requires less energy to produce to one that is more energy-intensive to make but whose life-long energy cost may be lower. The LCA of America's ubiquitous short-sleeved industrial work shirt (65% PE, 35% cotton) laundered 52 times and then landfilled, showed that the shirt's production accounted for 36% of all energy use, 26% of CO_2 equivalents, and 24% of water (Cartwright *et al.*, 2011).

Comparison of pure cotton, cotton/polyester blends, and pure polyester fabric by van Winkle *et al.* (1978) was one of the pioneering energy analysis studies embracing the LCA. The study found that polyester production required twice as much energy as producing cotton lint (clear advantage cotton) but because of a higher cost of cloth manufacturing, the total energy cost of a cotton shirt was about 20% higher than that of its pure polyester counterpart (slight advantage polyester), and after including the energy costs of maintenance (washing, drying, ironing) the cotton shirt was about 3.6 time more energy intensive (clear advantage polyester). Similarly, life-cycle analysis by Kalliala and Nousiainen (1999) found that sheets made from pure cotton and laundered 100 times require 72% more energy than 50/50 cotton/polyester blend, because two 100% cotton sheets are needed to cover the life-time of one blended sheet.

Cotton appears even more disadvantaged once nonenergy impacts are compared: the water requirements of the blend are less than a third of those for pure cotton, and global warming and acidification potentials are 38% lower than those for producing and laundering two pure cotton sheets. Going even further, we can consider the long-term cost of excessive soil erosion in cotton fields, soil quality decline due to salinization in irrigated cotton fields in arid regions, and the presence of pesticide residues in soil and water (Smil, 2008). All of these are avoided by using a synthetic fiber – but its production depends on a nonrenewable feedstock. But so does the cultivation of cotton: PE synthesis consumes about 1.5 kg of hydrocarbons per kilogram of fiber but growing a kilogram of cotton requires nearly 500 g of fertilizers and 15 g of pesticides made from hydrocarbon feedstocks, as well as liquid fossil fuels for farm machinery. And one element still missing in all of these comparisons is consumer preference: should the superior wearing qualities of cotton shirt count for nothing?

Also, when using LCAs in order to compare environmental impacts of different products, their weight may be a misleading common denominator. That is why Shadman and McManus (2004, 1915) objected to an analysis of microchip energy and material cost by Williams *et al.* (2002), calling the use of weight as the basis for comparison "arbitrary, nonscientific, and inaccurate" because "it does not account for performance, utility, and benefits" of compared products. Of course, this is fundamentally just another instance of a missing qualitative perspective, an inherent weakness of LCA comparisons based on mass.

But what LCAs have done is to allow comparisons within the same category of environmental consequences (when a choice of materials is possible, which one will have the lowest effect on water use or water pollution?) as well as more comprehensive rankings of materials according to several categories of environmental impact. This is important because material use is one of the three dominant ways that humans have been changing the biosphere: food production and energy supply (dominated by extraction and combustion of fossil fuels) are the other two great interventions. And when considered in its entirety, the intricate system of extraction, processing, transportation, use, reuse, and disposal of materials encompasses every major environmental interference, from land use changes (ranging from deforestation due to lumber and pulp production to destruction of plant cover and disruption of water cycle due to massive surface ore mines) to atmospheric emissions (ranging from acidifying gases to being a major contributor to anthropogenic warming).

I should stress that because of their different action modes and different qualitative consequences (including many physical changes, chemical transformations, and interferences in the structure and function of ecosystems), different spatial effects (ranging from local to global), and different durations (from ephemeral to consequences felt for millennia) there can be no ranking of anthropogenic environmental impacts. There is no unifying metric that would allow us to conclude that soil erosion should be a greater concern than photochemical smog, or that tropical deforestation is more worrisome than the enormous water demand of modern irrigated agriculture. And, obviously, it would be impractical to perform LCAs that would attempt to quantify every known environmental impact: that is why most analyses settle on a small number of leading indicators of environmental change or degradation.

Calculations of overall global warming potential – GWP, a relative indicator of the intensity with which specific GHGs absorb the outgoing terrestrial infrared radiation and of the life-time of those gases in the atmosphere – have been the most common component of recent LCAs. All GHGs are scored against CO_2 whose GWP value is 1 (IPCC, 2007). Values for the two other leading GHGs are 21 for CH_4 and 310 for N_2O and the highest equivalents are 23 900 for SF_6 (sulfur hexafluoride, used as an insulating medium in electric industry) and 11 900 for C_2F_6 (perfluoroethane, used in electronic manufacturing). Obviously, omitting or undercounting such emissions would have a huge impact on the final value. GWPs of materials are usually expressed in equivalents of CO_2 per unit mass of a material (kg CO_2-equiv./kg) but for buildings the values can be prorated per unit area (kg CO_2-equiv./m^2).

Calculations begin with specific emission rates expressed in t CO_2/TJ of energy; IPCC default rates are 56.1 for natural gas, 73.3 for crude oil, 94.6 for bituminous coal, and 101.02 for lignite, but actual nationwide or regional values can differ by a few percent for hydrocarbons and by more than 10% for solid fuels, with European lignites ranging from 90.7 to 124.7 t CO_2/TJ (Herold, 2003). With CO_2 usually being the leading contributor, this indicator will have a very high correlation with the mass of fossil fuel used in the initial production. For example, production of a kilogram of cement will release about 0.5 kg CO_2 per kg of but making a kilogram of hot-rolled sheet steel will release about 2.25 kg of CO_2 liberated from fossil fuels (NREL, 2013b).

The other two commonly assessed variables are acidification and eutrophication impacts. Emissions of sulfur and nitrogen oxides from combustion of fossil fuels, smelting of ores, and other industrial processes are the precursors of atmospheric sulfates and nitrates whose wet and dry deposition acidifies waters and soils. In LCAs this acidifying potential of products or processes is calculated in terms of grams of hydrogen ions per square meter (g H^+/m^2) or per volume of water (g H^+/l). Eutrophication is the process of nutrient enrichment of fresh and coastal waters (most often by releases and leaching of nitrates and phosphates) that leads to excessive growth of algae whose decay deprives waters of dissolved oxygen and creates anoxic zones that either kill or impoverish heterotrophic life in affected waters (Harper, 1992). In LCAs this impact is usually expressed in kilograms of PO_4^{3-} equivalents. Perhaps the least frequent variable included in LCAs is the smog-forming potential of products or processes, expressed as grams of NO_x per square meter.

As just explained in the previous section, production of materials claims roughly a quarter of the world's primary energy supply. Because the combustion of fossil fuels provides most of this energy – in global terms about 87% in 2010, with the rest coming from primary, that is mostly hydro and nuclear, electricity (BP, 2013) – production of materials is a leading source of emissions of particulate matter (including black carbon), SO_x and NO_x (whose conversion to sulfates and nitrates is the primary cause of acidifying precipitation), and GHGs. The water needed for processing, reaction, and cooling ends up contaminated or is released at elevated temperatures by many industries, and most of them also share the necessity of disposing of relatively large volumes of solid waste and small but potentially worrisome volumes of hazardous waste. Fortunately, these common environmental challenges have many common (generic would be the correct adjective here) solutions. At the same time, there are many specific challenges arising from unique physical or chemical properties or from exceptionally demanding industrial processes.

Many studies offer either fairly comprehensive environmental profiles of materials – including, as applicable, assessments of their GWP, acidification potential, water consumption, and water pollution burden – or comparisons of specific consequences (with GWP being, not surprisingly, a recent favorite). Here are brief environmental impact summaries of some common materials. Production of a kilogram of typical construction steel sections generates about 1.5 kg of CO_2 equivalent, 50 g of SO_2 equivalent of acidification potential, a negligible amount of eutrophication potential (0.36 g), and just 0.8 g of the photochemical smog-inducing ethane (WSA, 2011b). Production of a kilogram of PVC (consuming around 60 MJ) requires about 10 kg of water (excluding cooling demand), produces 1.9–2.5 kg of CO_2 equivalent, 5–7 g of acidification potential (as SO_2), 0.6–0.9 g of nitrification potential (as PO_4), nearly 0.5 of ethane (measure of photochemical ozone creation), 0.4–0.8 g of total particulate waste, and 5–8 g of hazardous waste (Sevenster, 2008).

What I like best about LCAs is that they repeatedly offer interesting and counterintuitive results. Which paving has the lower impact: natural granite-slab or a concrete sidewalk? Mendoza *et al.* (2012) concluded that – largely because of the energy needed to cut and move the stone – the granite sidewalk has a 25–140% higher impact than the concrete one. Which transport packaging of fruits and vegetables carries the lowest CO_2 equivalent: traditional light (less than 1 kg) one-way wooden boxes made from an obviously renewable material whose energy content can be recovered by incineration; even lighter one-way cardboard boxes that could be either incinerated or recycled; or much heavier (2 kg) multi-way plastic crates that must be washed between uses? When the impact is scaled to 10 annual rotations per person, the GWP of plastic crates is about 10% lower than that of wooden boxes and less than half of that of cardboard packaging; when scaled to 1 million annual rotations plastic crates generate about 332 kg of CO_2 equivalent, wooden boxes about 367 kg, and cardboard boxes 708 kg (University of Stuttgart, 2007).

Another counterintuitive result comes from a recent Thai LCA of the single-use thermoform boxes that are widely used for food packaging (Suwanmanee *et al.*, 2013). The study compared boxes made from hydrocarbon-derived polystyrene (PS) with those made from polylactic acid (PLA) derived from corn, cassava, or sugarcane by using country-specific material and energy inputs to calculate three impact variables, GWP (including direct GHG emission and indirect emissions from land-use changes), acidification potential, and photochemical ozone formation. PS boxes have a considerably lower overall environmental impact, above all due to the considerable indirect impacts caused by corn and cassava cultivation. And although plastics are relatively energy intensive, an LCA of energy consumption and GWP in Europe made it clear that their substitution by other materials would require an additional 53 Mt of crude oil and increase emissions by 124 Mt of CO_2 equivalent (Pilz *et al.*, 2010).

But natural materials can be a better choice, and often resoundingly so in the case of biomaterials. For example, a comprehensive LCA by Bolin and Smith (2011) showed that a wood/plastic composite decking results in 14 times higher fossil fuel use, 3 times more GHG, almost 3 times more water use, 4 times higher acidification potential, and about 2 times more smog potential and ecotoxicity than a wooden deck built with lumber treated with alkaline copper. The advantages of biomaterials matter most in long-lasting structures, and many studies have shown environmental advantages of wooden buildings and wooden house components.

For example, glulam beams embody about 500 MJ/m^2 of roof, compared to at least 1 GJ/m^3 for steel beams made from recycled metal and 1.6 GJ/m^3 for steel made from blast-furnace iron (Petersen and Solberg, 2002). Similarly, wooden floors are much less energy intensive than the common alternatives: the total energy per square meter of flooring per year of service was put at 1.6 MJ for wood (usually oak or maple) compared to 2.3 MJ for linoleum and 2.8 MJ for vinyl (Jönsson *et al.*, 1997). Another study found solid oak boards superior not only to linoleum and vinyl, but also to polyamide carpet and wool carpet (Petersen and Solberg, 2004).

LCAs indicate that wooden structures have significantly lower (as much as 65%) GHG emissions when compared to reinforced concrete buildings (pre-cast or cast-in-place), and similarly large savings for flooring or furniture. Eriksson's (2004) comparisons of Swedish houses show major energy and GWP savings. A four-story wood-frame residential house required 960 MJ/m^2, but after subtracting the recyclable energy of 1460 MJ/m^2 there was an actual credit of 530 MJ/m^2 compared to the net energy requirement of 1770 MJ/m^2 for a concrete house, and the GWPs (both excluding usage phase) were, respectively, just 30 and 400 kg CO$_2$-equiv./m^2. Additionally, increasing the timber content of typical residential buildings in Scotland could save as much as 75% of life-cycle GHG emissions for a small semi-detached house and up to 86% for a larger detached structure (Edinburgh Centre for Carbon Management, 2006). Moreover, wood recovered from demolished houses or discarded furniture could be burned, releasing more, or much more, energy than went into its production.

An additional advantage of managed forests and tree plantations are their contributions to oxygen generation and carbon sequestration and, if trees are harvested in rotation and promptly replanted, such plantations can be long-lasting stores of carbon. Even more importantly, proper forestry management can increase the phytomass storage and annual productivity of natural forests, as shown by a nearly century-long perspective from Finland. In a periodically studied area in the southern part of the country, growing stock in essentially identical areas increased 2.15 times between 1912 and 2005 and annual carbon sequestration averaged about 30 t/km^2 (Kauppi *et al.*, 2010). In addition, unless the wood cut in a natural forest is used directly as fuel or to make charcoal, its carbon sequestration effect can be prolonged by many decades once it is converted into structural lumber or furniture (Berge, 2009). Not surprisingly, typical life-spans of wood products remaining in place vary widely, with the greatest range for construction lumber (Lauk *et al.*, 2012).

Sawnwood remains in place for less than two decades in mobile homes, or it can last easily 80–100 years in well-built wooden houses in the USA and Canada. Life-spans of wooden furniture are generally much shorter, 10–30 years, with two decades being a good average. Wood used in packaging and shipping (boxes, crates, pallets) may be discarded immediately or reused for a few years, and packaging paper and paperboard have similar durability (one to six years, an average of about two to three years). Other persistent uses include finishing components of houses (stairs, doors, door and window frames, floors), barrels and casks for wine and spirits, specialty woods for musical instruments and for sport implements, and (a small but rewarding niche) wooden toys and puzzles.

But, as with any complex analysis, users should beware: LCA has plenty of problems and weaknesses and too often it still falls short of really comprehensive cradle-to-grave accounting. An obvious problem involves comparisons of materials made from feedstocks with a high energy content, with those whose embedded energy comes only from fuels and electricity used in their production. For example, LCAs show that the energy embodied in a flexible asphalt concrete roadway is roughly twice as high as that in rigid Portland cement concrete – but the difference is almost entirely due to the energy content of feedstock (Cement Association of Canada, 2006). Asphalt (bitumen) is a mixture of heavy hydrocarbons with a high energy value (40 GJ/t), but it is an unsuitable energy source and is obtained as a by-product of oil refining with an expenditure of only 0.5–6 GJ/t of process energy, in most cases no more than the production of

Portland cement. Consequently, a more logical procedure would be to compare only the energies needed to construct the two kinds of pavements, and that would yield a very similar value.

Remarkably, many life-cycle assessments take a cavalier approach to life-spans and just assume what seems to be a reasonable length or use (often unexplained) from previous studies. Perhaps most notably, this has been the case with buildings, whose longevity means that arbitrary assumptions may end up with particularly inaccurate appraisals of life-cycle impacts. Modern houses have life-spans mostly between 50 and 100 years, modern houses and commercial buildings last between just 12 and about 50 years, but many of them (particularly stores) undergo relatively frequent and extensive remodeling. An inquiry by Aktas and Bilec (2012) put the average life-time of American residential buildings at 61 years with a standard deviation of 25 years. That study established mean life-times of major house components: paint at about 7 years, carpets 10 years, linoleum and vinyl floors 22 years, hardwood floors and finishes 42 years, and ceramic 48 years.

Another category of errors arises from assuming generic rates rather than ascertaining specific flows. This problem is perhaps best illustrated by the examples of highly inefficient industrial processes that abound in most modernizing countries and that are still common in China. The LCA of GHG emissions of China's primary aluminum production for the year 2003 showed a rate of 21.6 t CO_2 equivalent per ton of metal, 70% higher than the average global level in 2000 (Gao *et al.*, 2009). But perhaps the most important shortcoming of LCAs is the fact that in many cases they still present only truncated and incomplete appraisals, and perhaps no other category of materials illustrates this reality as well as the plastics.

This may be hard to believe after checking the life-cycle inventory databases for major synthetic compounds: for example, NREL's (2013a) output flows for HDPE quantify about 220 chemicals including organics (from acetone to phthalates), heavy metals, radionuclides, and GHGs. And some published LCAs of specific plastics in specific settings, such as the cited work by Suwanmanee *et al.* (2013), account even for indirect effects resulting from land-use changes caused by cultivation of plants producing renewable feedstocks. But there are no LCAs taking into account the great longevity of some plastics (on the order of hundreds of years) and their now ubiquitous, and clearly highly disruptive, distribution in aquatic environments.

Their buoyancy, their breakdown into progressively smaller particles, and their eventual sinking through the water column to the sea bottom combine to make them a truly global and omnipresent environmental risk to marine biota: they are now found on the remotest islands as well as in the abyss, but their highest concentrations are in surface water and on beaches (Moore, 2008; Barnes *et al.*, 2009). The extent of this new kind of environmental degradation became clear with the discovery of plastic refuse trapped within the North Pacific Gyre: this was encountered for the first time in 1997 by Charles Moore and his crew, returning from the trans-Pacific sailing race, and it soon became known as the Great Pacific Garbage Patch (Moore and Phillips, 2011). Later studies estimated that at least 6.4 Mt of plastic litter enters the oceans every year; that some 8 million pieces are discarded every day; that the floating plastic debris averages more than 13 000 pieces per km^2 of ocean surface; and that some 60% of all marine litter stems from shoreline activities (UNEP, 2009).

Additionally, the latest summaries show that, despite many efforts to limit this, now planet-wide, degradation the accumulation is still increasing (STAP, 2011). Dangers to ocean life are posed by every size unit of discarded plastics: among the largest items are abandoned or damaged fishing nets that can ensnare fishes, dolphins, and often even whales; while aquatic birds often mistake small-size pieces of plastic for small fish or invertebrates and regurgitate them to their fledglings: the stomachs of many species show a distressing collection of such objects. And microplastics – the smallest pieces (sizes less than 5, 2, or 1 mm) that are manufactured for cosmetics, drugs, and industrial uses and that arise from abrasion and photodegradation of larger pieces – can be ingested by marine biota and can have serious metabolic and toxic effects (Cole *et al.*, 2011). Inevitably, masses of plastic microparticles have been also accumulating on shorelines where they endanger more organisms (Browne *et al.*, 2011).

This important example shows how incomplete and uncertain are even our best analytical procedures tracing the requirements and consequences of material production, use, and abandonment; it also provides a strong argument for much better management of materials and, obviously, recycling should be a key component of these efforts. A closer look at the practice reveals enormous, and inexplicably ignored, opportunities as well as expected complexities (including some counterintuitive realities): recycling is both a practical and effective tool that should be deployed to the greatest extent possible, but that extent must be carefully determined in order to avoid an even greater overall waste of materials.

4.7 Recycling

Recycling is the flow that closes a circle and that turns those overwhelmingly unidirectional sequences of modern material extraction–production–use–abandonment into cycles. This is, of course, how the biosphere works: provision of water, carbon, and nitrogen to the Earth's biota could not be sustained without biogeochemical cycles. In a civilization using billions of tons of materials every year, recycling is an exceedingly desirable flow that can greatly extend the life-span of known mineral reserves, lessen the need for new biomaterials, and prolong the utility of materials turned into manufactured products. No less beneficial is the associated reduction in the use of embedded energy (differences in overall energy demand per unit of product can be as high as an order of magnitude) and, consequently, often substantial alleviation of many negative environmental impacts that result from production and thoughtless abandonment of materials.

The term covers several outcomes. Material recycling may be a matter of reusing the entire product or component, recovering the original constituents by separating, crushing, or melting the discarded products, or burning the recovered materials to generate heat or electricity. Materials are reused either as salvaged (bricks are perhaps the most common example of this practice) or after minimal refurbishment (massive timbers salvaged from old buildings are often resawn). Most of the recycling is done in order to recover the original ingredients that can be used by recyclers or sold to other industries as raw materials. Metals could be recycled with a minimum degree of waste to produce materials that are qualitatively indistinguishable, or very close, to the original (primary) inputs, but with most materials recycling is more accurately described as down-cycling: high-quality paper becomes packaging stock or cardboard, expensive plastics are turned into cheap items.

Energy savings and reduced environmental impacts are among the most appealing, and the best studied, advantages of recycling. Production of recycled steel will require roughly 75% less energy and the LCA of common steel products (sections, hot-rolled coil, and hot-dip galvanized steel) showed recycling benefits of up to about 50% for both GWP and acidification potential (WSA, 2011b). Benefits of aluminum recycling, although not as high as commonly claimed, are even greater than for steel. A standard claim is that aluminum recycling saves of up to 96% of energy compared to production from bauxite that needs about 200 GJ/t. But that claim would be true only for melting old metal. Its delacquering (removal of any coatings) needs about 7 GJ/t, melting consumes 7 GJ/t, addition of pure aluminum to adjust the alloy composition takes about 8 GJ/t, and the production of containers (casting, rolling, blanketing, forming) adds 30 GJ/t for the total of about 52 GJ/t, a saving of 74% rather than 96% (Luo and Soria, 2008). Because of their ubiquity and their high rate of recycling, aluminum cans have been the subject of detailed life-cycle analyses in Japan (Mitsubishi, 2005) and the USA (PE Americas, 2010) and a comparative European LCA looks at aluminum cans and beer bottles (Green Delta, 2013).

But the benefits go beyond energy savings, as recycling lowers GWP, claims less water, and generates less water and air pollution: Grimes *et al.* (2008) provide many comparisons for steel, aluminum, and paper production. Recycled paper can be made with 40% less energy while generating 45% less waste water and 50% less solid waste (EPN, 2007). Energy savings for recycled high-density polyethylene are 45–50% and 40–45% for PVC. As for the emissions of GHGs, producing a kilogram of hot-rolled steel coil will generate 2.2 kg of CO_2 equivalent, while recycling will claim only 1.8 kg, preventing emissions of 0.4 kg CO_2; savings with the recycling of plastics are much higher, close to 2 kg CO_2/kg of both PP and PVC; and aluminum recycling offers the greatest GWP reduction with at least 10 kg CO_2/kg of Al scrap (GHK, 2006).

But recycling is also a quest that is often very difficult to pursue (be it because of logistic challenges, excessive costs, or negligible energy savings) and one that, unlike the flows of carbon or nitrogen atoms in grand biogeochemical cycles, often amounts to a fairly rapid down-cycling as the reused materials appear in less valuable guises. Some challenges are fairly universal, others are material-specific. Recycling of waste predictably generated during large-scale industrial processing (be it smelting, casting, or machining of metals, or production of plastics, glass, or paper) is obviously easiest to do, and it is the cheapest and the most rewarding way to reuse valuable waste. In some industries (above all in steelmaking, aluminum smelting, and manufacturing of plastic products) this is a major source of recycled material.

In contrast, expensive collection costs burden recycling of all wastes that are produced with low spatial densities, with fluctuating quality and varying rates: paper and plastic waste generated by urban households are the prime examples in this difficult category. But the greatest challenge is to recycle the increasing amount of electronic waste that is, in mass terms, dominated by a few plastics, a few metals, and screen glass, but that also contains small, but in aggregate quite substantial, amounts of more than a dozen elements.

An overall verdict about modern civilization's recycling performance is easy: so far its recycling efforts have been quite inadequate even when compared to what might be labeled as a barely satisfactory effort. For example, of the 60 metals and metalloids examined by a UN report, more than half (including all rare earths, as well as germanium, selenium, indium, and tellurium) have recovery rates of less than 1%, for only five elements is the rate between 1 and 25%, and 18 common (or relatively common) metals have end-of-life recycling rates above 50%, but rarely above 60% (Graedel *et al.*, 2011).

Another way to look at the current extent of recycling is to express the recovery of materials as a share of the latest annual outputs. For metals these rates cluster in the 30s: according to the Bureau of International Recycling (2011) the figures are 30% for zinc, 33% for aluminum, 35% for lead, 37% for steel, and 40% for copper. America's metal recycling rates calculated for the years 1998–2004 range widely, from highs of more than 90% for gold and silver (both about 97%), lead (95%), titanium (94%), and titanium (91%) to lows of 19% for zinc and 15% for cadmium, with iron and steel averaging 52%, copper 43%, and aluminum 42% (Sibley, 2011).

Calculating the amounts of metals that will become available for recycling depends on estimates of the average life-time of products. For example, for aluminum these range from discarding the items in a production year (beverage cans, other containers, and packaging foils) to a decade in consumer durables, and to more than 30 years in structural elements, machinery, and some equipment (Hatayama *et al.*, 2009). Unfortunately, some toxic heavy metals, whose release into the environment is particularly undesirable, have very low recycling rates. The only relatively common way of recycling cadmium is by returned Ni-Cd batteries, but their collection rate remains low. The recycling rate of fluorescent lights containing mercury is also too low.

I will devote the rest of this section (and of this chapter) to brief appraisals of the recycling efforts for four materials – two key metals (steel and aluminum) and plastics and paper – and of electronic waste, a category of discarded material that would most benefit from much enhanced rates of recycling. Resmelting of steel is by far the largest recycling effort in overall mass terms, an achievement made easier by the ubiquitous availability of the material, by the ease of its separation (magnetic sorting), and by its perfect recyclability: resmelted material retains all physical properties of the original metal. In 2010 worldwide use of steel scrap reached 530 Mt, an equivalent of more than 37% of all crude steel.

About 190 Mt (36% of all scrap) were in the category labeled by the industry as "own arisings," that is circulating scrap produced by mills and foundries during smelting, casting, rolling, and further processing of the metal (WSA, 2011a). The rest (64%) of steel scrap was purchased by steelworks and it originated not only from many domestic sources but from imports, including those from different continents, with the USA (about 20 Mt) and EU-27 (about 19 Mt) being the largest exporters in 2010. About a third of all purchased scrap (amounting to 110 Mt in 2010) was new steel (so called process scrap), the rest were old products (so called capital scrap).

National steel scrap/crude steel ratios cover a considerable range, from just 14% in China to 100% in Luxembourg; the ratios for major producers range from 43% in Germany and 56% for the EU-27, to 68% in Poland, and 79% in Italy; the US, Japanese, and Russian ratios in 2010 were, respectively, 75, 35, and 32%, reflecting to a large extent domestic availability of old and new steel scrap, whose supply has been boosted in the

USA by generations of high rates of car and major appliance ownership made largely of the metal. Recycling has become particularly important for making stainless steel (alloyed with Mo, Ti, V, and W) that is in increasingly high demand.

High energy pay-off has made aluminum the world's most widely recycled material: about three-quarters of the 700 Mt of the metal that have been produced worldwide since the 1880s are still in circulation as secondary, tertiary, and so on, reincarnations (AA, 2013). Global recycled aluminum production (refining and remelt) topped 8 Mt in 2007 and was 7.7 Mt in 2010, of which 4.3 Mt were made in the EU, nearly as much as the primary output of 4.4 Mt (EAA, 2013). But there are clear limits to the metal's recycling. Hatayama *et al.* (2009) calculated that, in the coming decades Japan, the USA, the EU, and China could reduce their primary aluminum consumption to, respectively, 60, 65, 30, and 85% of the levels prevailing between 2003 and 2005 – but also that by 2050 there will be 12.4 Mt of obsolete aluminum scrap that cannot be recycled due to high concentrations of alloying elements.

Global aluminum recycling rates are about 90% for transportation equipment (particularly airplanes) and construction materials (OEA, 2010), but the most often recycled item made of the metal is the aluminum beverage can. America's recycling of aluminum cans, closely monitored by the Aluminum Association (2013), fluctuated between 50 and 58% during the first decade of the twenty-first century and in 2011 it reached about 800 000 t of the metal with a nationwide collection rate of 58%. But that high rate comes from counting all domestic and imported scrap cans in the numerator but only domestically produced cans in the denominator; with can imports now in millions the proper calculation, advocated by the Container Research Institute (2013) should count only those recycled cans that were made in the USA, and the recycling rate then drops substantially, from 58.1 to 49.6% in 2010.

But the rates are much higher in those US states that levy charges on container sales, and international comparison of used beverage can recycling rates in 2010 shows that the three global leaders – Germany, Finland, and Norway with rates, respectively, of 96, 95, and 93% – have deposit schemes (EAA, 2012). But even recycling rates between 50 and 70% (the most common values for affluent countries) are far higher than the returns for PET bottles. In the USA, the PET bottle recycling rate peaked in 1995 at about 37% and fell to 25% by the year 2000, and even slightly lower afterwards, a poor performance that resulted in billions of bottles landfilled or discarded in cities, along roads, in forests, and on shorelines. The lightness and dispersion of these containers will always make their near-perfect recycling challenging: it takes 20 000 PET bottles to make 1 t of recyclable waste.

Recycling of plastics brings the expected benefits of lower energy use – typically about 55–60 MJ/kg of PE, or savings of 60–80% (Vlachopoulos, 2009) – and a commensurate reduction of environmental pollution, but collection and processing of plastics is much more challenging than the reuse of metals, or paper, because household (or institutional) wastes contain mixtures of five major polymers (sometimes with two compounds forming parts of a single container, often in highly fluctuating ratios) that must, after time-intensive collection, be manually separated. Correct separation is difficult but imperative: a single PVC bottle in a load of 10 000 PET bottles can ruin the entire melt (ImpEE Project, 2013). Sorting is followed by cleaning (but removing print and labels cannot be

100% successful) and reduction to uniform pellet sizes ready for reuse, still mostly only in lower-grade applications such as cheap carpets, garbage cans, or park benches.

Even in Europe, with its profusion of national rules and EU directives, recycling rates remain very low. Collection rates of post-consumer plastic waste are impressive, more than 90% in eight EU countries and in Switzerland, and an average of 43% (25.1 Mt) in 2011 for the EU-27; but more than 40% (10.3 Mt) of collected plastics were then landfilled and of the 14.9 Mt (60%) that were recovered nearly 9 Mt were incinerated (energy recovery) and only about 6 Mt (or close to 10% of annual production) were recycled (Plastics Europe, 2012). Of course, plastics have a high energy content (mostly between 41 and 45 GJ/t) and for this reason the recovery of feedstock energy is preferable to landfilling, but that option has been mostly avoided in the USA where, in 2010, waste-to-energy facilities had the capacity to incinerate only 14% of the generated municipal solid waste (USEPA, 2011).

Opportunities for much higher rates of plastic recycling are ubiquitous. Beverage bottles (now a truly universal plastic waste, an eyesore, and both a land and marine pollutant) continue to have low recycling rates: in 2010 just 28% for HDPE bottles and 29% for PET bottles and jars (USEPA, 2011). The EU's collection rates range from just 5% for plastic bags and sacks, to 10% for plastic trays, films, and stretch and shrink wraps, and 25% for bottles and containers. In 2011, 1.7 Mt of used carpets were discarded in the USA, and despite a fairly strong growth of carpet recycling (with mechanically separated fibers that are ground, cleaned, and remelted) only 9% of old carpeting ended up in new products (CARE, 2012). The overall record is no better: in 2010 only 8% of all plastics present in municipal solid waste in the USA were recycled (USEPA, 2011). Similarly, in the EU nearly 50% of all plastics go to landfill, roughly an equivalent 12 Mt of crude oil a year (Potočnik, 2012).

All affluent countries, as well as China, have done much and, increasingly, better with recycling paper. Global consumption of recycled fibers rose from 152 Mt in the year 2000 to nearly 224 Mt in 2010; this mass is about 20% larger than virgin fiber production and it means that roughly 55% of the world's paper and paperboard were made from recycled material in 2010 (Magnaghi, 2011). China's traditional shortage of wood led to common use of nonwood fibers (mainly straw) whose pulping is less energy expensive, and to high reliance on recycled paper: nonwood fibers and imported waste paper provide raw material for nearly 60% of pulp. Between 2000 and 2010 China's use of waste paper rose from 18 to 66 Mt and its imports quadrupled (BIR, 2011).

Collection of household waste paper is expensive, and a thorough processing of the material is needed to produce clean fibers for reuse. This includes defibering of paper, cleaning and removal of all nonfiber ingredients (most often adhesive tapes, plastics, and staples), and de-inking is needed if the fibers are to be reprocessed into white paper. Recovered fibers can be used to make 100% recycled products, but more often they are blended with virgin fibers. Reprocessing shortens the cellulose fibers and this means that paper can be recycled no more than four to seven times. Japan has been the most assiduous collector and reuser of waste paper.

Japan's recovery rate went from 34% in 1960, to 50% by 1985, and 78% in 2010 (PRCP, 2012). Given the fact that the denominator used to calculate this rate includes a great deal of nonrecoverable or nonrecyclable paper (sanitary paper, tissues, treated paper) the country's Paper Recycling Promotion Center (2009) was right to call the rate of nearly 80% "a very significant achievement." For comparison, the US recovery rate has been also fairly high, reaching almost 65% in 2010. But Japan's recent utilization rate of collected waste paper has trailed the recovery rate by 12–13%, whereas the gap was only about 5% in the year 2000.

Recycling of e-waste is particularly challenging, because the devices commingle many compounds and elements that must be separated by a sequence of mechanical and chemical operations. Silicon is the core of modern computers, but transistors and microchips could not function without the presence of a multitude of materials, including many heavy metals. As already noted, elements used to dope silicon include arsenic and phosphorus and also boron and gallium. Printed wire boards, disk drives, expansion cards, electrical supplies, and connections add up to a hoard of materials, small (and for many elements even miniature) per unit but highly consequential in the aggregate. Most of the mass in electronic devices is steel, glass, plastics, copper, and aluminum (in covers, cases, screens, connections) but more than dozen other metals are also present in tiny amounts, eight of them (As, Cd, Cr, Co, Hg, Pb, Sb, and Se) classified as hazardous (USGS 2001; Williams, 2011).

Lead is of the greatest concern. Li *et al.* (2006) used toxicity characteristic leaching procedure (TCLP) to show that the element's concentration in TCLP extracts made from computer wire boards had 30–100 times the level of Pb (5 mg/l) that classifies waste as hazardous, while other heavy metals had essentially no potential to cause toxicity through leaching. Another lead-related concern is due to batteries used to power computers, which hardware and software companies in cooperation with governments and philanthropic organizations promised to donate to low-income countries where many people are not connected to the grid. Cherry and Gottesfeld (2009) concluded that between 3.5 and 4.5 kg of lead could be lost to the environment for every 12-kg battery, even with very high (90–100%) rates of recycling.

But even if they were Pb-free, computers should not be discarded as their concentrations of some precious metals are equal, or higher, than those present in metallic ores. This is, most notably, the case with gold: there is as much as 1500 g AU/t of circuit boards, 40 times the metal's concentration in ores mined in the USA (USGS, 2001). Concentrations of copper in discarded computers are also high, and the challenges of heavy metal contamination, recycling of rare elements, plastics, and glass extend to the still increasing range of electronic products that contain printed board, microprocessors, and connections (and the portable ones also power sources): this category of e-waste now extends from photocopiers and facsimile machines to electronic games and cellphones (Robinson, 2009).

What happens to these devices is only poorly known. Surprisingly, many people never throw the smaller devices (especially computers) away and keep them in basements or storage, too many units are simply thrown into the garbage and end up in landfills, some are stripped of some materials and a substantial share of e-waste is exported to Asia (above all to China and India) for dismantling and recycling, a practice associated with some truly horrid labor conditions (working without any protection while handling toxic substances) and severe environmental problems (acids and heavy metals seeping into soils and entering streams and ponds), particularly in the infamous recycling center in Guiyu, in China's Guangdong province.

The uncertain fate of e-discards is illustrated by the fact that, in their material flow analysis of used computers in the USA, Kahhat and Williams (2012) could do no better than to offer rather wide ranges concerning the fate of 40 million used and scrap devices, concluding that 6–29% of them were exported, 17–21% were landfilled, and 20–47% were collected for domestic recycling (but the actual recycling rate was much lower). Problems with "traditional" e-waste (TVs, CD players, desk computers, laptops, game consoles) have been compounded with the rise of cellphones. Since their commercial introduction during the 1990s (I am beginning the count only with the rise of digital cellular networks, leaving out pre-1991 heavy and inconvenient devices with limited connectivity) these devices have seen continued growth of ownership.

More than 100 million cellphones were sold for the first time in1997, more than 1 billion a year by 2009, and by the end of 2012 there were more than 6.5 billion devices in use (more than 1 billion in China) and total shipments of mobile devices (including tablets, notebooks, and netbooks) surpassed 1.9 billion units including 1.7 billion cellphones (dotMobi 2013). With an average life-span of just 18–24 months, this implies an annual discard rate of 800–1100 million cellphones a year. The best data for the USA, published by the EPA, show dismally low recycling rates for all mobile devices (just 8% in 2009, or fewer than 12 million units, compared to 129 million discarded units in 2009), and rates of 17% for TVs and 38% for computers (USEPA, 2012).

Although cellphones are the lightest e-devices, the aggregate amount of metals in annually discarded units is considerable. With about 15 g Cu, 0.3 g Ag, and 0.03 g Au per unit (USGS, 2006), 130 million devices annually discarded in the USA contain about 2000 t of copper, 45 t of silver, and 4 t of gold, besides nearly 9000 t of plastics (they make up 55–60% of the total mass) and 2500 t of ceramics and glass – and worldwide totals of these materials incorporated in discarded cellphones will be at least six times larger.

5

Are We Dematerializing?

The *Oxford English Dictionary* (OED) defines the verb *dematerialize* as becoming free of physical substance. This definition has obvious supernatural and spiritual connotations: it has been used in science fiction (in the USA most famously in *Star Trek*) where objects and persons disappear by means of an unexplained process; dematerialization of Christ's body is said to explain the apparent imprint of torso, limbs, and wounds on the fabric of the shroud of Turin (Berard, 1991); and a famous Yoghi asserted that a man will be free only once he "is able to dematerialize... his human body... and then materialize it again" (Yogananda, 1946).

Fortunately, my concern is with more mundane matters, and OED's example of real-world dematerialization refers to replacing physical records or certificates "with a paperless computerized system." This is an example of dematerialization that appears to be factually correct and that has been universally accomplished. Printed stock certificates (including many large and ornately designed versions) have become just strings of ones and zeros in electronic memories – but a closer look shows that this has been only an apparent dematerialization, just one of many immaterial e-replacements that have either resulted in no substantial overall material savings (because of the need for less obvious deployment of other materials in the new infrastructures needed to provide the substitutes) or, more often, have actually led to an increased consumption of paper (a trend clearly correlating with large-scale computerization).

But a much more common category of dematerialization is not the complete elimination of a particular material, but a substantial reduction of material used, be it a finished item, a service rendered (per unit of power delivered by a prime mover, per unit of computer performance) or a unit of national economic product. This **relative dematerialization**, has been one of the key trends in modern manufacturing whose quest for higher productivity and lower prices has brought reduced use of materials (be they traditional and inexpensive, or modern and costly) for products ranging from beverage cans to jetliners, and from simple structural elements in buildings to computers.

Making the Modern World: Materials and Dematerialization, First Edition. Vaclav Smil.
© 2014 John Wiley & Sons, Ltd. Published 2014 by John Wiley & Sons, Ltd.

And there are still other categories of dematerialization to consider. Herman *et al.* (1990, p. 334) argued that from an environmental point of view it is not the reduced unit mass of products that matters (as smaller and lighter items could be of inferior quality and their more frequent replacement could generate more waste) but that dematerialization "should perhaps be defined as the change in the amount of waste generated per unit of industrial products." This might be a desirable and revealing choice, but collecting the necessary information about all possible waste-streams and coming up with relative valuations of different emissions or effluents (as mass reduction of one waste-stream can be replaced by a much smaller mass release of a compound that is more harmful or more durable) is obviously challenging. LCA's (life-cycle assessment) goal is to provide precisely that kind of information but, as shown in the preceding chapter, such assessments should be used with caution.

Similarly, two other measures of dematerialization – declining consumption of goods or lower use of energy per unit of economic product – face a number of intractable data challenges relating to the accounting for overall economic output; while adjustments for inflation can be done fairly easily and with acceptable uncertainty, conversions to purchasing power parities (PPPs) are far more questionable. Moreover, energy use per unit of GDP may be a common measure of an economy's overall energy intensity but (even when setting aside the uncertainties inherent in converting various energies to a common denominator) a closer look shows that it is fundamentally flawed, that its narrow interpretation gives very limited insights and that it is only a poor proxy for tracing both historical and recent changes of material consumption in growing economies. I will review, deconstruct, and assess all of these dematerialization measures.

5.1   Apparent Dematerializations

When explaining dematerialization, the OED should have used the conversion from blueprints to computer-assisted design (CAD) as a far more consequential example of disappearing uses of paper than the replacement of printed stock certificates by electronic versions. That apparent dematerialization eliminated roomfuls of workers at their slanted drafting boards and replaced large numbers of paper blueprints filed in heavy storage steel cabinets with electronic graphics displayed on screens and saved initially on tapes and then on magnetic devices, hard drives, and various portable storage media. CAD applications, beginning with two-dimensional designs and progressing to complex three-dimensional capabilities, were used first in military projects, GM began its CAD development during the late 1950s, commercial aircraft makers joined during the 1960s (Weisberg, 2008).

The total number of blueprints needed to build more complicated machines steadily increased, first between the 1890s and 1920s as internal combustion engines and steam turbines began to replace steam engines, and then (between the 1900s and 1930s) as gasoline and diesel engines were energizing more advanced car and truck designs. And the complexity of airplane design increased the number of blueprints by another order of magnitude. A World War II bomber required about 8000 drawings – and the adoption of jet propulsion (starting in the 1940s for military designs and a decade later for civilian aircraft) and the introduction of wide-bodied jetliners during the 1960s brought another order of magnitude increase.

The Boeing 747, conceived in 1965 and launched in 1969, required about 75 000 drawings adding up to nearly 8 t (Boeing, 2012). This is why, during the 1970s, Boeing developed a mainframe-based CAD system that eventually contained more than 1 million drawings and that was replaced, after more than 25 years of service, by a server-based Sun Solaris operating system (Nielsen, 2003). During the 1990s such airplanes as the Boeing 767 and Airbus 340 became the most impressive embodiment of pure CAD. Even the earliest CAD systems often halved the time spent on a particular task, and later advances reduced labor requirements to between 10 and 20% of the original need.

No less important have been the paper savings brought about by automated electronic redesign, as previously laborious adjustment and changes needed for redrawing entire blueprints were turned into just handfuls of mouse clicks. For example, during the design of the B-29 (Stratofortress, America's largest and most complex long-distance bomber) more than 4000 test drafts were thrown away for every final draft making up the plane's final production version (Herman, 2012). In contrast, CAD makes it easy to try out a large number of variants and adjustments, as appropriate software follows any ripple changes through the entire design and amends everything. As CAD became the norm it replaced experimental and trial drafts and proposal and final blueprints in virtually all categories of construction, manufacturing, and network design with disembodied electronic versions. In less than two generations advanced countries experienced a nearly universal dematerialization of design.

Of course, the dematerialization claim is correct only when looking at what has been eliminated, and that is why I have used the qualifier "apparent." Thanks to CAD, large and small sheets of paper are gone, as are drafting tables, chairs, and utensils, and large steel storage cabinets – but creating and preserving dematerialized blueprints requires extensive infrastructures of modern electronic computing, redundant mass data storage, and communication, ranging from specialized software (written by using other computers) to large flat screens to the roomfuls of servers required to process Internet traffic (globalization of design and the reliance on construction subcontractors on different continents, be it for making airplanes or cellphones, call for routine international sharing of design proposals and specifications) – to say nothing about the increased demand for electricity and hence for the requisite infrastructure of its reliable supply (generating by combustion of fossil fuels or harnessing of renewable flows, transforming the current and transmitting it along HV and distribution lines to final consumers).

Consequently, even in a case that appears to be the perfect example of dematerialization, the reality is nothing but a complex form of material substitution. Diffusion of CAD has reduced wood harvesting, production of pulp, paper, and drafting implements, and steel storage – but it has generated additional demands for the complex (and energy-intensive) material infrastructures of the modern electronic world and, secondarily, demands for the materials needed to energize those systems by reliable generation of electricity. Given the intricacy of modern electrical and electronic infrastructures, their multiple uses and their very different longevities (software or screens may be replaced in a matter of months, thermal power stations will stay put for decades) it would be very difficult to come up with a net assessment of the material balances involved in such complex substitutions. Even if such exercises were much easier to conduct, there is little doubt that their results would have a wide range of outcomes.

Microprocessors have helped to dematerialize an increasing assortment of products and services, from books (e-books accessible on computers and on numerous e-readers) and educational materials (on-line language lessons and tutorials with teachers in a different continent) to banking statements and supplier invoices, and from correspondence (e-mail) and transportation tickets (e-tickets on cellphone screens) to phone directories (even though fat yellow page books are still available) and maps. And new forms of e-art do not require canvass, paint, stone, or clay, e-palettes offer millions of hues; and while the world's great museums will not go the way of stock certificates, visiting many of them can be now done entirely (room by room) without any travel or any physical presence via virtual e-tours (for a list of museums offering these tours see http://www.virtualfreesites.com/museums.museums.html).

Obviously, electronic infrastructures (including material and labor back-ups) will be very different for creating a virtual tour of a small museum than for designing and maintaining a system for reservation and e-ticketing by a large international airline. And it would be even trickier to account for the consequences of these dematerializations. Access to on-line reservations for every airline flying a particular route has created more competition, reduced prices, and contributed to increased frequency of flying: paper tickets have vanished but more materials are needed to support the complex infrastructures of international air travel. Similarly, an e-tour of a museum may actually prompt the viewer to take a trip in order to see the real thing, but we will never know the net impact of virtual activities on the overall material requirements of the museum itself: the only certain conclusion is that they will not go away.

5.2   Relative Dematerializations: Specific Weight Reductions

Reduction of material inputs in production can be accomplished in four principal ways: by gradual improvements that do not involve new materials; by substitutions (often relatively rapid) of constituent materials with lighter or more durable alternatives; by intensified recycling, particularly effective in those cases where reusing brings major energy savings; and by the introduction of entirely new devices that perform the desired functions with only a fraction of the mass needed for their predecessors. The first approach usually requires nothing more than a smarter design that reduces specific material input without compromising the desired function: lighter steel beams or engineered timber used in modern construction are good examples of these design changes, but the most ubiquitous instance of this development are lighter beverage cans.

Beverage cans are also a good example to illustrate the second and the third category of change, relatively rapid substitution of materials and intensive recycling. Replacement of steel by aluminum reduced the mass of soft-drink cans but, given the high energy intensity of primary aluminum production, that substitution would have led to higher energy costs. However, more common use of recycled aluminum has also lowered the need for new materials required to produce primary inputs and has resulted in substantial overall energy savings. In mass terms, steel recycling offers the best global example of this welcome consequence, as the practice obviates the large-scale deployment of large amounts of materials needed to produce the metal from its ores.

As for the last category, perhaps the best example of adopting entirely new solutions has been the displacement of heavy and bulky vacuum tubes, first by tiny transistors and then by transistors crowded onto silicon chips to make microprocessors. In reality, these processes of relative dematerialization are not mutually exclusive and, as in the just described case of beverage containers, strategies that combine two or more approaches have been common. But this does not mean that the quest for dematerialization has been the paramount factor of product evolution. The history of virtually any kind of manufactured product could be presented as a story of declining costs and improving performances, be they measured by convenience of use, greater power, longer durability, or higher reliability (with their course following expected logistic curves as additional gains bring diminishing returns).

In some cases, reduced material use has spearheaded these trends or has been a critical factor driving the improvements: mass reduction, together with safer handling, has been an explicit design goal in replacing glass by metals in distributing beer and soft drinks. First came steel beverage cans, then aluminum cans, and both kinds of containers have been progressively redesigned in order to use thinner body sheets and smaller tops. In other cases, relative dematerialization has been a welcome consequence of innovations motivated by other goals. The substitution of tiles by plastic flooring simulating tile design was not driven by mass considerations but by the affordable cost of materials and installation; and the heavy glass, cast iron, and ceramic objects (bowls, pots) present in traditional kitchens were not replaced by plastics and thinner metals primarily because of their weight, but because they, too, were more expensive and often less safe to handle.

The evolution of soft drink cans also illustrates both an often protracted process of substitution and a surprisingly large magnitude of eventual improvements. Steel cans for storing many kinds of food have a long history that goes back to the beginning of the nineteenth century, to the first commercial steps to produce canned food by Nicolas Appert in France and Peter Durand in the UK (Shephard, 2000). But steel containers began to be used for beer and nonalcoholic beverages only during the 1930s and reached their greatest popularity during the 1950s. The first aluminum cans were introduced by Coors in 1959, Royal Crown Cola was the first soft drink maker to use them in 1964, Coca Cola and Pepsi Cola followed in 1967 (Aluminum Association, 2013).

A decade later steel cans were on the way out, and none of them have been used for beer since 1994 and for soft drinks since 1996, but they have retained their dominant position in preserving fruits, vegetables, and meats (CMI, 2013). Aluminum cans are made from alloys (containing about 5% Mg, less than 2% Mn, and traces of Fe, Si, and Cu), their insides are often protectively coated, and they have crimped-on tops opened by pull tabs. At 85 g the first aluminum cans were surprisingly heavy; by 1972 the weight of a two-piece can dropped to just below 21 g, by 1988 it was less than 16 g, a decade later it average 13.6 g, and by 2011 it was reduced to 12.75 g (CMI, 2012).

This means – assuming average weights of 19 g in 1980 (when 41.6 billion cans were shipped) and 13 g in 2010 (97.3 billion cans shipped) – that over the three decades the 2.3-fold increase in the consumption of cans was accompanied by only a 60% increase in the mass of aluminum. Another way to illustrate this relative dematerialization is to note that the mass of 2010 cans made as heavy as they were in 1980 would have required an additional 580 000 t of aluminum, or as much metal as is needed to build eight Boeing 747s or (because of their lower Al content, only 20% of empty weight) more than 25 Boeing 787s.

Moreover, as already noted in the preceding chapter, the aluminum can is the most frequently recycled container (indeed the most recycled small object) with more than half of the annual US output (no matter how the rate is measured) now coming from recycled metal as relative dematerialization and reuse combine to moderate the energy cost and environmental impact of aluminum cans. And the mass of 13 g per can is not the end of container light-weighting: in 2012 a European maker introduced its latest 330 ml can (US cans are slightly larger, containing 7.5% more volume or 355 ml) that weighs just 9.5 g (Ball Packaging Europe, 2012). But this often cited example of relative dematerialization is dwarfed by a much more consequential and yet (curiously) overlooked evolution of prime movers, the shift from animate to inanimate energies and the subsequent advances of engines and motors. This neglected but fundamental development deserves a closer look.

Muscles, converting chemical energy in food and feed into kinetic energy of human and animal labor, were the dominant, animate, prime movers in all pre-industrial civilizations. Mechanical (inanimate) prime movers had limited roles in traditional societies, but became the enablers of the modern economic development characterized by industrialization, mechanization of agriculture, and urbanization and resulting in unprecedented gains in quality of life. The first mechanical prime movers were simple wooden machines, water wheels (first introduced during antiquity), and windmills (used since the early Middle Ages) converting the energies of flowing water and wind into reciprocating or circular motion that was used to mill grains, press oil from seeds, saw wood, pump water, or blow air.

Combustion engines convert the chemical energy of fossil or biomass fuels into the mechanical energy of reciprocating or rotary motion. External combustion in steam engines (heating water in boilers to supply steam, the working medium, to the machines moving parts) came first, very inefficiently, during the first half of the eighteenth century, and then with increasing efficiencies after 1760, until the machines reached their performance plateau by the end of the nineteenth century. Internal combustion engines were commercialized after 1860, starting with inefficient and heavy stationary engines burning coal gas, proceeding to much lighter gasoline-fueled automotive engines (pioneered during the late 1880s) and to relatively heavy but inherently more efficient engines designed first by Rudolf Diesel during the 1890s.

The twentieth century brought two other categories of internal combustion engines: gas turbines powered either by liquid fuels (kerosene for jet planes) or by gases (by natural gas in large stationary gas turbines), and rocket engines using a variety of liquid or solid propellants. The third class of modern prime movers has been used to convert the kinetic energy of water, wind, and steam into electricity: steam turbines date back to the 1880s, their modern high-capacity designs receive superheated steam from large coal or natural gas-fired boilers and they have been the dominant means of generating the world's electricity for more than a century. Large hydro turbines have supplied substantial shares of electricity in many countries with rich water resources, and during the past decade modern triple-bladed giant wind turbines have become important electricity providers, particularly in some European nations.

The declining specific mass/power (g/W) ratio has been, besides the growth of their unit capacities, perhaps the most remarkable long-term trend affecting all categories of modern prime movers. The rate's inverse (W/g), commonly known as power-to-weight ratio (or specific power), is one of the most revealing characteristics used when assessing and comparing the performance of engines or machines; and its variant, thrust-to-weight ratio, is used for gas turbines and rocket engines deployed in commercial and military jet propulsion and in launching payloads to space. The baseline for comparison is the performance of the first fuel-powered mechanical prime movers, steam engines of the eighteenth century. By 1750, after decades of marginal improvements, a standard-sized, Newcomen, inefficient steam engine developed about 15 kW, weighed nearly 9.6 t, and had a mass/power ratio of about 640 g/W (Smil, 2008).

Three decades later even Watt's famous improved design (patented in 1765) remained a very heavy machine (9.2 t, 15 kW) rated at just over 600 g/W. This was the same order of magnitude as the specific mass for the two most important animate prime movers, hard-working men and large draft horses. Although the power of brief (anaerobic) human exertions can be as high as 400 W (prorating to less than 200 g/W), men can work steadily at rates between 60 and 80 W, averaging about 1000 g/W of useful labor. The heaviest working horses (English Clydesdales, French Percherons) weigh nearly 1 t and develop more than 1 hp (745 W), that is, again, specific power close to 1000 g/W during prolonged pulls (Smil, 1994).

By the mid nineteenth century the specific mass of many steam engines was below 500 g/W and by the late nineteenth century the best design (triple-expansion engines) rated less than 100 g/W, still too heavy to energize affordable road transport. Internal combustion engines changed that. In 1874, Otto's first (noncompression and stationary) internal combustion engine had a nearly 4-m tall cylinder and was very heavy: at about 900 g/W its mass/power ratio was several times that of the best contemporary steam engines. Compression lowered the engine's size and weight and reduced the ratio to about 270 g/W by 1890, less than that of the small stationary steam engines used in workshops and factories.

The first order-of-magnitude improvement came with the introduction of gasoline-powered engines, designed during the 1880s by Karl Benz, Gottlieb Daimler, and Wilhelm Maybach. By 1890 Daimler and Maybach had introduced their first four-cylinder engine whose mass/power ratio was less than 30 g/W (Beaumont, 1902). In 1901 Maybach revealed Mercedes 35, the prototype of all modern vehicles whose four-cylinder engine had power of 26 kW but whose aluminum block and honeycomb radiator lowered its weight to just 230 kg, yielding a mass/power ratio of 8.8 g/W. The second order-of-magnitude improvement came during the twentieth century with the transformation of cars from expensive oddities into affordable machines.

The world's first mass-produced car, Henry Ford's Model T launched in 1908, was initially powered by a 2.9-l four-cylinder engine with a mass/power ratio of about 15 g/W but by the time the car's production ended (in 1927, although the production of replacement engines continued until August 1941) the ratio had declined to less than 5 g/W. By the early 1930s, even such powerful engines as Pierce Arrow's powerful V-12 rated less than 4 g/W, and gradual changes before and after World War II eventually brought the mass/power ratio of commonly used automotive engines to less than 2 g/W by the 1950s and to less than 1.5 g/W by the mid 1960s: even Ford's 298 V8 for the 1965 Mustang rated just 1.4 g/W. During the first decade of the twenty-first century most passenger cars had engines with a mass/power ratio between 1 and 1.5 g/W, and the most powerful turbocharged engines rated well below 1 g/W.

Similarly low ratings were necessary to transform early airplanes carrying just one or two people into progressively heavier passenger and military aircraft. Wilbur and Orville Wright designed the engine that powered their first successful flight on December 17,

1903 (it was built by skilled bicycle mechanic Charles Taylor): its weight was 91 kg and it eventually produced 12 kW, resulting in a mass/power ratio of 7.6 g/W. The two world wars brought new record ratings, the first one with Liberty, America's first mass-produced aero engine with 1.34 g/W in 1917, the second one with the Wright R-3350 radial engines powering the B-29 (Superfortress) bombers with just 0.74 g/W (Gunston, 1986). Its successor, Wright R-1820-82A, was deployed just before the conversion of large-scale commercial flight to jetliners during the 1960s and weighed only 0.59 g/W.

Inherently heavier diesel engines – that have conquered heavy road, rail, and ocean transport (and have also made important inroads into the passenger car market, especially in Europe) – have seen similar weight reductions. Rudolf Diesel's third prototype, a massive (nearly 4.5 t) upright stationary machine with a piston stroke of 4 m that was used for specification tests in 1897, had a maximum power of 13.5 kW and a mass/power ratio of 333 g/W, three times higher than the best contemporary large steam engines and an order of magnitude above the improving gasoline-fueled designs (Diesel, 1913). The first stationary commercial unit sold in 1898 was still very heavy (296 g/W), but weights began to decline with mobile applications. By 1910, marine diesels were around 120 g/W and during the 1930s German engineers brought the mass/power ratio of the Jumo Junkers 205 aircraft engine (595 kg, 647 kW) to a remarkably low level of 0.92 (Wilkinson, 1936).

Today's diesel engines in passenger cars are only slightly heavier than their gasoline counterparts: two versions of the Mercedes Benz OM 651, introduced in 2008, rate, respectively, just 1.5 and 1.35 g/W (Daimler, 2008), and in heavy road service (buses, trucks) the ratios are as low as 3 g/W (Smil, 2010). Much heavier machines that are the prime movers on nonelectrified railways weigh between 3 and 10 g/W, and the world's most powerful marine diesel engines that propel massive container ships rate nearly 30 g/W: the 80.1 MW 14-cylinder turbocharged Wärtsilä's RT-flex96C, whose stroke length is 2.5 m, has a specific mass of 28.72 g/W (Wärtsilä, 2013).

Starting in the 1940s, jet engines (gas turbines) reduced the specific mass of the lightest prime movers by yet another order of magnitude, to less than 0.1 g/W. Whittle's pioneering W.1 turbojet of the early 1940s needed 0.38 g/W, early commercial turbojets of the mid-1950s rated about 0.25 g/W, and the latest turbofan engines weigh less than 0.1 g/W (Smil, 2010). The standard way to relate the mass of aero engines to their performance is by calculating the thrust-to-weight ratio: it increased from 1.6 for Whittle's prototypes of the early 1940s to nearly 4.0 for turbojet engines of the late 1950s, to 5.5 for large turbofans of the 1990s, and it reached 6.3 for GE90-11B, the largest turbofan introduced in 2003. This means that the latest turbofan engines need only a quarter of the mass to deliver a unit of propulsive power.

But even such low ratios would not be enough to escape the Earth's gravity, and only rocket engines have mass/power ratios lower than that. In terms of thrust/weight ratio, rocket engines have advanced from less than 20 for the PGM-11 that propelled Redstone, America's first ballistic missile in 1958, to nearly 140 by the early 1970s when the Kuznetsov group designed the NK-33, the world's most powerful liquid oxygen/kerosene engine that never carried out its intended mission to the Moon. To compare these specific masses by using the same units as in the case of car and airplane engines, when the Saturn C 5 rocket propelled Apollo 11 on its path to the Moon (in July 1969) its mass/power ratio (even after counting all the fuel) was merely 0.001 g/W.

Reduction of specific mass has also been impressive in the case of the largest stationary prime movers: the mass/power ratios of steam turbogenerators installed in thermal power plants are 2 orders of magnitude lower than the rates for the best late-nineteenth century steam engines. Those machines, rating nearly 100 g/W, were used in all early coal-fired electricity generating stations but, by 1891, just seven years after its patent, Charles Parson's steam turbine rated no more than 40 g/W; by 1914 that ratio was reduced to 10 g/W, and modern large turbogenerators have a specific mass only a bit above 1 g/W while the largest stationary gas turbines are only a bit heavier at around 2 g/W.

Finally, a few paragraphs about the relative dematerialization of electronic devices in general and computers in particular. These improvements are among the most appreciated instances of the dematerialization process, and in the case of computers this trend has been even more significant because their declining mass has been inversely related to their performance: the machines are not only smaller, their power has increased as they have become lighter, a result of the remarkably long duration of Gordon Moore's law, formulated in 1965 at a time when the most complex integrated circuit had 50 transistors. Moore's original forecast was for the doubling of transistor density per integrated circuit every year, but a decade later he extended the doubling period to two years (Moore, 1965, 1975). Comparisons of declining computer weight often start with the Apollo 11 guidance system that was used to steer and then to land the capsule on the Moon in August 1969. That computer had just 2 kB RAM and with mass of 32 kg that prorated to 16 g/B (Hall, 1996). In 1981 IBM's first personal computer had 16 kB RAM and mass of 11.3 kg or just 0.7 g/B.

In 2011, my new Dell Studio laptop had 4 GB RAM and weighed 3.6 kg, resulting in a specific mass of mere 0.0000009 g/B. This means that between 1969 and 2012, in 42 years, the mass/RAM ratio was reduced by 7 orders of magnitude, a truly stunning speed of relative dematerialization. In quantifying the overall effect of this relative dematerialization, we should start in 1981 with the IBM PC, because the Apollo computer was not a commercial system and was not really portable. In 1981, the aggregate RAM of some 2 million computers sold in that year was just over 30 GB and their mass added up to about 22 000 t; 30 years later the aggregate RAM of more than 300 million computers sold in 2011 was on the order of 1.2 Eb (10^{18} b) but their mass was about 1.2 Mt: consequently, an aggregate memory gain of 7 orders of magnitude (40 million times) was accompanied by only a 60-fold increase of total mass.

This is an exceptional example of relative dematerialization associated with an enormous operational improvement, and performance/mass gains of an order of magnitude are limited to devices that are functionally dominated by microprocessors. Modern cameras, whose design was transformed by the introduction of microchips and the subsequent rise of digital photography, are an order of magnitude lighter than their predecessors of just 30 years ago (my comparison is in terms of image quality and hence it ignores the small Instamatic-type devices that have been available since the 1960s). Scores of new digital designs weigh just around 100 g while standard SLR cameras of the 1970s and 1980s weighed about 1 kg and considerably more with their telephoto or macro lenses and with the bags needed to protect the lenses and to carry rolls of film.

Cellphones offer an even better example, as they have combined an impressive rate of relative dematerialization with no less impressive gains in communication quality and overall functionality. In 1973, Motorola's first mobile phone weighed 1135 g; in 1984 the first commercial device, Motorola's DynaTAC 8000X, weighed 900 g; in 1994 Nokia's units still had mass around 600 g; by 1998 the company had devices weighing only 170 g; and in 2000 the average phone weighed 120 g and the lightest design just 96.35 (GSM Arena, 2013). Since 2005, the average weighted mass has remained fairly steady at 110–120 g, because new phones are thinner (average 22 mm in 2000, 14 mm in 2010) but their displays are larger, 3.75 cm in 2000 and 7.5 cm in 2010 (GSM Arena, 2013). Apple's popular iPhones reflect that change: the original design released in 2007 weighed 135 g with a 6.2 cm-wide display, iPhone 5 released in 2012 had a mass of 112 g and display width of 7.6 cm (Apple, 2012).

Tupy (2012) made a list of 16 devices or functions that are replaced by applications installed on an iPhone: camera, PC used to receive e-mail, radio, fixed phone, alarm clock, newspaper, photo album, video recorder, stereo, map, white-noise generator, DVD player for movies, rolodex, TV, voice recorder, and compass. Even when leaving out all those functions that are far from being true equivalents of what they have replaced (truncated versions of the news do not replicate a real newspaper and a tiny screen cannot match viewing large-screen HD TVs) we are left with an array of equivalent, or even superior, substitutions: a cellphone can be a perfect watch and alarm clock, it works fine as a Sirius radio, its portable list of contacts neatly replaces the old-fashioned rolodex, GPS is a superior locator and direction finder compared to a compass and a sheaf of maps, it will do as a voice recorder, and many cellphones take better-quality pictures than most standard SLR cameras, to say nothing about eliminating purchases and development of film and printing pictures.

Even when adding just the weight of modern, that is mostly electronic, versions of replaced devices – a watch, a portable alarm clock, a small portable radio, a pocket voice recorder, and a digital camera, each of them weighing on the order of 100 g – we end up with total mass of 500 g compared to a bit more than 100 g for a cellphone; more liberal replacement criteria could double the mass of replaced devices, and adding up the weight of their mechanical predecessors from the 1950s would raise the aggregate to several kilograms and bring the relative dematerialization gain to (or very close to) 2 orders of magnitude. Keep this impressive comparison in mind: I will return to it in the penultimate section of the last chapter!

But where microchips are not the dominant component of the total design, there has been no even remotely similar mass decline, and in some cases microprocessor-driven dematerialization has been actually accompanied by substantial increases of overall mass. Passenger cars (and other two-axle four-wheel vehicles) are perhaps the best example of the countervailing trend. Although they remain complex assemblages of mechanical components (containing on the order of 30 000 parts) all of their key functions (above all, timing of engine operation) and many subsidiary tasks (deployment of airbags being perhaps the most critical one) are controlled by microprocessors using software that is more complex than that onboard fighter jets or jetliners (Charette, 2009). Modern cars are mechatronic hybrids with electronics accounting for up to 40% of the cost of the most expensive vehicles. But I will show in the following section that the total mass of average American vehicles used for passenger transport has been steadily increasing, driven not by engineering necessities but by personal preference.

And among the enormous number of quotidian consumer products that still contain no microprocessors (from apparel to cookware, from basic tools to hand-built furniture) there has not been a single example where mass per unit of product or per an indicator of performance has improved by several orders of magnitude, even reductions on the order of 1 magnitude (resulting in identical product performance while reducing its mass to around 10% of its original value) are extremely rare, and in most cases relative dematerialization has amounted to less than 30% compared to the same types of products available a generation ago.

5.3   Consequences of Dematerialization

Before taking a closer look at the material consequences of dematerialization, I should stress that this important indicator gives a necessarily limited and imperfect insight. A more realistic measure that would also consider the environmental impacts of weight reduction (land use, carbon emissions, aquatic eco-toxicity) might show that weight loss may not be the most important consideration: small amounts of heavy metals in electronics may have greater impacts than the larger mass of plastics (van der Voet *et al.*, 2005). In order to appraise these impacts we need specific comparative LCAs that, in some cases, would have to be repeated frequently as new materials and devices become parts of dematerialization processes.

Secular declines of specific mass/power ratios of products and prime movers could be used to make many revealing comparisons similar to that made in the last section when I equated mass of aluminum saved because of lighter beverage cans with a number of new Boeing airplanes. None of these comparisons is more stunning than a calculation of total computer mass when assuming that the average mass/RAM ratio of more than 300 million machines sold worldwide in 2011 remained at the early 1980s level: the total weight of materials (metals, silicon, plastics, glass) used to make different kinds of those personal computers would have been nearly 850 Gt, 2 orders of magnitude greater than the total mass of carbon extracted worldwide in that year in all fossil fuels and more than 200 times the mass of all materials consumed in that year by the US economy.

And, to introduce just a few common examples, everything else being equal, engines of today's intercontinental jetliners would be at least 50% heavier if their thrust/power ratio had remained at the early 1960s level; the engines in today's cars would have to be at least 3 times as heavy if they were built as the best designs of the early 1930s; if massive container ships were powered by the best steam engines the weight of these prime movers would be 3–4 times heavier than the weight of the latest, highly efficient, marine diesels; and today's thermal electricity generation powered by the most efficient and the largest steam engines available at the beginning of the twentieth century would require 50–100 times more material (mainly iron and steel) than does today's production energized by steam produced in large boilers in order to power steam and gas turbines.

These contrasts also make it clear that relative dematerialization has been a key reason for falling costs, greater affordability, and mass diffusion of successive generations of products, as well as for preventing some truly unimaginable levels of environmental pollution and degradation. Obviously, it would have been physically impossible for the global sales of personal computers to reach more than 300 million units in 2011, indeed not even 30 million machines, if their mass/RAM ratio had remained at the early 1990s level: even the most affluent and the most innovative society would not be able to acquire and handle the material throughput created by such a demand.

Similarly, today's passenger cars would not be so powerful if their mass/power ratio had stuck at the level of the 1930s: unlike the case of personal computers – whose mass/RAM ratio would have stuck in a not-so-distant past – we could build and drive large numbers of very heavy cars, but their prohibitively high capital and operating costs would prevent any mass ownership. And electricity generation powered by steam engines could never match the low cost of supply from massive turbogenerators with very low mass/power ratios: continuing reliance on steam engine-powered plants would have greatly restricted average consumption of electricity, would have prevented a truly mass ownership of electric and electronic appliances, and could have supported only limited household and public lighting and a relatively modest deployment of industrial electrochemical processes.

Of course, lower specific mass of products and lower mass per unit of performance of prime movers have not been the only major reasons behind falling costs and more widespread ownership of products and greater affordability of service. Technical innovation usually works in complex, interactive (albeit often unanticipated) ways and it becomes difficult, if not impossible, to assess the relative contributions of individual factors. The lower costs of energy needed to produce, process, and transport materials have been a key contributor and other principal factors include the declining cost of progressively more mechanized manufacturing, frequent (and in some cases almost continuous) redesign of commercially successful products, and introduction of new techniques and processes, some of incremental nature, others representing radical departures.

But lower energy costs would have been impossible without new products and processes: just think of the machinery needed for the cheapest surface coal mining, or of the impact of modern rotary rigs as they replaced the old percussion tools used in hydrocarbon extraction, or of the possibilities for economic long-distance electricity transmission brought by direct current high-voltage lines. In turn, the new materials needed for these products often required much higher energy inputs than did the production of their predecessors: aluminum vs. steel and plastics vs. glass are the two most obvious examples. And while many innovations that lowered operation costs were deliberately developed to fill a niche, others had originally nothing to do with their eventually widespread commercial applications: Czochralski's invention of pure silicon crystals growth, the material foundation of modern solid-state electronics, is a perfect example (Scheel and Fukuda, 2004).

In an overwhelming majority of cases, these complex, dynamic interactions of cheaper energy, less expensive raw materials, and cheaper manufacture have resulted in such ubiquitous ownership of an increasing range of products and more frequent use of a widening array of services that even the most impressive relative weight reductions accompanying these consumption increases could not be translated into any absolute cuts in the overall use of materials. Indeed, there can be no doubt that relative dematerialization has been a key (and not infrequently the dominant) factor promoting often massive expansion of total material consumption. Less has thus been an enabling agent of more.

This counterintuitive phenomenon was first described in relation to energy consumption by William Stanley Jevons, an English economist, one and half centuries ago:

It is wholly a confusion of ideas to suppose that the economical use of fuels is equivalent to a diminished consumption. The very contrary is the truth. As a rule, new modes of economy will lead to an increase of consumption according to a principle recognised in many parallel instances (Jevons, 1865, p. 140; emphasis is in the original).

Savings arising from the increasing efficiency of energy conversions have been a factor in promoting their more frequent use and in driving up the overall use of fuels and electricity.

Examples abound: just compare the number of lights and electronic gadgets in an average household in 2010 with the total of their much less efficient predecessors in 1930. In 1930, even in the richest countries, a typical household had just a few (six to eight) lights, a radio, and perhaps a small electric stove; 80 years later it has more than two dozen lights, an array of electrical appliances (including such major energy consumers as refrigerators, stoves, toasters, washing machines, dishwashers, clothes dryers, and air conditioners) and a still growing range of electronic gadgets ranging from TVs and CD players to game boxes, personal computers, and cellphones. Clearly, relative dematerialization is decidedly one of those "many parallel instances" noted by Jevons as less means more.

Progressively lower mass (and hence decreased cost) of individual products, be they common consumer items or powerful prime movers, has contributed to their increased use as well as to their deployment in heavier (more powerful, larger, more comfortable) machines. Inevitably, this has resulted in a very steep rise in demand for the constituent materials – even after making an adjustment for increased populations or for a greater numbers of businesses. Mobile phones offer the most obvious recent example of this increase. Their truly explosive diffusion has more than made up for their impressive evolution from brick-like units to slim designs (some of them thinner than a pencil) and the net outcome has been a much expanded demand for energy-intensive materials. With an average mass of 600 g in 1990 (when about 11 million units were in use) and 118 g in 2011 (when there were nearly 6 billion subscribers), the aggregate mass of cellphones rose from less than 7000 t in 1990 to more than 700 000 t, a gain of 2 orders of magnitude.

Evolution of American passenger cars provides another well-documented example of an impressive relative dematerialization, in this case that of the machine's prime mover, that has not been accompanied by any aggregate decline in consumed materials. Improved power/mass ratios of American gasoline engines (in Europe Diesel's engines are common but in the USA their share rose from 0.2% of new models in 1975 to just 0.6% by 2011) slowed down the growth of demand for materials needed to make those prime movers, but this effect was more than negated by the combination of increased rate of ownership, higher average power of those relatively dematerialized engines, and a greater mass of machines.

Here is a comprehensive comparison of the situation in 1920 and 2010. In 1920 the market was dominated by Ford's Model T whose 15 kW (20 hp) engine (with a low compression ratio of 4.5:1) had a mass of 230 kg and hence a mass/power ratio of 15.3 g/W. By 2010, the company's bestselling passenger car, Ford Fusion, was powered by the Duratec 25 I-4 engine rated at 130 kW (with a high compression ratio of 9.7:1) and weighing only about 120 kg, resulting in a mass/power ratio of just 0.92 g/W. Taking a conservative average of 1 g/W for all light-duty vehicles registered in 2010 would mean that the typical mass/power ratio of gasoline engines commonly deployed in passenger cars was reduced by 93% in 90 years.

But these notable improvements of the typical mass/power ratio of gasoline car engines were accompanied by substantial increases of average engine power. The Ford Fusion had a 8.7 times more powerful engine than did the Model T, but that multiple underestimates the overall rise, as pick-up trucks and SUVs gained a larger share of the market over the past 30 years. The average power of a new cars sold in 1975 was 102 kW and then (as a result of rapid crude oil price increases) it declined to around 80 kW during the early 1980s, but soon after the crude oil price fell it began rising again. The average power of new vehicles has been above 150 kW since 2003, and after a slight drop in 2009 reached a new record at 170 kW in 2011 (USEPA, 2012). The typical power of new US cars thus rose more than 11-fold between 1920 and 2011. Even when assuming a conservative average of 150 kW for all vehicles registered in 2011, the increase in average power would have been almost exactly 10-fold since 1920.

Not surprisingly, there has been a high degree of correlation between average power and average car mass: during the 1920s and 1930s cars became heavier as they acquired fully enclosed metal bodies and a growing array of amenities, and this necessitated the use of more powerful, and heavier, engines. In 1908 Ford's Model T weighed just 540 kg, 30 years later Ford's popular Model 74 sedan weighed 1090 kg. Since 1950, light-duty vehicles have become heavier because they have become both larger and equipped with more features. This trend was reinforced by the rapid adoption of SUVs and by the growing habit of using vans and pick-ups as passenger cars carrying usually a single person. Automatic transmissions (now in about 95% of all new American vehicles), air conditioning, servomotors operating windows and mirrors, audio systems, and better insulation add to an average car's mass. But, as long as crude oil prices remained low, America's car designers were quite unconcerned about vehicle weight, only about their (often rather dubious) appearance.

As a result, in 1975 (the year when the USEPA began to monitor the mass of newly sold cars) when the typical mass of European cars was between 750 and 900 kg, the mean weight of US cars and light trucks was 1.84 t. The two rounds of rapid oil price increase during the 1970s favored smaller vehicles, and the average mass of new US cars declined to 1.45 t by 1981, but once the world oil price collapsed in 1985 larger cars and light trucks became again more popular and the introduction of SUVs has worsened this trend. SUVs had less than 2% of the new car market in 1975, just over 5% a decade later, but by 1995 their share had reached 12%, and by 2004 new passenger vehicles were heavier than in 1975 as they reached a record average weight of 1.86 t. By 2005, the SUV share was nearly 26% (with the heaviest designs weighing nearly 4.7 t) and by 2010 it had declined only marginally to about 25% (USDOE, 2005; Davis *et al.*, 2011).

Because, by 2011, America's automotive market was splintered among scores of brands and hundreds of models, the best-selling vehicle in 2010 was not actually a passenger car but a heavy (2130 kg) pick-up, the Ford F-150. High shares of light trucks and SUVs have pushed the weight of an average new vehicle above 1.7 t since the year 2000 and, after a small dip in 2009 and 2010, the mean in 2011 was very close to the record high at 1815 kg, and hence virtually unchanged since 2004 (USEPA, 2012b). This means that vehicle weight increased 3.4-fold between 1920 and 2011. These ratings,

Table 5.1 *Weight, engine power, and mass/power ratios of US cars, 1920–2011.*

Year	Car	Engine		
	Weight (kg)	Weight (kg)	Power (kW)	Mass/power (g/W)
1920	540	230	15	15.3
2011	1850	170	170	1.0
Multiples	3.4	0.74	11.3	0.06

summarized in Table 5.1 show that while the typical mass/power ratio declined by 93% between 1920 and 2011, average engine power increased more than 11-fold: this means that the increased engine power erased about 75% of the material savings resulting from the vastly improved mass/power ratio of these prime movers, and that these gains were completely negated due to more than trebled average curb weight.

This failure to reduce car mass is even worse when assessed in per capita terms, because the rate of vehicle ownership did not remain constant between 1920 and 2010: passenger car registrations increased from fewer than 8 cars/100 people in 1920 to 27/100 people cars in 1950. After this, it became common for a family to own two or more vehicles: in 1950 only about 3% of US households had a second car, by 1965 that share reached 24%, and by 2011 it was nearly 40%, with about 20% of all families having three or more cars at just over 80 vehicles/100 people. Between 1920 and 2011 relative registrations thus increased almost exactly 10-fold, and even after completely eliminating the effects of the intervening population growth (from 106 to 308 million people) which would have, everything else being equal, resulted in nearly three times as many vehicles on the road, the combined effect of additional power, higher weights, and higher rate of ownership had increased the average per capita mass of materials deployed in passenger cars registered in 2011 nearly 35-fold when compared to their mass in 1920.

And similar weight and ownership trends have been evident even for small Japanese and European vehicles. In 1973, the first Honda Civic imported to North America weighed just 697 kg while the 2012 Civic (LX) with automatic transmission (and with air conditioning as standard) is, at 1198 kg, more than 70% heavier, and its engine is 2.8 times more powerful (105 vs. 37 kW). For the company's other bestselling design, the Honda Accord, the curb weight increased by more than 50% between 1980 and 2012 and engine power nearly tripled. And while some small archetypal post-World War II European cars weighed less than 600 kg (Citroen 2 CV: 510 kg, Fiat Topolino: 550 kg), the average mass of European compacts rose to about 800 kg by 1970 and to 1200 kg by 2002 (WBCSD, 2004). Since 1995, car weight has increased substantially for nearly every vehicle segment as the average European production added 100 kg every five years, with many models now weighing more than 1500 kg (Cuenot, 2009).

Car ownership in Japan rose from less than 1 car/100 people in the early 1960s to nearly 50 cars/100 people by 2011, and during the same period the increase in Germany was nearly 6-fold, to more than 50 cars/100 people. Higher rates of ownership and greater vehicle weight thus combined to increase the mass of European or Japanese car fleets by the same order of magnitude (10^1) as in the USA. So much for any dematerialization of that most ubiquitous of all large modern machines! And new designs – hybrid drives and electric cars – will not result in much lighter vehicles as they have to accommodate more complicated power trains or heavy batteries. For example, GM's 2012 Volt weighs more than 1.7 t and the electric Ford Focus weighs 1.67 t. And there is yet another variable that must be considered in assessing material consumption related to passenger vehicles, the average fuel efficiency: obviously, its steady improvement would have significant effects on the materials needed for extraction, transportation, and refining of crude oil and for the distribution of refined products.

Just the opposite happened in the USA, where the average performance of new vehicles declined for more than four decades, and then, after a decade of improvements, stagnated for another two decades. In 1936 (the first year for which the mean fuel consumption can be reliably calculated) the average performance of new vehicles was more than 15 mpg and by 1973 it had declined to only 13.4 mpg (Sivak and Tsimhoni, 2009), a remarkable case of technical retrogression during an era of rapid technical advances. The trend changed only thanks to the OPEC-driven rise of crude oil prices: in 1975 the US Congress adopted the Corporate Average Fuel Economy (CAFE) regulations and the first mandated efficiency increase took place in 1978.

By 1985, the pre-1975 performance had more than doubled to 27.5 mpg. But then, although crude oil prices fell (from nearly $40/bbl in 1980 to only about $15/bbl by 1986) the CAFE standard was left unchanged for more than two decades; with the influx of SUVs, exempt from the standard, the adjusted fuel economy of new vehicles (reflecting actual driving experiences rather than laboratory testing) was merely 21 mpg in 2008, 5% lower than 22 mpg in1987, and by 2011 the rate had risen to just 22.8 mpg, an astonishingly poor performance and an incredible lack of progress during a quarter century that witnessed so many important technical advances.

Finally, there has not been any progressive automotive dematerialization in the USA because, until 2004, the average distance driven annually per capita increased, making higher demands on fuel production and distribution, on car depreciation and maintenance, and on road repairs. While the average distances driven annually by Europeans and by the Japanese leveled off at relatively low levels, in the USA that same per capita mean more than trebled between 1950 and 2000 (from about 4800 km to more than 15 500 km) and it peaked at 16 165 km in 2004 before declining by about 5% by the year 2010 (BTS, 2013). The conclusion is clear: relative dematerialization of the internal combustion engine and the use of lighter elements and compounds in vehicle construction has only slowed down the rate of overall material consumption claimed by making and operating vehicles. By the year 2010 (after adjusting for population growth) the combination of expanded ownership of heavier cars, light trucks, and SUVs operating with inexcusably poor typical efficiency and driven longer distances had increased the average per capita mass of American vehicles more than 30-fold compared to 1920.

An analogical calculation involving the relative dematerialization of prime movers and the massively rising aggregate consumption of materials could be made for the modern airline industry. The thrust-to-weight ratio of the largest turbofan engines (such as those used to power the Boeing 747 or Airbus 380) is now roughly 60% higher than that of the turbojets used by the first jetliners of the late 1950s – but this significant dematerialization of performance has been accompanied by the introduction of more powerful engines needed to propel many larger airplanes. Boeing's pioneering 707 was powered by Pratt & Whitney's JT3D engine, whose rated thrust was about 80 kN, while the most powerful engine of the first decade of the twenty-first century was GE's GE0-115 B rated at 512 kN, and engines used on the mid-size jetliners that dominate intra-continental flights (Boeing 737 and Airbus 320 models) have thrusts ranging mostly between 100 and 150 kN, that is typically 25–50% more powerful than the first commercial turbojets (Smil, 2010).

The aircraft industry has been the foremost pioneer of material substitutions using light-weight metals and compounds, a trend that has now culminated in increasing reliance on carbon composites. But this relative dematerialization has not resulted in lighter commercial fleets as their composition shifted toward larger aircraft. This trend is best illustrated by the evolution of Boeing 737, the most successful jetliner in history: since its introduction in 1967 the company has sold more than 7000 units with each successive model weighing more than its predecessor. In 1967 the maximum take-off weight (MTOW) of the Boeing 737-100 was just short of 50 t, by 1981 the 300 series had a MTOW of 56.5 t, planes of the 800 series – in service since 1998 – weigh 79 t, and the latest design – 900ER flying since 2006 – weighs 85.13 t, a 70% increase of MTOW in 40 years. Larger capacities (100 passengers in the 100 series, 203 in the 900 series) have brought relative – per-passenger – dematerialization for Boeing 737 models, as the MTOW declined from 494 kg/person to 419 kg, a 15% decrease in 39 years.

But this has not been a universal experience. For example, the MTOW of the Boeing 747 rose from 333 t for the 100 series launched in 1969 to 442.25 t for the latest version 747-8i – but (because of a longer maximum range) so did the relative weight, from as low as 605 kg/passenger in 1969 to nearly 950 kg/passenger for the 747-8. And, contrary to the impression created by the unprecedented extent of composite materials in that new plane, the zero-fuel weight of the Boeing 787 is actually higher than that of the Boeing 747-8: when the comparison is made for three-class seating (467 people in the 747, 242 people in the 787) the smaller plane rates at 665 kg/passenger, the larger one at 623 kg/passenger (Boeing, 2012). In any case, a large increase in the total number of operating aircraft and an enormous increase in the frequency of flying makes a much greater difference.

The US commercial jetliner fleet and the total of passenger-kilometers flown would have grown by no more than 70% if they had just matched the expansion of the country's population; in reality, between 1960 and 2010, the total fleet of commercial jetliners increased 3.5 times and passenger-kilometers logged on just domestic flights multiplied more than 18-fold. Obviously, the relative dematerialization of jet engines, and of some aircraft designs, has had only a mildly retarding impact on the overall growth of material demand on America's aviation industry that expanded not only directly due to more and larger planes with more powerful engines, but also indirectly due to an enormous increase of fuel consumption for more frequent flights and associated aircraft services (cargo handling, catering, airplane maintenance) and owing to continuous expansion of airports and their infrastructures. And the process is now being replicated in China, where expansion of aviation is taking place at rates that have already marginalized any material savings resulting from a reliance on the latest engines to power advanced jetliner designs.

The second example of tracing long-term trends in consumption of a product affected by relative dematerialization also shows that even the combination of reduced unit mass with relatively high rates of recycling is insufficient to make up for increased per capita consumption. Data on actual shipments of US-made beverage cans by category are available only since 1979: between that year and 2011 the weight of a can was reduced by 25%, but per capita consumption of aluminum cans doubled from 149 to 296 a year (AA, 2013): once again, relative dematerialization has been a contributing factor to an absolute increase in material consumption.

And, counterintuitively, even the rising recycling rate of America's aluminum beverage cans has not resulted in overall energy savings. As already noted, that recycling rate reached 65% in 2011 (AA, 2012). This makes a great difference to the overall energy cost, as the cans made from recycled metal cost only about 5% of the energy needed for new cans. But even when assuming that only 30% of all cans were recycled in 1979 and that the energy cost of aluminum production was 10% higher than in 2011, the overall per capita energy cost of new and recycled aluminum cans was almost exactly the same in 2011 as it was in 1979, and hence even the second-order effect of relative dematerialization (its energy cost) has not been translated into absolute savings. To be sure, the combination of lower unit mass and higher recycling rates was able to produce twice as many cans with the same amount of energy, but the total material throughput has increased.

Finally, I must note those reductions of specific mass that have been accompanied not only by an increased overall consumption of materials but also by increased specific energy cost of substitutes. By far the most common instance in this category has been the substitution of steel by aluminum. Steels have a specific density between 7.7 and 8.0 g/cm^3 compared to 2.6–2.8 g/cm^3 for aluminum alloys. Again, the US automobile industry makes it possible to quantify the consequences of this shift that began on a large scale during the late 1970s when aluminum began to replace cast iron in intake manifolds. That substitution was accomplished in a matter of years, and the next wave brought cylinder heads, pistons, blocks, transmission cases, and wheels cast from aluminum alloys.

As a result, by the year 2000, an average American car had about twice as much of its weight (some 120 kg) in aluminum than in 1980 and four times as much as it had in 1960 (AA, 2003). Assuming that iron and steel components fill the same volume (in reality the two volumes would not be identical but the difference does not substantially affect the calculation), 120 kg of aluminum replaced about 350 kg of steel. But while steel is nearly three times heavier than aluminum, its energy cost is much less than a third of the energy needed to produce the lighter metal. The best US average of overall energy cost of production is 20 MJ/kg for steel and about 218 MJ/kg of aluminum, an order of magnitude difference (Worrell *et al.*, 2008). Some specialty steels may require as much 50 MJ/kg, but even in these cases the difference remains more than 4-fold. Producing 120 kg of aluminum components for an average US car would thus require about 24 GJ of energy, producing 350 kg of cast iron and steel costs no more than about 8 GJ, a 4-fold difference.

Consequently, all weight reductions associated with replacing steel by aluminum result in a substantially higher overall energy cost of materials. At the same time, it must be pointed out that this difference could be reduced by lower operating costs of lighter automobiles, railcars, and trains and that this reduction will increase with the degree of substitution: aluminum is still a minor constituent of cars but not in rail transport: wheels and driving assemblies are still made of steel but both large freight cars – used to carry bulk minerals and other materials – and high-speed passenger trains (Japan's *shinkansen*, the French TGV, or Spanish AVE that travel at velocities close to, or even above, 300 km/h) have their hoppers and bodies made of the lighter alloys. And aluminum hoppers have yet another advantage that reduces their life-time energy cost: even after 20 years of wear they have about 25% lower floor and wall metal loss than do steel cars (AA, 2003).

Replacement of cast iron and steel by plastics brings the same combination of advantages (lighter components, reduced energy cost for machines where these components make a substantial part of the total mass) and disadvantages (invariably higher energy requirements for production, with most plastics costing anywhere between four and six times as much per unit weight than common steel alloys). Again, the substitution makes the greatest difference in the case of the life-time operation cost of energy-intensive means of transport. This is particularly true in the case of jetliners, whose interiors now include (for larger planes) more than 5 t of advanced thermoplastics and other synthetic and composite materials in the floors, ceiling panels, overhead baggage bins, window frames, seats, food and drink galleys and trolleys, lavatory modules, and class dividers (Composites World, 2006).

I could cite many other instances where substitute materials, chosen for their lighter specific weight as well as for other superior or unique properties, have lowered the mass of finished products but required substantially higher energy inputs. Common examples range from plastic flooring replacing wooden boards or tiles in buildings, to titanium taking the place of steel in aircraft components, and from polyethylene bags replacing sturdy canvas to plastic garbage bins (also made from polyethylene or from polypropylene) taking place of previously ubiquitous much heavier galvanized (zinc-coated) steel metal trash containers. An obvious question to ask is: what effects have these relative product or sectoral dematerializations had on the national scale?

5.4 Relative Dematerialization in Modern Economies

The earliest stage of intensive modernization (the latter half of the nineteenth century in most countries of Atlantic Europe and in North America, and post-1950 China) is marked by a rising demand for construction materials – dominated by traditional bulk inputs of timber, stone, sand, and mineral aggregates and by expanding production of cement and crude-oil derived paving materials – due to the rapid and extensive development of material-intensive transportation, industrial, and residential infrastructures. This stage is also characterized by rapid increases in consumption of primary metals due to unprecedented rates of urbanization, industrialization, and the construction of railways and (in maritime nations) of new steel-hulled ship fleets.

The next phase of material consumption experienced rising material demand due to the continuing expansion of industrialization and the next phase of urbanization driven by electrification and mass adoption of automobiles; in turn, those industries required materials for increased fossil fuel extraction, electricity generation, car industries, and road construction. Another major factor is the growing use of agricultural inputs (fertilizers, machinery) as higher disposable incomes boost demand for food in general and for animal products in particular and accelerate the process of the sector's mechanization and chemization. The United States of the 1920s, leading European economies and Japan of the 1950s, and post-1980 China are excellent examples of this demand stage. Inevitably, use of some traditional materials will eventually reach saturation point and this development will either result in temporary consumption plateaus or, in many instances, there will be reversals of long-lasting upward trends as specific material intensities will decline (Herman, *et al.*, 1990; Bernardini and Galli, 1993).

This secular decline of traditional materials consumed per unit of economic product that appears during the second stage of material consumption has been perhaps the most often noted category of relative dematerialization. Mass reduction would be perhaps the most preferable term, and the process can be measured as a declining mass of a specific product, as a lower annual use of a commodity per capita (for example, average US annual per capita consumption of wood declined during the twentieth century from about 791 to 282 kg), or as decreasing mass of a particular material or of all materials used annually in a national economy per unit of economic product. A notable example in this latter category would be the US consumption of all metals, that declined from about 27 kg/$ of GDP in the year 1900 to less than 15 kg/$ by the year 2000, a 45% reduction in a century (Matos, 2009).

A common corollary of this process is an intensification of international trade that increases the share of material imports in general and reduces (or even eliminates) domestic output of energy-intensive materials in affluent economies. These shifts are an expected outcome in the evolution of material uses in modern economies as they move into the post-industrial era – and they should be seen as a part of more general relative reordering of consumption that accompanies rising affluence. More than 150 years ago, Engel (1857) observed that the share of disposable income spent on food declines as income rises, an inevitable outcome given the simple fact that average daily per capita food intakes compatible with a healthy and long life should remain below 9 MJ (less than 2200 kcal), a total that (even when including plenty of more expensive animal protein) can be now purchased with less than 10% of America's (and about 15% of the EU's) average income (USDA, 2012).

Similar saturation levels and inversions have become evident in many other instances of specific consumption. Even in the absence of technical advances, average household expenditures on energy for cooking, lighting, and heating would eventually become saturated, and with continuing efficiency improvements they have been falling: in 2010 only about 12% of an average disposable income in the USA paid for household energy consumption, including gasoline for cars (USCB, 2012). As for material consumption, rising incomes obviously do not produce commensurate increases in demand for a wide assortment of manufactured products ranging from appliances to furniture. At the same time, widespread possession of a widening range of consumer goods and the deliberately engineered rapid obsolescence of many products are two notable factors that militate against dematerialization even in the most affluent societies already suffused with goods, and the net outcome can be determined only by taking a longer look at aggregate demand in modern economies.

How closely does it follow the described pattern of vigorous rise followed by saturation and relative decline? Or, to rephrase this question in macroeconomic terms, what has been the extent of decoupling between GDP growth and overall consumption of materials? Has it been a weak decoupling when material consumption grows at a slower rate than GDP, or a strong one when material consumption declines in absolute terms even as inflation-adjusted GDP keeps on growing? Data limitations do not make it easy to answer this question. GDP series in constant monies are readily available, and affluent countries have historical series of products going back commonly to the last decades of the nineteenth century. But (a few notable exceptions aside) the trade data needed to establish actual domestic consumption are much less complete (or are available only in value terms), totals for major material categories (construction aggregates, metals, other minerals) are either unavailable or have to be reconstructed from incomplete series, and until recently no country besides the USA had long-term series of aggregate material consumption.

As already explained, comparative reconstructions by Adriaanse *et al.* (1997) and Matthews *et al.* (2000) offered only fairly short-term (1975–94 and 1975–96) aggregates for domestic processed output, domestic hidden flows, and total domestic output of all materials (including and excluding oxygen flows) in Germany, Japan, Netherlands, and the USA, with the second report adding Austria. Any general conclusions based on these studies were limited by the brevity of the time span as well as by some obvious national peculiarities. Germany's specific total domestic output (kg/Deutschmark) declined by about 20% between 1975 and 1990, but the reunification with the East (and resulting inclusion of much less efficient East German industries and, even more so, of massive hidden flows associated with DDR's lignite mining) brought a temporary reversal of that trend. And Japan's impressive decline of total domestic output per constant yen ended after 1990 as the country's economy entered its first lost decade.

The same two limitations apply to the EU's new material accounts, whose systematic publication began in the year 2000. This data series is the most detailed set of material flows quantified according to common rules for a large group of countries, but the brevity of the available data (with less than a decade available as of 2012), the reversals suffered by all of those economies since 2008 due to the greatest post-World War II economic crisis, and many national specificities preclude a simplistic conclusion regarding any possibly long-term trends. Moreover, the EU's material aggregates also include all fossil fuels – an analytical choice that inevitably boosts the totals of both direct and hidden flows for countries that are highly dependent on coal – but exclude fertilizers, materials whose inputs made a critical difference to the prosperity of Europe's agricultural sector. Keeping these realities in mind, these statistics show that between 2000 and 2009 resource productivities – reverse indicators of relative dematerialization measured in €/kg and including all fossil fuels – have improved in most EU countries.

These productivity gains rose by 17% for the EU-27, with the German rise at 23% and the French one nearly identical at 22%, and with the highest gains in Latvia (up by 56%) and the Czech Republic (up by 40%, mostly due to reduced lignite consumption). And average annual change for the period shows a small (−0.3%) decline in total domestic material consumption for the EU-27, even as the Union's gross economic product rose by nearly 17% during that period, with the highest reduction in Italy (−2.8%/year), the Netherlands (−2.1%), and the UK (−1.8%). Consequently, for the EU as a whole, the experience during the first decade of the twenty-first century can be labeled as a weak variety of strong decoupling, while the shifts in Italy, the Netherlands, and the UK were clear cases of strong decoupling, as was the German achievement, with GDP up about 9% and total material consumption down by nearly 14%.

In contrast, 15 countries (including Poland, Cyprus, Malta, Slovakia, Ireland, and Bulgaria) experienced only weak decoupling as their material requirements rose at a slower rate than their GDP: for example, for Poland the overall material demand (including the country's large coal extraction) rose by 12% between 2000 and 2009, but GDP grew by nearly 50% (a case of strong weak decoupling); while for Slovakia the difference was much smaller with 35% more materials and 55% higher GDP during the same period. Romania was the greatest exception with GDP up strongly by about 55% but with the country's material consumption expanding 2.5 times during the decade, a clear case of rising inefficiency of material use and increasing dependence of economic growth on higher material inputs.

Disaggregations show some particularly large declines in material requirements for metal ores and products (decadal decline of 3.3% for the EU 27, 3.4% for Germany, 9.4% for France, and 6.3% for Italy) but gains for nonmetallic minerals (1.4% annual increase for the EU-27) and for waste products. Several exceptional rates (such as a 40% increase in Dutch sand and gravel requirements between 2000 and 2009 or a 13.3% decline in Romania's metal ore consumption during the same period) reflect national peculiarities (continuing conversion of areas below sea level to farmland and residential areas and reinforcement of coastal defenses by the Dutch, closures of unprofitable deep mines in Romania).

Longer perspectives are available for several countries thanks to recently published reconstructions of total material requirements: I have already noted a study of Japan's 1887–2005 flows by Krausmann *et al.*, (2011); Schandl and Schulz (2002) traced UK consumption between 1850 and 1997, and Kovanda and Hak (2011) did the same for Bohemia and Moravia (since 1993 the Czech Republic) between 1855 and 2007. All of these series include fossil fuels (whose extraction dominated the mass of domestically produced materials during the latter half of the nineteenth and the first half of the twentieth century) and food crops, materials excluded from this book, and show the flows only in highly aggregate forms.

Consequently, the best choice for examining the long-term relationship of material use and economic development is to rely on more than a century of disaggregated and reliable American data. As already explained, this series of material inputs into a national economy is maintained by the US Geological Survey: its first version ranged from 1900 to 1995 (Matos and Wagner, 1998) and it was subsequently extended to 2006 and 2010 (Matos, 2009; Kelly and Matos, 2011). The series contains data for agricultural and forestry (wood, paper and paperboard, and recycled paper) products, metals, and minerals (subdivided into primary and recycled metals, and industrial and construction materials) that were used to make products in the United States: it leaves out food and fossil fuels (as large shares of these materials are consumed directly) but it includes all phytomass that is used in manufacturing (ranging from cotton to wool, and from timber to recycled paper).

In order to quantify the dematerialization of the US economy during the twentieth century, I used inflation-adjusted US GDP in constant (2005) dollars and the total material consumption series prepared by Matos (2009). The material intensity of the US economy first shows a rise from 340 g/$ in 1900 to 500 g/$ in1929, followed by four decades of fluctuations (a low of 260 g in 1945, a high of 540 g in 1960) and then a steady decline from 490 g in 1970 to 300 g in 2000 and 290 g/$ in the year 2005, that is a reduction of about 40% in 35 years. The post-World War II doubling of average material intensity is not surprising given the fact that the total input is dominated by minerals used in construction and that the first two post-war decades saw an unprecedented expansion of their use for new housing and industrial and transportation infrastructures. Their mass is dominated by stone and sand and gravel, and their share of the US material input rose from 54% of the total in 1945 to 72% in 1960.

After excluding construction materials, the trend becomes much clearer, with maximum material intensities peaking during the second decade of the twentieth century at 230–280 g/$ and then, except for a brief post-World War II bounce, declining steadily to about 150 g by 1960, 130 g by 1980, and 75 g by the year 2000, resulting in an overall decline of nearly 75% in 90 years. This rate was almost perfectly matched by the decline in metal intensity (primary and recycled inputs) of the American GDP that was reduced from the peak level of about 44 g/$ during 1920 to just 13 g by the year 2000 (and 11 g in 2005). The only greater drop was for wood intensity of GDP, from about 150 g/$ in 1910 to less than 30 g by 1950 and to just 7 g in the year 2008. Use of construction materials per unit of GDP rose, in a fluctuating manner, from about 130 g/$ in 1900 to a peak of roughly 400 g in the early 1960s, and the subsequent decline brought it to less than 230 g in the first years of the twenty-first century.

This overall reduction contained trends ranging from a continuous and large decline (wood: down from 157 g/$ in 1900 to just 8 g/$ in 2000) to slower and less pronounced reductions (primary metals going from 27 g/$ in 1900 to 25 g/$ in 1950 and then to 8 g/$ by 2000), and to a temporary rise and subsequent fall (evident for both paper and industrial minerals). Secular reduction of material intensity of GDP in modern societies reflects the changing nature of modern economic products (with services accounting for a steadily increasing share of the total) and the ways we quantify it (with intellectual inputs valued far above the manual labor). But a different metric reveals a very different result. In per capita terms, America's usage of all material inputs rose from about 1.9 t in 1900 to 5.7 t in 1950 and to just over 12 t in the year 2000, a 6.4-fold increase in a century. Per capita rates for all nonconstruction minerals rose from almost 1.2 t in 1900 to 3 t in 2000, a 2.6-fold rise; and the consumption of metals grew from about 135 kg/capita in 1900 to 510 kg/capita in the year 2000, a nearly 4-fold increase.

And the real increase in material throughput has been even greater than the USGS compilations indicate because of the growing imports of manufactured goods, including an influx of material-intensive goods: just think of the metals and plastics in machinery and cars, the aluminum and plastics in jetliners, the heavy metals, rare earths, and lithium in electronics, magnets, and batteries. A conservative conclusion would be that the country's average annual per capita consumption of inputs into manufacturing at least quadrupled during the twentieth century, and houses and their contents have been the key component.

In 1900, the typical size of a new American home was 90 m² (1000 ft²), it changed little during the century's first half, but since the end of World War II the average size of a newly built US home has increased substantially, with the mean area rising from 100 m² (1100 ft²) in 1950 to about 220 m² in 2005, even as the average household size declined during those years from 3.7 to 2.6 members (USGS, 2008). Consequently, in 1950, the average built area was about 27 m²/capita compared to more than 80 m²/capita in the year 2000, a 3-fold gain (Wilson and Boehland, 2005). And by 2005 the mean for custom-built houses had surpassed 450 m², nearly five times the size of an average Japanese home, while "mansions" in excess of 600 m² had become fairly common.

Average per capita increase in the total house mass was even higher than indicated by the growth of built area because the houses have better insulation, double windows, larger garages (by 2005 about 85% of them had double-garage), larger driveways, and often extensive associated structures (decks, patios, storage sheds) and are crammed with more furnishings and with more consumer products and finished with heavier materials (marble bathrooms, stone floors, granite kitchen counters). Per capita mass gain has thus been at least 3 to 3.5-fold in just 50 years! Only the Great Recession checked the growth of average US house, with new structures averaging less than 190 m^2 in 2009 (NAHB, 2010).

Changes in qualitative terms have been no less important, as many lighter but more energy-intensive materials displaced traditional choices: synthetic wall-to-wall carpets or plastic floors became used instead of wood boards or parquets; aluminum siding replaced boards or mortar; steel, aluminum, and plastics displaced wood in window and door frames and in outside and garage doors widely used for doors and windows; and plastics, rather than metals, became the dominant choice in pipes.

The first wave of mass acquisition of household items that began during the 1920s was virtually complete by 1960 when some 95% of all US households had an electric or gas stove, a refrigerator, and a radio – but still only 40% had a washing machine, clothes dryers were only in every fifth home, and only about 10% of households had central air conditioning and a dishwasher (Felton, 2008). By 1980, a large majority of households had these items and they began to acquire electronic goods (the third wave): CD players, computers, video games, flat-screen and HD TVs, and cellphones, all possible only thanks to tiny microprocessors and other highly energy-intensive electronic components.

Hence, even as America's declining infrastructural expansion has reduced demand for cement (whose consumption fell from 23 g/\$ in 1950, to about 15 g/\$ in 1975 and just over 11 g/\$ in the year 2000), the electronic era created new demand for ultra-pure silicon, whose typical energy cost is more than 3000 times the energy cost of cement and 1000 times the energy cost of crude steel made from iron ore. As a result, even small quantities of highly energy-intensive products can claw back substantial shares of fuel and electricity savings made by reducing the use of more massive but less energy-intensive inputs (3300 kg of cement = 1 kg of Si).

Clearly, there is no recent evidence of any widespread and substantial dematerialization – be it in absolute or relative (per capita) terms – even among the world's richest economies. Undoubtedly, they have seen a more subdued growth of raw material inputs, and trends explain this moderation: affluent societies have already put in place extensive and highly material-intensive infrastructures; ongoing outsourcing of material-intensive (and often also polluting) industries to foreign low-cost producers has lowered the direct domestic consumption of primary inputs; and relative dematerialization has slowed down the growth of demand.

In a few countries with reliable data the overall material inputs have stabilized or have even slightly declined (most notably in Germany and the UK) while some of their specific inputs continued to rise. Other high-income economies, including the USA, France, and Spain, have seen slowly, but continuously, rising material needs and, as a group, affluent countries have not experienced any aggregate dematerialization but rather a substantial increase in the total demand for materials during the latter half of the last and during the first decade of this century. At the same time, the modernizing countries of Asia, Latin America, and Africa, in absolute terms led by China and India, have been experiencing unprecedented rates of material demand.

This new Asian material surge has been the main cause of rapid demand increases of the past generation: between 1990 and 2010 the increments were 180% for steel, 210% for aluminum, 170% for copper, and 320% for cement (USGS, 2013). Obviously, most of these increases are attributable to domestic development, but given the worldwide integration of economic activities quantifications a significant share of this new material demand has been embedded in exports to affluent countries. In this new world economy, the global scale is the most appropriate way to demonstrate that modern civilization has become more, not less, massive even in per capita terms.

Yet another factor helping to explain this continuing rise has been the change in energy consumption that, as in the case of material use, has been determined by two opposite processes. On the one hand is the declining energy intensity of products, industrial sectors, and entire national economies: given high material intensities of modern energy extraction, conversion, and distribution, this decline has been a major contributor to reducing the relative material intensity of modern societies. On the other hand are increases in overall fuel and (even more so) electricity consumption that have been driven by a combination of continued population growth and increasing per capita consumption engendered by higher standards of living and higher mobility. Per capita energy consumption is still unacceptably low in most of Africa and among the low-income groups of Asia and Latin America, it has been rapidly advancing in a number of modernizing countries (above all in China) and, until quite recently, it has been slowly rising even in affluent nations with already very high levels of fuel and electricity demand.

5.5 Declining Energy Intensities

Declining specific use of energy – be it per unit of extracted raw material or finished product – has been one of the key markers of modernization. In this book I have introduced many examples of such efficiency gains, and recounting just three cases of leading materials reiterates those impressive reductions achieved between 1900 and 2000 and continued during the first decade of the twenty-first century: in the year 2010 a ton of steel was made with only about 20% of the energy needed in 1900, a ton of aluminum needed about 30% (USDOE, 2007), and the respective reductions for ammonia synthesis and cement production were 70% and more than 80%. Similarly impressive, or only slightly lower, long-term efficiency gains could be cited for virtually all leading materials.

And given the intensive global competition and the relatively rapid diffusion of new techniques, it is not surprising that specific fuel and electricity requirements for principal energy-intensive products, be they metals or fertilizers, are very similar in all affluent countries. Readers familiar with modern energy development may also already appreciate that impressive efficiency gains have also accompanied large-scale diffusion of virtually all modern energy conversions, be it heating a house to a desired temperature or transporting goods between continents. Historical efficiency gradients for common converters have been fairly steep: wood-burning fireplaces converted less than 10% of the fuel's chemical energy to useful heat, the rate for good late nineteenth to early twentieth century hand-stoked coal stoves had efficiencies no better than 20–25% (often much lower), oil-fired furnaces were rated up to 50%, a generation ago it was common to burn natural gas in home furnaces with efficiencies of 70–75%, and now the best performance is 95–97%.

The heating sequence becomes even more impressive when we consider automatic and clean gas combustion compared to the chores and dirt associated with low-efficiency coal combustion. In the second case, technical evolution moved from small (less than 5000 dwt) freight ships powered by steam engines (best efficiencies less than 15%) to massive (commonly in excess of 100 000 dwt) bulk carriers (for ores, fuels, and other loose materials) and container vessels (the largest ones one carrying in excess of 10 000 units) powered by diesel engines whose best efficiencies are very close or even slightly above 50% (Smil, 2010).

Inevitably, these improvements had to add up to significant aggregate changes, to gradually lower energy intensity of modern economies, the variable measured in units of energy (preferably joules) per unit of national currency or, for international comparisons, per constant US$ (that is monies adjusted for inflation). The most obvious advantage of tracing this indirect measure of relative dematerialization is that historical data series for total primary energy supply – TPES, an aggregate of all fuels and primary (that is hydro, nuclear, wind, and solar) electricity – are much more readily available than those for aggregate material inputs or for specific material uses. The secular decline of energy intensity indicates adoption of more efficient, and hence almost always less material-intensive, extraction, processing, and manufacturing techniques, as well as the diffusion of less wasteful ways of household energy consumption.

The heuristic or exhortatory value of comparing energy intensities is undeniable, but the EI ratio must be approached with caution. If the measure is interpreted in a naïve, ahistorical, and abstract fashion its use only reinforces some inaccurate notions, and it misleads more than it enlightens. Deconstruction of the measure offers a deeper under-standing of underlying realities, uncovers a number of serious data limitations, leads to a careful interpretation of the differences in levels and trends, and helps to avoid sim-plistic, and hence potentially counterproductive, conclusions. The first concern is about the measure's constituents. National TPES rates must be calculated by converting fuels and primary electricity to a common denominator: joules in scientific writings, ton of oil equivalent (toe) in energy publications. As fossil fuel flows are among the best monitored variables in modern societies, such conversion errors will be relatively minor.

Electricity generation is measured even more accurately, but the greatest disparities in national TPES values arise due to differences in converting primary electricity (that is mostly hydro generation, but increasingly also wind and solar, with geothermal energy being still marginal). The first conversion choice is to use a thermal equivalent, whereby 1 kW h is obviously equal to 3.6 MJ (1 J being 1 Ws); the alternative is to convert primary electricity by using an average heat content of the fossil fuels used to produce a nation's thermal electricity. Because thermal generation has typical efficiencies of just 33% (a good national average) to 38% (average for the best modern plants) such an equivalent will raise the conversion rate to 9.5–10.9 MJ/kW h. Each of these choices carries a bias when comparing aggregate energy use in a country that relies only (or overwhelmingly) on fossil fuels to generate its electricity, with a nation that derives most of its electricity from hydro generation.

With the first conversion, the aggregate energy use in a country relying heavily on water will be lowered, and hence its energy use will appear to be more efficient than in a nation with marginal reliance on hydro. The second conversion choice will eliminate that apparent undercount by creating the opposite problem. But those hydro-dependent countries that generate much more electricity per capita than the countries relying heavily on thermal generation do so precisely because they can tap abundant water resources, and converting their primary electricity generation at roughly triple the rate of electricity's thermal equivalent will, inevitably, and misleadingly inflate their TPES. An additional complication is due to the fact that the average efficiency of thermal generating stations keeps improving, and hence the average conversion ratio must be adjusted to reflect these gains. For example, the average US conversion rate was 14.8 MJ/kW h in 1950, about 11 MJ/kW h in 1975, 10.8 MJ/kW h in the year 2000, and 10.3 MJ/kW h in 2010 (USEIA, 2013). Where reliable historical data are not available estimates must suffice.

There are now four major sources of worldwide energy statistics: the United Nations (UN, 2013), the International Energy Agency's balance sheets (IEA, 2013), the US Energy Information Agency (USEIA, 2013), and British Petroleum's annual *World Energy Review* (BP, 2013). The UN and IEA use thermal equivalent (3.6 MJ/kW h) for hydroelectricity and the prevailing efficiency rate (about 33%) of fossil-fueled generation for nuclear electricity; BP converts hydroelectricity by assuming an average 38% conversion efficiency (an equivalent of 9.47 MJ/kW h); and the USEIA converts hydroelectricity by using the prevailing national efficiency rate for thermal generation (recently about 35% or 10.3 MJ/kW h).

The consequences of these choices are as to be expected. In 2009 (the last year for which all of these sources could be compared) the TPES of Norway, the country with the world's highest hydroelectric generation per capita, was just 27.48 Mtoe according to the UN and 28.25 Mtoe according to the IEA, but 43.4 Mtoe according to BP and the USEIA conversion put it at 47.75 Mtoe, a 73% difference between the lowest and the highest value! Similarly, Canada, with its high dependence on hydro generation, consumed only 254 Mtoe according to the IEA, but the USEIA total was 328 Mtoe, a 29% difference. There is no reconciliation of these two approaches and the least that must be done for all international comparisons of energy intensities is to define explicitly the conversions used to calculate TPES. And, naturally, any long-term comparisons of energy intensities must avoid splicing data from different sources and use only identical conversion for secular comparability.

As problematic as the TPES aggregates can be, GDP values are even more problematic, and that remains true even after accepting the fundamental limitations of GDP: after all, it can grow even as a nation's average quality of life may be declining, as the educational achievements of its citizens may be stagnating, and as its irreplaceable resources are being plundered for ephemeral profit. Even for economies with good historical statistics, all pre-World War II GDP estimates are less reliable than their post-1950 counterparts, and for many modernizing economies they are simply unavailable, or amount to nothing but rough estimates: these realities make reliable long-term international comparisons questionable. Moreover, recent GDPs, calculated according to a UN-recommended System of National Accounts, exclude all black market (underground economy) transactions whose addition would boost the total by 10–15% even in the most law-abiding countries, and could double the economy's size in the most lawless settings.

But, once again, the most important bias comes from conversion. In order to make international comparisons, national currencies must be converted to a common denominator (usually US$). This was done for decades by using nominal exchange rates, a practice that exaggerates the disparity between high- and low-income countries and biases any comparisons between nations with floating and fixed exchange rates. But a more recent use of conversions based on PPPs is not without its problems; its adoption lifts the real GDPs of modernizing countries but, in the absence of a universally valid consumption basket, there can be no single, unequivocal PPP value for countries whose economies, everyday habits, income allocations, and expectations differ greatly from the reference (US) model. And, of course, for any longer-term comparisons all GDP values must be expressed in constant (inflation-adjusted) terms and these terms must be clearly specified.

Here are just a few revealing examples of what you get depending on what you choose to work with. When using the USEIA energy conversions (resulting in the highest national aggregates among the four data series) and the national GDPs in terms of standard exchange rates expressed in constant (2005) US$, the 2009 values for the world's leading economies range from 4.9 MJ/$ for Japan, 5.0 MJ/$ for Germany, and 7.7 MJ/$ for the USA, to 27.4 MJ/$ for China. The Chinese economy is clearly not that inefficient, and when the intensities are recalculated in terms of PPP (again, for 2005 US$) they change to 5.6 MJ/$ in Germany, 5.8 in Japan, remain 7.7 in the USA, and are reduced by nearly 60% to 11.4 MJ/$ for China.

When using the lowest energy aggregates (according to the UN's conversions), energy intensities using exchange-rated GDPs (in 2005 US$) go to as low as 2.8 MJ/$ for Japan, are about 4 MJ/$ for Germany, 6.5 MJ/$ for the US, and about 23 MJ/$ for China; when expressed in terms of PPP, UN-based energy intensities go up to about 4.3 MJ/$ for Japan and 4.4 MJ/$ for Germany, remain at 6.5 MJ/$ for the USA, and decline to 9.6 MJ/$ for China. Depending on the choice of energy and GDP values the 2009 rate for the USA (in 2005 $) is thus as low as 6.5 and as high as 7.7 (a difference of about 15%), but for China the range is nearly 3-fold (2.85 times), from just 9.6 to 27.4. The difference for Japan is more than 2-fold (5.8 vs. 2.8), for Germany it is about 40% (5.6 vs. 4.0).

These comparisons reaffirm some widely held perceptions about the performance of economies: efficient Japan and Germany are doing much better than energy-wasting USA and Canada, while even the considerable magnitude of China's recent economic advances has not been able to erase the country's Maoist legacy of inefficient resource use. Other impressions concerning the economic prowess of nations could be confirmed by a quick check of the latest values of energy intensities: in 2009 the EIA-based, PPP-adjusted rate was 14.7 MJ/$ for Russia, and the Saudi rate was just above 15 MJ/$, reflecting energy wasted by these two hydrocarbon-exporting energy superpowers.

But are these comparisons of energy intensities, done in a uniform manner, truly revealing indicators of economic efficiency and, in obverse, of waste and environmental degradation? Not at all: as other aggregate measures, they hide as much as they reveal and hence it is necessary, before praising or condemning a particular performance, to take a closer look at what they really convey. Reasons for substantial differences in energy intensities are complex, and I have identified six key variables that explain national peculiarities (Smil, 2003, 2008). These include fundamental physical attributes (country size and climate); composition of energy supply and the degree of energy self-sufficiency; differences in final (sectoral) energy use; and the level of discretionary (personal or household) energy demand. These factors must be considered in combination, because no single variable can be a revealing indicator of a nation's expected energy intensity.

Everything else being equal, a large territory will require more extensive, and hence more expensive, infrastructures and more energy devoted to transportation and communication. This makes the greatest difference for flying, the most energy-intensive form of transportation. Aircraft performance is almost entirely eliminated from consideration, because commercial airlines all over the world use the same large jets produced by the duopoly of Airbus and Boeing and the same regional jets made by another duopoly, that of Bombardier and Embraer; their fleets differ only as far as the shares of specific models are concerned.

Intercity distances will be thus a major factor in determining the frequency of flying (and energy use) in societies with similar disposable income, and major North American cities are far further apart than their counterparts in Japan or the EU. A simple way to demonstrate this is to measure the distance connecting the three largest metropolitan areas: in Japan the distance from Tōkyō to Ōsaka to Kyōto adds up to less than 800 km; when the EU is considered as a single country it is 800 km from London to Paris to Frankfurt; but it is more than 3000 km both in Canada (Toronto–Montreal–Vancouver) and the USA (New York–Los Angeles–Chicago).

As result, most of the trips between Tōkyō and Ōsaka or London and Paris are by train, while North American intercity travel is dominated by flying, often by shuttle services departing every 1 or 2 h. Country size is also a key factor in average distances traveled by car, particularly during the past 20 years when car ownership in most European countries and Japan rose to levels almost identical to those in the USA and Canada while the gap between the performance of North American and other vehicles shrank due to higher efficiencies. But the US annual driving mean is still around 20 000 km a year, nearly twice as much as in Japan or in major EU countries.

Before the advent of mass ownership of air conditioning, the energy–climate link was overwhelmingly a matter of heating, but now every US utility has a summer peak due to high AC demand while high electricity prices have slowed down the rate of AC adoption in the EU. Still, in colder climates the average number of heating degree days remains the best predictor of household energy consumption, and Canadians (5700 heating degree days per year in my city, Winnipeg) will always use more energy to keep warm than Germans; their capital has 2900 degree heating days (EEA, 2013). The matter becomes complicated by differences in expected comfort levels: in winter Americans would not tolerate British indoor temperatures, in summer many rooms are now cooled to excess in newly affluent Asian households.

But it is the composition of energy supply and its origins that are perhaps the two most important determinants of the overall level of energy intensity. A high degree of energy self-sufficiency may be great for the balance of payments, but extraction and conversion of fossil fuels are energy-intensive activities – while high levels of energy imports tend to promote higher conversion efficiencies. A large proportion of the energy intensity gaps between Saudi Arabia and Japan, or between Australia and South Korea, can be explained by the combination of these two realities, and the composition of the primary energy supply has even greater consequences because the conversions of liquid and gaseous fuels are more efficient, more flexible, and have lower environmental impacts than the burning of hard coals and lignites.

Although finely milled coal blown into large boilers can be converted into electricity with efficiencies of up to 42%, matching the combustion of fuel oil or natural gas, coal-fired plants have to consume a higher portion of the generated electricity on air pollution control (capturing fly-ash, desulfurization, reducing NO_x emissions) and solid waste disposal (a 1 GW plant will generate tens of thousands of tons of fly-ash a year). And there is now a superior alternative to burning natural gas in large boilers: up to 60% of the fuel used in combined-cycle generation (gas turbine followed by a steam turbine) can be converted to electricity. Efficiency disparities are much larger when the fuels are used directly for space and water heating, with coal stove efficiencies typically well below 50%, while high-efficiency household furnaces now convert 95–97% of fuel's energy into useful heat.

And because virtually all modern industrial processes depend on incessant flows of electricity, nations with high shares of primary electricity can avoid inevitable losses inherent in thermal electricity generation and, everything else being equal, can have lower energy intensities. At the same time, low-cost hydroelectricity attracts electricity-intensive industries, and such countries will have disproportionate shares of such activities. Canada is a perfect example of this reality. The country is the world's second largest generator of hydroelectricity and also the world's third largest producer of aluminum, and ranks among the world's top 10 in other energy-intensive metals and products, including ammonia, copper, lead, and nickel (USGS, 2013). Not surprisingly, Canada's relatively high share of energy use in ore mining, nonferrous metal smelting, and ammonia synthesis will boost the overall energy intensity of its economy, while the near-absence of these industries in Japan or Italy will lower their overall energy intensity rates.

The point has already been made. Energy intensity is far from being a close proxy of material consumption. Its most notable disadvantages are that national rates of TPES are strongly influenced by the structure of energy industries as well as by climate and the country's size and population density; moreover, a specific national intensity reflects not only more efficient (and thus most likely less material-intensive) uses of energy but also more common ownership of many energy converters and their more intensive use. Even so, the long-term trends of energy intensity are broadly comparable with long-term changes in material intensity of a national economy, although the measure should not be used as a proxy for international comparison of material use because of the inherent diversity of the just noted factors. What can it tell us about long-term changes?

The history of global energy intensity cannot be traced with accuracy: while twentieth-century energy statistics are fairly accurate, and even the flows between 1850 and 1900 can be satisfactorily reconstructed, all published estimates of worldwide economic product are questionable. Using Maddison's (2007) estimates of global economic product (in constant 1990 US$) results in very similar energy intensity rates for 1900 and 2000, about 11 MJ/$, with the intervening rise from 1900 to 1970 and subsequent decline equal to about 20% of the peak level. These shifts reflect combinations of complex trends dominated by large consumers with very different energy trajectories (USA vs. China, Germany or Japan vs. India or Indonesia) but it seems fairly certain that a further decline is most likely as affluent nations become gradually even more efficient and as populous modernizing countries continue to rationalize their still relatively wasteful energy uses.

National historical perspectives are more amenable to revealing interpretations, and data needed to calculate long-term series of many national energy intensities are readily available in statistical compendia published by the United Nations (UN, 2013), the International Energy Agency (IEA, 2013), the US Energy Information Agency (USEIA, 2013), and British Petroleum (BP, 2013). The UN series goes back to 1950, other series begin at different times during the 1960s or 1970s, and while the UN statistics list TPES in terms of coal and oil equivalents and also in joules, other sources have toe as their common denominator. As already explained, conversion to these common denominators introduces inevitable errors due to changing qualities of fuel, and there is also no uniform way to convert primary electricity to heat equivalents.

National GDP figures are also readily accessible in compendia published by the UN (2013), the UNDP (2013), the International Monetary Fund (IMF, 2013), and the World Bank (2013a), with some series going back to the 1950s. All values must be expressed in constant monies and meaningful international comparisons should use conversions to PPP (rather than to official exchange rates) in US$. The easiest way to do this is to consult already calculated energy intensity series published regularly by the OECD (2013; in Mtoe/1000 constant US$) and the USEIA (2013; in Btu/constant US$). Long-term perspectives show similar patterns of national energy intensities in Western countries that have experienced transitions from noncommercial biofuels to coal, and then to hydrocarbons and primary electricity.

The energy intensity of their economies rose appreciably during the early stages of their industrialization but the peaks were usually pointed, without any extended plateaus, and they were followed by substantial declines as maturing economies began to use their energy (and hence material) inputs much more efficiently. In the UK the peak had already taken place by 1850, in Canada in 1910, and a sharp American peak was reached around 1920 (Smil, 1994). By the end of the twentieth century the intensities of these three countries were, respectively, about 80, 65, and 60% below their respective peaks. In contrast, the energy intensity of Japan's economy peaked only in the early 1970s. Differences in slopes and peaks are due to specific conditions of individual economies, that is, to the onsets of industrialization, its progress, the primary structure of economies, the rate of technical innovation, and dependence on energy imports.

More recent changes – all calculated from data using GDP expressed in PPP terms (USEIA, 2013) for the three decades between 1980 and 2010 – have shown some impressive declines, with China recording a globally unprecedented drop of 67%, the overall energy intensity of the UK economy falling by 50%, the US rate decreasing by 44%, post-unification Germany using 23% less in 2010 than in 1991, and the global mean down by 20% in the three decades since 1980, the rate almost exactly matched by the Japanese decline. Each of these shifts has arisen from a number of specific changes, with some of them having potential for further continuation while others have been unique developments.

In China, the most decisive factor has been the transition from an extraordinarily wasteful, low-productivity Maoist economy to a more efficient, modern configuration of productive forces increasingly relying on advanced industrial processes. During the 30 years between 1980 and 2010, the average energy intensity of China's steel production declined by 43% (Sun *et al.*, 2013) and that of cement production fell by 55% (Lei, 2011). Given the country's still unfavorable fuel mix (with a high dependence on coal) and far from exhausted opportunities for higher industrial, commercial, and household efficiencies, this downward trend of average energy intensities should continue, albeit at a slower rate, for decades to come. In the UK, the decline has been led by the country's continuing shift from coal to hydrocarbons and by further deindustrialization; the first process has now gone about as far as it can go, while the fate of the country's once globally dominant industrial sector remains in question.

Deindustrialization has been also an important contributing factor in the American decline of energy intensity and, as I will demonstrate, that decline is less impressive once the energies needed to produce the increasing mass of imports are accounted for. Energy use in post-unification Germany was lowered by the closure of many inefficient industries in the former DDR, but the entire country continued on its steady path of gradual efficiency improvements even while deriving a relatively large share of its GDP from energy-intensive manufacturing.

With all major energy consumers lowering the energy intensities of their economies it is inevitable that the global rate will also decline. Its annual decrease averaged 0.75% between 1980 and 2010 and there is no reason why more determined action should not keep it falling by about 1%/year for the next few decades.

I went into such detail in order to show that energy intensity is a peculiar measure that is not easily interpreted. The long-term decline of energy intensity may be a laudable result of determined efforts to improve energy conversion efficiencies in industries, household, and transportation, and it may be a good proxy indicator of gradual dematerialization. But it can also be the not so laudable outcome of a large-scale deindustrialization that has seen energy-intensive activities (metallurgy, chemical syntheses, heavy manufacturing) moving abroad even as an increasing share of all kinds of consumer product is imported. And a prolonged plateau, very slow decline, or even temporary rise in energy intensity may not reflect any deterioration of conversion efficiencies in major consumption sectors, but either an inherently energy-intensive stage in a country's economic development (establishing new industries, extending transportation infrastructures, coping with rapid urbanization) or a higher level of personal (discretionary) energy use by a more affluent population that negates some, or most, of the gains in industrial and service sector efficiency.

Post-Franco Spain is an excellent example of the first case: its energy intensity barely changed between 1980 and 2000 as the country belatedly built its modern infrastructures; and Japan's rising energy intensity during the 1990s was largely due to the fact that the Japanese began to save less and to spend more on themselves (Smil, 2007). Consequently, there is no simple verdict that rapidly declining intensities always indicate a desirable trend and that a slow rate of improvement is an indicator of poor performance. The measure is too indiscriminating to allow such judgments. What I wrote a decade ago still stands: "… those simple quotients are products of complex natural, structural, technical, historical, and cultural peculiarities. The lower the better may be a fine general proposition, but not one to be applied to national energy intensities in absolute terms and on a worldwide scale" (Smil, 2003, p. 81).

And we must remember that the Jevons effect can have economy-wide repercussions as a significant share of specific energy savings may be used, directly or indirectly, to increase energy use elsewhere. Direct rebounds (efficiency gains resulting in lower costs that lead to higher energy use) are easier to quantify, but no generalization is possible. Studies show their effects ranging from negligible (less than 5% increase) to substantial, with more than a 50% rise in specific energy consumption – and there is even greater uncertainty about indirect rebound, that is, on spending the income freed by savings on products or services that are equally or even more energy-intensive, particularly on a nationwide scale (IRGC, 2013).

5.6 Decarbonization and Desulfurization

Decarbonization of national and global energy supply – a gradual shift toward burning fuels with lower carbon content and generating more primary electricity – can be seen as a form of dematerialization that is particularly welcome from the environmental point of view because it reduces specific emissions of CO_2, the most important greenhouse gas, as well as the generation of acidifying sulfur and nitrogen oxides and particulate matter. The process is an organic result of gradual energy transition from solid to liquid to gaseous fossil fuels and of rising shares of primary electricity (hydro, nuclear, wind, solar, geothermal) in the total energy supply. In turn, these transitions have been driven by the need for increasingly higher power densities of final energy uses that are required to support urbanization, industrialization, higher intensity of transportation, and more affluent lifestyles.

Wood (and charcoal) were the dominant sources of thermal energy in pre-industrial villages, and in the temperate European climate the power density of their combustion would have been as low as $10 \, W/m^2$ of living area in a small single-story house; burning coal in a four-story apartment building required more than $200 \, W/m^2$ and modern high-rise buildings (heated in northern latitudes, air conditioned in the subtropics and tropics) consume more than $1000 \, W/m^2$ of their foundations. Even when including streets and parks, densely populated modern cities require constant energy supply at rates from 50 to $100 \, W/m^2$. This is why Ausubel (2003, p. 2) concluded that "the strongly preferred configuration for very dense spatial consumption of energy is a grid that can be fed and bled continuously at variable rates."

Obviously, natural gas – safely and reliably delivered by pipelines, burned in modern high-efficiency furnaces, with its temperature effortlessly controlled by resetting or programing a thermostat, and whose combustion generates just H_2O and CO_2 – is a superior source to liquid fuels that must be stored in tanks and distributed by trucks to final consumers; and it is an incomparably better choice than coal, a low-energy density, bulky, and dirty solid whose combustion emits particulate matter and SO_2. The correlation is unmistakable: modern high-energy societies rely on energies that can deliver high power densities thanks to the combination of high energy densities of better fuels converted with relatively high efficiencies.

The energy density gradient for fuels shows a roughly a 3-fold difference between the extremes: dry wood has $18-20 \, GJ/t$, most steam coals rate between 22 and $26 \, GJ/t$, refined oil products are within a tight range of $40-44 \, GJ/t$, and a ton of natural gas contains $53.6 \, GJ/t$. Higher hydrogen content explains this progression of higher energy density. Combustion of pure carbon releases $32.8 \, GJ/t$, combustion of pure hydrogen liberates $142 \, GJ/t$: fuels with higher hydrogen shares will have higher energy densities, and decarbonization can be seen either as the decline of carbon content or the rise of hydrogen content, or can be expressed as a ratio of hydrogen and carbon atoms present in specific fuels and in aggregate national or global energy consumption.

Wood, composed mostly of cellulose, hemicellulose, and lignin, is usually about 50% C and its H content is only about 6%, typical bituminous coals contain 65% C and 5% H by mass, common fuels refined from crude oil (gasoline and kerosene) have 86% C and 13% H, while methane's (CH_4) has, obviously, 75% C and 25% H by mass. When assuming $19 \, GJ/t$ as a mean energy density of absolutely dry wood, the fuel's complete oxidation would release about 30 kg C/GJ. Combustion of good bituminous coal will emit roughly 25 kg C/GJ (which means that the substitution of coal for wood will release 10–15% less carbon per unit of energy); the typical emission factor for refined liquid fuels is about 20 kg C/GJ (or 20–25% lower than for coals); and complete combustion of methane will liberate only 15.3 kg C/GJ, nearly 20% less than gasoline, 40% less than typical bituminous coal, and 50% less than wood.

When using averages of 50% C and 6% H, wood would have atomic H:C ratio of about 1.4. But that is not the ratio of atoms whose combustion is converted to thermal energy: a large part of wood hydrogen is never oxidized because hydroxyls (OH·) radicals that are part of cellulose and hemicellulose escape to the air during the early stages of combustion (Shafizadeh, 1981). The actual H:C ratio of wood is variable but no higher than 0.5 compared to coal's 1.0, 1.8 for common fuels refined from crude oil (gasoline an kerosene), and 4.0 for methane (the actual ratio for natural gas is marginally lower due to the presence of higher homologs, C_2H_6 and C_3H_8, besides CH_4). For comparison, when Ausubel (2003) charted the historic shift toward decarbonization, he worked with the following H:C average: 0.1 for wood (too low, I would argue), 1 for coal, 2 for crude oil and, naturally, 4 for methane.

Decarbonization of the modern energy supply began first in the USA (already during the second half of the nineteenth century) but the process became widespread and proceeded at accelerated rates only after World War II. When comparisons are made on the basis of energy content (not on mass), high-carbon fuels (wood and coal) accounted for 94% of the global total in 1900 and still 73% in 1950, but by the year 2000 their share was down to about 38%. Decarbonization was also helped by rising shares of primary (hydro and nuclear) electricity: its share in the global primary energy supply rose from just 2% in 1950 to about 10% half a century later. But integrations of individual fuel components show that during the twentieth century coal supplied significantly more energy than crude oil, roughly 5.3 vs. 4.0 YJ (Smil, 2010).

Thanks to China's rapid post-1990 expansion of coal extraction, coal's share in the global primary energy supply has actually been slightly higher in 2010 than it was in 1990, and natural gas, the fuel with the least carbon content, still supplied no more than 25% of all fuel energy in the year 2000 and just over 27% in 2010 despite the fact that its global extraction had more than doubled between 1980 and 2010, from about 1.4 to nearly 3.2 Tm^3 (BP, 2013). As a result, the overall carbon intensity of the world's fossil fuel supply declined only moderately during the course of the twentieth century: in 1900 it was about 25 kg C/GJ of fossil fuels, by 1950 it was about 22.5 kg C/GJ, and in the year 2000 it was basically unchanged at about 22.3 kg C/GJ, altogether only a 10% decline during the twentieth century. Thereafter, higher natural gas extraction negated the effect of Asia's rising coal production and the global rate fell by about 6% in a decade, to 20.9 kg C/kg by 2010.

The progress of decarbonization becomes more pronounced when the calculation is done for the world's TPES, including biofuels and carbon-free primary (hydro, nuclear, wind, solar) electricity: the denominator (TPES) becomes larger, but so does the nominator due to emissions of CO_2 from the combustion of wood and crop residues (in 1900 their combustion emitted more CO_2 than the burning of coal, by 2000 they still contributed about 17% of the total). With all energies included, the post-1900 decarbonization trend is clearly more pronounced, with global rates declining from about 27.5 kg C/GJ in 1900 to 21.2 kg C/GJ in 2010, a 23% drop in 100 years. The global desulfurization trend appears even more pronounced with a drop of about 60% in sulfur intensity of TPES in 110 years.

When expressed in terms of the H:C ratio of fossil fuels (leaving out wood), the global indicator of decarbonization rose from 1 in 1900 (when coal supplied all but 5% of all fossil energies) to about 1.6 by 1950, and 1.8 by 1980, but during the next two decades it leveled off at roughly 1.9 and by the year 2010 it had declined a bit to 1.83. This plateau, followed by a slight decline, was due to China's extraordinarily large rise in coal extraction, from about 600 Mt in 1980 to 3.25 Gt in 2010; this kept China's fossil fuel H:C ratio at 1.15 between 1980 and 2000 and then, even with large increases in consumption of hydrocarbons (crude oil combustion nearly doubling in a decade, natural gas use more than quadrupling), it allowed it to rise only marginally to 1.17 by the year 2010.

When Marchetti (1985) plotted the weighted H:C ratio of global fuel consumption, he found that the long-term trends indicated a global mean of about 3 by 2010, when it was actually only 1.83. He foresaw the arrival of a global methane economy (H:C of 4) shortly after 2030, and of a hydrogen-dominated economy (predicated on massive generation of H_2 without fossil fuels) in the closing decades of the twenty-first and the early decades of the twenty-second century. Two decades later Ausubel still expected, according to the historical trend in decarbonization, "methane to provide perhaps 70% of primary energy soon after the year 2030" and that "large-scale production of carbon-free hydrogen should begin about the year 2020" (Ausubel, 2003).

In reality, the decarbonization process has slowed down since 1990, the global H:C fuel ratio has been stagnating, and natural gas rose from 26% of the world's primary fossil fuel supply to 29% by 2000, before declining a bit to 28% by 2010. This was a gain of 8% in two decades – while a gain of 250% would be required to lift the gas to 70% of the global fossil fuel supply by shortly after 2030, clearly a most unlikely shift. Ausubel was right to foresee a much more important role for natural gas than most people thought possible in the early 1980s – but his anticipated rise of hydrogen, the good element, has clearly been proceeding at a slower pace.

In any case, what is true about relative and absolute dematerialization of global material consumption is true about relative and absolute decarbonization of the global energy supply: in both cases relative declines have been unmistakable, impressive, and will continue, but as there have been no absolute global declines in materials consumed, so there have been no worldwide reductions in carbon extracted and CO_2 emitted. Just the opposite has been true, as global carbon emissions from fossil fuel combustion rose from less than 550 Mt C in 1900 to about 1.6 Gt C in 1950, to almost 6.8 Gt in the year 2000 and, after a 35% increase in a decade, to a new record of 9.13 Gt C (or 33.5 Gt CO_2) in 2010 (CDIAC, 2013).

And this contrast is also true as far as all of the world's major economies are concerned: multipliers for the twentieth century carbon emissions were 8.6 for the USA, 62.8 for Japan, and (the following figure is correct!) 35 700 for China. All Western countries have recently seen a clear moderation in the growth of CO_2 emissions: during the 25 years between 1985 and 2010, US carbon emissions rose by 22% (or at most by 24% after adding emissions generated by net imports of goods produced abroad, compared to 63% during the 25 years between 1960 and 1985), those of Japan increased by 18%, while Germany's emissions recorded a 27% decline, largely due to the post-unification collapse of the former East German economy. In contrast, Chinese CO_2 emissions quadrupled, and Indian releases of fossil carbon rose by nearly 430% (CDIAC, 2013).

The verdict is clear: although the generations-old process of relative decarbonization of the world's fuel supply experienced a significant slow-down during the second half of the twentieth century, it is bound to continue over the coming decades, above all thanks to substantial gains in natural gas production; in terms of the overall primary energy supply, it should be helped by further gains in hydro, wind, and solar electricity. Even so, there are no imminent prospects for any major reductions in absolute emissions of CO_2 – and, of course, the atmospheric behavior and the future levels of absorbed outgoing radiation are determined by the absolute atmospheric levels of the gas, not by reductions of the relative burdens per unit of product or service.

Except for some natural gases that are almost pure methane, or mixtures of CH_4 and higher homologs of the straight-chained alkanes series (mostly C_2H_6 and C_3H_8), all fossil fuels contain some sulfur, with shares ranging from less than 0.5% in the lightest (sweet) crude oils to more than 4% in low-quality lignites: 1.5–2% is the most common range for bituminous coals and globally traded crude oils (Smil, 2008). The combustion of fossil fuels generates more than 90% of all anthropogenic sulfur, with the rest originating largely in the smelting of copper, zinc, and lead. Emitted SO_2 oxidizes to sulfates and the resulting acid precipitation (and also acidifying dry deposition) affects soils, plants, and both aquatic and terrestrial ecosystems, and damages exposed metals as well as limestone and marble. Its effects have become particularly noticeable across parts of Western and Central Europe, Eastern North America, and China.

Rising combustion of fossil fuels has been accompanied by increased emissions of SO_2, from about 20 Mt in 1900 to just over 58 Mt by 1950, and the peak level of about 150 Mt was reached in 1979–80 (Smith *et al.*, 2010). But, unlike in the case of carbon, this absolute rise of sulfur emissions has been accompanied by pronounced relative desulfurization: I have calculated its rates as declining from about 465 g S/GJ in 1900 to about 400 g S/GJ in 1950, and to 270 g S/GJ by 1980. Higher shares of hydrocarbons in the global energy supply have, on average, a much lower sulfur content than coal, and high-S crudes and natural gases rich in H_2S are routinely desulfurized before combustion. After 1980 this desulfurization trend was amplified due to the adoption of flue gas desulfurization (FGD) in large coal-fired power plants.

FGD removes SO_2 by reactions with basic compounds (mostly lime or ground limestone); $CaSO_4$, its main product, is either land-filled or can be used in wallboard manufacture. FGD was first commercialized during the 1970s and by the year 2000 nearly 40% of the world's coal-generating capacity had some level of FGD (Smil, 2008). Emission declines were impressive: in the USA they were cut nearly by half between 1970 (the first year the USEPA began to made consistent calculations) and 2000 (USEPA, 2013); globally, relative SO_2 emissions were reduced to about 175 g S/GJ by the year 2005 as the absolute emissions declined to about 107 Mt (Smith *et al.*, 2010).

As emissions were cut in Europe and North America, they were rising rapidly in China and India: between 1990 and 2005 both China's and India's SO_2 emissions doubled, respectively to more than 32 Mt and nearly 10 Mt/year; the combination of deep cuts in the West and large increases in the East resulted in a slight global lowering of SO_2 releases, from 121 Mt in 1990 to 112 Mt in 2005. But the rise of Asian emissions introduced a greater uncertainty into the global estimates. For example, a comparison of half a dozen estimates of SO_2 emission by China, (the world's worst polluter) shows that the difference between the lowest and the highest value was 38% for the year 2000 and 35% for the year 2005 with, predictably, the lowest totals claimed by China's Ministry of Electric Power (Lu *et al.*, 2010).

Whatever the actual total might be, it seems that even the Chinese emissions finally began to decline after 2007 because of FGD installed at many of China's new (and some older) coal-fired power plants, a reduction that has been confirmed by the Ozone Monitoring Instrument aboard NASA's Aura satellite (Li *et al.*, 2010). Another post-2007 event that helped to reduce the global SO_2 emissions was the worst post-World War II economic crisis, and Klimont *et al.* (2013) put the 2010 total as low as 103 Mt, nearly a third lower than the global record in 1980, and a rate that implies specific emissions of only about 118 g S/GJ, a third lower than the 2000 value.

Accounting uncertainties cannot change this conclusion: unlike in the case of CO_2 emissions, global dematerialization of sulfur emissions has taken place not only in relative, but also in absolute, terms and it has done so despite the rising combustion of coal because of an effective technical fix (FGD). Unfortunately, decarbonization of flue gases is a much greater technical challenge than their desulfurization, and recent years have seen too many exaggerated promises regarding its near-term performance and too many unrealistic forecasts (Metz *et al.*, 2005; Muradov and Veziroğlu, 2012). During the next two decades it should become clear if effective (and affordable) decarbonization of large stationary sources is a practical solution, or if the best way to decarbonize is to keep changing the make-up of the primary energy supply.

6

Material Outlook

Those readers who have persevered (and have, along the way, complained about too many numbers) have now reached the point where they should be impressed by the magnitude and complexity of the global material edifice erected by modern civilization since the middle of the nineteenth century, and no less so by the magnitude of incessant material flows required to operate and maintain it. Although there was no shortage of admirable extraction, construction, and consumption feats in pre-1850 history, only the creation, transformation, and expansion of modern civilization made human societies dependent on enormous, incessant, and now also truly global flows of materials. After millennia of reliance on a limited range of materials – wood and other phytomass, stone, gravel, sand, and clays, and a dozen metals – that were processed in relatively simple ways, we now use thousands of materials whose qualities are tailored to a still widening array of specific requirements.

This growing diversity and dependence have been accompanied by some remarkable efficiency gains that can be impressively demonstrated by long-term tracing of specific energy and material uses, that is by declining energy intensities and by relative dematerialization, as well as by moderation, or even near-elimination, of typical environmental impacts. At the same time, growing populations and improving quality of life have translated into a steadily rising demand for virtually all traditional materials: the only notable exceptions have been such ancient biomaterials as flax or hemp, but in aggregate every modern nation now needs more of the wood, stone, sand, metals, and nonmetallic minerals that have been used since antiquity than it did 100, 50, or 25 years ago. And to these traditional materials we have to add the greatly expanded production of new materials – including synthetic fertilizers, polymers (plastics), and metals and nonmetallic elements not previously exploited by pre-industrial societies. In all of these cases there has been no dematerialization in absolute terms, not on the national level for any major economy and not on the global level.

These realities lead to many obvious questions. How long will these now so well-entrenched trends continue? How long can they continue? What are the natural resource

---

Making the Modern World: Materials and Dematerialization, First Edition. Vaclav Smil.
© 2014 John Wiley & Sons, Ltd. Published 2014 by John Wiley & Sons, Ltd.

limits to further multiplications of mineral extraction and to modern syntheses based on nonrenewable raw materials? Are there any insurmountable economic obstacles that will force us to moderate our consumption of materials? Even more fundamentally, are there environmental constraints that will make it impossible to replicate the twentieth-century material expansion during the twenty-first century? And then there are the most important questions of all: should we advocate some rational limits to material consumption? And, if so, how should we be going about it?

Some of these questions can be answered on the basis of indisputable physical evidence or by referring to the most likely course of realistic technical advances. Consequently, in the opening sections of this closing chapter, I will first contrast the best available evidence concerning the availability of principal material resources with the extent of possible future requirements, and then look at the opportunities for material substitution, enhanced recycling, and reuse. Other questions posed in the preceding paragraph entail fundamental value judgments and in all those instances it is much more difficult to find a common ground and to come up with recommendations that would be most likely to meet the twin criteria of widespread acceptance and true effectiveness.

Complex linkages of many dynamic determinants of future material growth, including variables subject to relatively sudden shifts, make all long-range forecasts inherently uncertain. More than two decades ago I decided to abstain from offering such prognoses and, true to this resolution, will not provide any estimates of the most likely levels of global or national demand for particular materials by the year 2025 or 2050, nor will I present any scenarios of future rates of dematerialization or reductions in environmental impacts attributable to material extraction and use. What I will do instead is to appraise the most probable trends of some fundamental variables that will determine the future rates and consequences of material use.

I will start with concerns about the availability and extraction of natural resources whose incessant flows have become indispensable for the maintenance and advancement of modern civilization. In the next section I will stress many encouragingly impressive opportunities for wasting less: potentials for both material and energy savings are nowhere near early exhaustion. These opportunities will be further enhanced by new materials and by the continuing advances in relative dematerialization, topics for a separate section. I will close the book (without normative prescriptions) by raising questions about some fundamental departures from the twentieth-century model of mass consumption, and about the chances of more rational arrangements.

6.1   Natural Resources

This is not the place to review the evolution of concerns about the availability of natural resources and their adequacy to maintain the expansion of modern civilization. In the USA, systematic interest in these matters began during the 1950s (The President's Materials Policy Commission, 1952), advanced greatly during the 1960s (Landsberg and Fischman, 1963; Landsberg, 1964; NAS, 1969) and has been intertwined with energy and environmental concerns since the 1970s. Boulding (1970) called for the end of the linear economy that extracts resources, transforms them into commodities, and spews them out into pollutable reservoirs, a process he considered inherently suicidal.

Global concerns about the exhaustion of mineral resources have been growing during the 1970s and 1980s (Meadows *et al.*, 1972; Barney, 1980; McLaren and Skinner, 1987; Ehrlich and Holdren, 1988), developed into arguments between catastrophist and cornucopian camps (Simon, 1981, 1995) and became combined with an even broader concern about the extent and rapidity of global environmental change. The latest phase of these concerns began with prediction of an imminent peak in global oil extraction that was to be followed by a steep decline in production (Campbell, 1997; Deffeyes, 2001; for a critique of peak oil see: Smil, 2006a).

That there has been no global oil production decline – and that it is unlikely to happen anytime soon – has not stopped the transference of such fears to other resources, particularly as China's seemingly insatiable demand for all minerals reversed a century of steadily declining cost of raw materials for the world's affluent consumers (Sullivan *et al.*, 2000; *The Economist*, 2011). Gold has become one of the latest additions to the list of about-to-peak extractions (Kerr, 2012). In fact, the "running out" meme has now been universally applied: Heinberg (Heinberg, 2010) declared that peak of everything is upon us and Klare (Klare, 2012), more moderately, sees "the end of easy everything." Klare does not see any imminent scarcity but high prices and ferocious competition as companies and nations fight among themselves for dwindling resources.

In contrast to the complexity and contentiousness of these matters seen from long-term perspectives, the near-term outlook for the supply of essential mineral resources is neither highly uncertain nor difficult to write about. The only prerequisite of such a review is to make clear the often misunderstood fundamental difference between reserves and resources (McKelvey, 1973). Resources denote the total mass of material (element, compound, mineral, ore) in the Earth's crust, and a distinction can be made between resources on land and under sea. Obviously, these aggregates are not known with a high degree of certainty, and their estimates tend to increase with more drilling and more extensive targeted exploration. Even if we were to know the exact amount of a particular resource, that information would not allow us to calculate the time of its ultimate exhaustion because no mineral could ever be completely exploited: long before reaching such a point the costs of removing it from excessive depths or isolating it from deposits where it is present in minuscule concentrations would make its recovery quite uneconomical.

This is why the reserve category is of much greater practical interest, as reserves are the shares of resources that could be extracted from known locations with known costs and by using available techniques. This means that resources in place are given and finite (but poorly known) while reserves are movable totals, accurately known but constantly changing: they are transferred from resources through investments and technical advances, through the continuous process of exploration, and through the deployment of better recovery, mining, and processing techniques. Specific reserve totals, for a country, a continent, or the world, are commonly divided by relevant annual production totals to calculate reserve/production (R/P) ratios. For example, according to the USGS the global R/P ratio for copper was 42.8 years in the year 2011 (USGS, 2013).

This ratio does not imply that there will be no copper left to mine by the end of 2054. Indeed, the metal's global R/P ratio was nearly identical in 1995 and in 1980 (Doggett, 2010), and the relative constancy of this (and most other mineral extraction ratios) means that industries successfully maintain acceptable levels of reserves relative to annual production. Put in another way, the translational effort of moving minerals from resource to reserve category assures the production rate for a given number of years: not too short to engender uncertainty, not too long as there is no advantage to verifying reserves whose extraction may not take place for a century. Unfortunately, too many poorly informed commentators interpret R/P ratios as the dates for running out of a particular resource. In reality, our civilization is no danger of running out of any major mineral, not imminently (in years), not in the near term (in one or two decades), and not on the scale of average human life-span (60–80 years).

Resources of common construction materials – sands, clays, stones – are immense, eliminating any concerns about their availability on a civilizational (10^3 years) time-scale. Resources of silicon, the material of electronic age, are obviously in the same super-abundant category. The survival of billions of people requires adequate fertilizer applications to produce enough food, and there is no danger that we are going to run out of the atmospheric nitrogen needed for the synthesis of ammonia. And the compound that serves as the most convenient source of hydrogen, natural gas, is being discovered and produced in increasing quantities. In recent years some catastrophists have begun to create a new concern about peak phosphorus (Beardsley, 2011), even about peak potash.

Perhaps most notably, this renewed fear (general and phosphate-specific) was raised in *Nature*, the world's leading science weekly, where a World View column concluded that

> Prices of global raw materials are now rising fast. This ... is a genuine paradigm shift, perhaps the most important economic change since the Industrial Revolution. Simply, we are running out ... There is now no safety margin ... there is the impending shortage of two fertilizers: phosphorus (phosphate) and potassium (potash). These two elements cannot be made, cannot be substituted, are necessary to grow all life forms, and are mined and depleted. It's a scary set of statements ... What happens when these fertilizers run out is a question I can't get satisfactorily answered and, believe me, I have tried. There seems to be only one conclusion: their use must be drastically reduced in the next 20–40 years or we will begin to starve (Grantham, 2012, p. 303).

I commented that Grantham could have tried just a bit harder (Smil, 2012). An Internet search would have led him to *World Phosphate Rock Reserves and Resources* published by the International Fertilizer Development Center (Van Kauwenbergh, 2010). This latest detailed assessment of the world's phosphate reserves found them to be adequate to meet fertilizer demand for the next 300–400 years. Similarly, the 2012 edition of the USGS's *Mineral Commodity Summaries* contains significant revisions of phosphate rock reserves in Morocco, Russia, Algeria, Senegal, and Syria and puts the worldwide R/P ratio at about 370 years (USGS, 2012). And the International Fertilizer Industry Association (whose members are most of the world's most prominent fertilizer producers and traders) has emphasized that it **"does not believe that peak phosphorus is a pressing issue, or that phosphate rock depletion is imminent"** (emphasis in original; IFIA, IFIA, International Fertilizer Industry Association, 2013).

But, appropriately, that statement also emphasized that all "efforts to minimize phosphorus losses to the environment and optimize phosphorus use should be encouraged" – not only because such efforts save money but also because excessive applications of phosphates are a major cause of aquatic eutrophication. Indeed, the call for minimized losses should be a universal admonition, not primarily because of any fears of imminent resource exhaustion but because all wasteful uses of materials have both economic and environmental costs. And (a point to which I will return in the closing section of this book) much larger gains in reducing applications of all fertilizers could be achieved by reducing average per capita meat consumption in all affluent countries where fertilizers are now used not to grow food but animal feed.

As for potash, the latest R/P ratio cited by the USGS is in excess of 250 years. Similarly, there are also no looming shortages of metallic ores. USGS puts global iron ore reserves at more than 80 Gt and resources at more than 230 Gt, copper ore reserves and resources at, respectively, 690 Mt and more than 3 Gt: these totals translate (for 2011 outputs) to R/P ratios of about 30 and 40 years: these are, as already noted, not much different from ratios 20 or 40 years ago, even though the process of translating resources into reserves has had to keep up with massively increased production. R/P ratios are similar for nickel and manganese, and they are considerably higher for titanium, vanadium, and silver. Nor are the rare earths actually rare, their crustal abundance is not particularly low but sites with extractable concentrations are less common.

China's near-monopolization of their supply and a temporary reduction of their exports created this impression of rarity. The response to these restrictions has been a surge in exploration in North America, Australia, and Africa and the opening of new mines will soon change that undesirable situation (CCC, 2012). There are also opportunities for substitution (Holliday *et al.*, 2012), and as soon as China began limiting its exports of rare earths, Japanese companies responded with a stepped-up research effort for substitutions and for recycling. Hitachi successfully developed machines that can automatically recover magnets from hard disk drives and compressors, and then use a dry process to abstract high purity neodymium and dysprosium from the material (Nemoto *et al.*, 2011).

A global assessment by Gordon *et al.*, (2006, p. 1213) concluded that "there is no immediate concern about the capacity of mineral resources to supply requirements for the geochemically scarce metals." Similarly, after reviewing the needs of a critical sector, Graedel (2011, p. 332) found that "at least for the foreseeable future, and from a global perspective, there are no significant supply challenges for any of the parent metals related to energy." Of course, these conclusions do not exclude the likelihood of higher energy requirements and hence higher production costs. The richness of many exploited metallic ores has been declining for generations, with relative degradation averaging about 1% a year.

Naturally, for some ores in some countries, the decline has been much faster; for example, Malaysian tin mining companies are now removing twice as much waste as two decades ago (Sims and Rusmana, 2011). As a result Norgate and Jahanshahi (2011) estimated that the combination of this declining quality (excluding Al and Fe whose ores are expected to remain unchanged) and higher demand might lead to a quadrupling of energy requirements for the worldwide production of principal metals (Al, Cu, Fe, Ni, Pb, Zn) by 2030, from about 2 EJ in 2010 to 8.7 EJ. On the other hand, such calculations do not factor in continuing technical advances or future shifts in demand.

And gold's use illustrates another adjustment shift that could make the metal available for essential applications, a huge opportunity for rationalizing its consumption: in 2011 demand for gold was dominated by jewelry (48%) and money (41% in bars, coins, medals), with only 11% (about 450 t) claimed by industry, with electronics consuming

320 t or less than 8% of the total demand (Savage, 2013). During the coming decades any conceivable increase in gold demand by the electronic industry could be met easily by a more rational division of the metal's final uses and by its near-complete recycling from gold-rich e-waste: surface gold mines yield 1–5 g Au/t of rock while computer circuit boards yield 250 g/t and mobile phones up to 350 g/t (Owens, 2013).

In any case, matters look different from a longer perspective, particularly when assuming that the rest of the world will eventually want to consume as much as the affluent nations in general, or the USA in particular, have been doing recently. For example, Gordon *et al.* (2006) estimate that the world's total copper resource is 1.6 Gt, but to provide every person of a future global population of 10 billion people with 170 kg of copper stock (the per capita mean for North America around the year 2000) would require 1.7 Gt Cu, more than the estimated resource in the crust. Of course, even more of the metal would be needed to replace inevitable losses during use and recycling. And the authors find that similar assumptions would also require the tapping of the lithosphere's entire stores of zinc and platinum.

Similar calculations demonstrating eventual resource limits can be made by assuming the continuation of the historic consumption growth rate. For copper (whose pre-1900 global consumption added up to about 10 Mt) the 50 year consumption averages since 1900 are as follows: 70 Mt until 1950, 340 Mt until the year 2000, and (assuming continued post-1950 growth at 3.3%/year) 1700 Mt until 2050, once again a cumulative total surpassing the estimate of ultimately recoverable metal. Such calculations can be interpreted either as incontestable proof of an approaching catastrophic collapse of modern material-intensive civilization – or as an interesting but simplistic exercise, because reasonable assumptions (based on known adjustments historically deployed by societies) could change those outcomes in fundamental ways.

The required stock of a material can be greatly reduced by lower per capita consumption, particularly as the price of the material will rise; better product design could maintain desired functionality with a fraction of deployed material; the need for new production could be lowered by assiduous recycling; these changes can greatly reduce the expected growth of demand; the population may never reach the expected count for a given year; and there is no reason why the North American or US per capita stock should be used as the most desirable model to follow. Besides the availability of natural resources and the level of average per capita demand (annual use or circulating stock) there are three other fundamental variables (or constraints) that will determine the extent of future material production: our capability to extract them at affordable cost; the degree of environmental impact accompanying material use; and the degree of material substitutability.

The last option has its inherent limit set by the evolution of the biosphere: hydrogen, oxygen, carbon, and nitrogen are the macro-constituents of every form of life and have no substitutes, but grand biogeochemical cycles assure their ready availability. Also required is a less massive presence of elements needed to build special organs or to enable biochemical processes: the list of these elements includes Ca, K, Na, Mg, Cl, S and, in trace amounts, Co, Cu Cr, F, Fe, I, Mn, Mo, P, Se, Si, Sn, and V. But the total mass of these elements that need to be mobilized to support the Earth's life is minuscule compared to the mass of nonmetallic minerals and metallic elements extracted by modern civilization, and so the concerns about substitutability apply mainly to those metallic elements that are needed in relatively large amounts but are present in the crust in relatively modest concentrations (unlike the enormous resources of Fe, Al, Ca, K, Mg, or Si) or in less accessible forms.

Copper is a perfect example of a metal belonging to this category: it is not needed in such enormous quantities as steel but its presence in common consumer products is substantial. For example, an average car contains 1.5 km of copper wire and the aggregate mass of the metal ranges from 20 kg in a small vehicle to 45 kg in large SUVs (USGS, 2009). What would we do without new copper? Turning to waste would be a limited option: by the end of the twentieth century the USA had extracted 101.5 Mt of copper and 16.5 Mt of metal remained in production wastes, mostly in some 12 Gt of tailings and in 212 Mt of slag (Lifset *et al.*, 2002). But, as Goeller and Weinberg (1976) pointed out decades ago, in the long run electrical copper is almost entirely replaceable by aluminum, and that element and steel, titanium, plastics, and composites can replace its structural uses. While the presence of elements in low concentrations, hence incurring high energy costs for production, will cease to be a problem once we master highly efficient conversion of renewable energy flows.

In reality, future material demands will be shaped by numerous dynamic linkages of the outlined variables. Examples of these intertwined effects abound. Even the near-exhaustion of high-quality mineral deposits would be no cause for concern if a new and affordable extraction technique could tap resources considered utterly uneconomical in the past. Even a severely limited resource potential (basically un-extractable at any cost) ceases to be a problem where a ready substitution becomes available. Consumption of a still abundant material may carry unacceptable environmental costs – but those, too, can be reduced (even eliminated) by a substitute or by a drastic shift in per capita demand.

During the last two to three decades we have already witnessed the initiation of some of these shifts and the formation of new trends. On the most fundamental level there have been some notable reductions in the average rates of economic growth in many affluent societies. Between 1990 and 2008 (the last year before a sharp decline due to the worst post-1945 economic downturn) the annual compound growth of GDP was 2.73% in the USA, 2.03% in Western Europe, and 1.24% in Japan (IMF, 2013), rates much lower than those for the period between 1950 and 1973 (the year of OPEC's sudden oil price rise) when Western European economies grew by 4.79%/year, the USA averaged 3.93%, and Japan 9.29%/year (Maddison, 2007).

This expected slow-down (characteristic of all large, maturing systems) has been accompanied by a leveling off, and even by some moderate declines, in per capita consumption rates of both energy and some principal materials. In 2010, average US per capita consumption of primary energy was about 8% below the 1980 level, in the UK the decline for the same period was 9%, while the rates have been stable (with slight fluctuations) since the mid-1990s in affluent countries as different as Italy and Sweden or France and Japan (USEIA, 2013). In material terms, saturation of per capita consumption, in some cases followed by decline, has been observed for such diverse commodities as steel and cement. After four decades of generally declining (but fluctuating) rates, per capita steel consumption in 2008 (before the drop caused by the global economic downturn) was considerably lower than in 1970 not only in the UK but also in France and the USA, and it was slightly lower in Canada and Japan (WSA, 2013).

And some English commentators made much of the UK's long-term absolute decline of material inputs, speculating that the country had passed its maximum level of material consumption (Clark, 2011). Total material requirements (direct and indirect flows) were down by about 21% between 1990 and 2010 and direct material inputs were 20% lower during the same time (ONS, 2012). The UK's material productivity more than doubled during those two decades as the mass of resources declined in relation to the rising economic activity. But those material aggregates contain all fuels, and hence they were heavily influenced by the demise of British coal mining and a strong shift to hydrocarbons, and the 2010 figure reflects the country's failure to recover from the economic crisis that began in 2008. Between 1970 and 2007 the UK's total material requirements had actually risen by 16%. Obviously, a longer period of time will be required to see a clear long-term trend.

But there is little doubt that the combination of relative dematerialization (per unit of GDP, per capita), minimal (or nonexistent) population growth and the rapid aging of affluent societies will be increasingly translated into weakening demand for materials. The average annual population growth rate of more developed regions was already down to just 0.4% between 2005 and 20101, and by 2050–55 it is forecast to be zero; aging is already much in evidence in Japan and in the EU, and by 2030 nearly 30% of people in more developed countries will be older than 60, compared to 20% in the year 2000 (UNDESA, 2013). As a result, future economic growth in affluent societies may not require substantial additional material inputs as any new requirements resulting from moderate GDP growth (1–2%/year) sufficient for already largely saturated markets with a slowly growing (and in many countries declining) and aging population could be largely or completely met by a combination of continued lowering of average energy intensities, of relative dematerialization of products and services, and by intensified recycling and reuse.

And the greatest contributor to the rising material demand of the past generation will not extend that record for another 20–30 years: Chinese growth rates of material consumption must come down. After all, in 2010 the country was already consuming 36% more steel per capita than the EU-27 and 50% more than the USA. And the recent peak per capita rates of China's cement consumption were 2.5 times the Japanese level, more than 3 times the German or the US rate, and in some of the country's regions they were as high, or even higher, than Spain's 2007 rate of 1300 kg/capita that preceded the collapse of Spain's construction industry; all countries whose annual cement consumption surpassed for a while 1 t/capita experienced, sooner or later, the burst of their construction bubble (Bell, 2012). These realities must be considered, yet again, in conjunction with the country's slowing rates of population growth and with the relatively rapid aging of its population: by 2040 China will have the same share of people above 60 years of age as Japan had in 2010 (UNDESA, 2013).

Of course, even a near-stabilization of aggregate material requirements in affluent economies (accompanied by steadily falling per capita consumption) and considerable moderation of growth in China will still leave us with decades of potentially huge demands for energy and for all kinds of commodities in India and other populous Asian countries as well as in Africa. This means (barring unprecedented natural catastrophes or prolonged economic downturns) that we will not see significant declines in the overall level of material extraction and production, and that reduced global rates of growth would count as a success during the next one or two generations. But within these confines of continuing (albeit moderated) growth there are many opportunities for change, for producing materials inputs much more efficiently, and for designing them for greater longevity and easier recyclability to last longer, in order to reuse all commodities much more effectively while creating fewer environmental impacts.

6.2 Wasting Less

Despite many recent improvements at various stages of production and use and reuse of materials, opportunities for wasting less, be it raw commodities or finished products, remain both substantial and ubiquitous. A systematic review of these opportunities would easily fill another book: here I will offer just some notable examples illustrating the diversity of realistic improvements and their significant rewards. I will do so by concentrating on four categories of reducing waste: by avoiding the use of a material through better design; by more rational, less wasteful, manufacturing; by more intensive and more efficacious recycling; and by rewarding material substitutions.

But before turning to these specifics I must emphasize that the most important indirect way to rationalize and to moderate modern material consumption is to keep on pursuing many outstanding opportunities for reducing energy cost all along the material chain, starting with mineral extraction and processing, through transportation and manufacturing, to distribution and reusing or recycling of unwanted components and products. The remaining differences between current best practices and energies of formation indicate the extent of potential savings: the difference between the two rates is nearly 2.5-fold for aluminum, 2-fold for steel, but only about 1.3 times for ammonia.

Of course, in practice it is impossible to come very close to the theoretical minima required to liberate metals from ores or to synthesize compounds from their constituent elements, but considerable savings are achievable without any stunning innovations. For example, a detailed appraisal of the US pulp and paper industry found the average energy cost – weighted by the type of production process (kraft, thermo-mechanical) and paper types (printing, linerboard, etc.) – of 27.4 GJ/t, while the application of the best available practices would reduce the rate by about 26% to 20.5 GJ/t (Jacobs and IPST, 2006). Savings of 20–30% with the application of the best available practices would be common in many other industries.

Avoiding the use of materials should be a high priority in all cases where such omissions do not result in products that are less safe, less convenient to use, or less durable. Opportunities for saving by design are ubiquitous and would have the greatest and longest-lasting effect in new construction: sub-optimal design of buildings means that a 30% reduction in overall material use is commonly possible. The easiest choices for avoiding material use are in packaging. Intercontinental transport requires adequate protection, but excessive packaging of consumer items (be it tools, food, or special gifts) has become such a ubiquitous attribute of modern affluent societies that people have become inured to this often gratuitous waste; they just remove the packaging and discard it.

This is a relatively recent development: even during the 1970s it was not the case that every small tool was encased in a blister pack glued to a paper backing (making it more difficult to recycle the two very different materials); paper bags, easily recyclable, dominated grocery shopping; and vegetables and fruits did not come pre-loaded in clam-shell containers or in small plastic pails. Japan is an unequaled paragon of this packaging excess, with items wrapped and placed in several layers of paper and plastic. The numbers involved are not trivial: in per capita terms the annual generation of plastic packaging waste alone is now close to 40 kg in Italy, Spain, and the Netherlands and the EU average is about 30 kg, with a total (including Switzerland) of 15.6 Mt in 2008 (EC, 2011). This shift to excessive packaging has been driven by the quest to reduce sales jobs and speed up check-outs, but its material, and environmental, consequences have been detrimental.

One of the best systematic reviews of possible material savings in production can be found in Allwood and Cullen's (2012) appraisal of waste reductions of metals, particularly steel and aluminum. The authors trace these opportunities in six different sequential categories: using less metal by design; reducing yield losses; diverting manufacturing scarp; reusing metal components; striving for longer-life products; and reducing final demand for metal services. Even material-sparing design, perhaps the most obvious pursuit that should be the aim of rational engineering, is a far from exhausted option. Allwood and Cullen (2012) estimate that light-weight design could reduce the mass of steel beams by 20–50%, that the savings could be as high as 30% for reinforcing bars and pipes and 40% for car bodies, and that in aggregate they could add up to as much as 100 Mt of steel, or nearly as much as the total pre-2008 US consumption. Additive manufacturing (three-dimensional printing) offers the least wasteful way to produce complex shapes with minimal waste (Gibson *et al.*, 2010; Gebbhardt, 2012).

Longer product life is another obvious material-sparing option, but some of the greatest rewards could come from reducing yield losses and diverting manufacturing scrap. Most people are surprised to learn that substantial shares of newly produced metals (more than 25% of all steel, almost 50% of all aluminum) never make it into final products: 90% of the initially produced steel may end up in a simple I-beam, but the share is just over 50% for an aluminum can and mere 10% for aluminum aircraft wing skin. Better design could raise these shares, and although these materials are recycled as internal scrap, their remelting consumes energy and generates atmospheric emissions.

Some of this internal waste is an inevitable by-product of manufacturing processes: waste comes from blanketing, cutting, and trimming of metal, turnings (swarf) are generated by drilling, milling, and grinding. Metals are also wasted due to surface and internal defects and the exigencies of handling and in some operations a great deal of scrap comes from excessive orders as new material in excellent condition is treated as waste and returned for recycling. Allwood and Cullen (2012) estimate that up to 30 Mt of steel blanketing skeletons and up to 20 Mt of aluminum swarf would not have to be recycled by melting but could be reused in even less energy-intensive ways. Similarly, loss of materials on construction sites is commonly around 10% of initially delivered mass and could be easily halved by appropriate management.

Opportunities for enhanced recycling remain great even in the case of paper and aluminum cans, the two materials whose recycling rates are the highest in all affluent countries (Japan's paper recycling may be the exception as it is already about as complete as is practical). Perhaps most notably, until 2008 paper was still the largest discarded material going into US landfills (almost 21% of the total mass, compared to nearly 17% for plastics), and although by 2010 it had fallen to just below plastic's share (16.2 vs

17.3%) the total mass of buried paper was still nearly 27 Mt/year (USEPA, 2011a): that is more than the annual production of all paper and paperboard in the same year in Germany (FAO, 2013).

And while the mass of paper landfilled in the USA in 2010 was half of the total in 1990 (26.7 vs 52.5 Mt), during the same two decades the mass of discarded plastics rose by 70% and the total of buried polymers, 28.5 Mt, was greater than the combined annual production in Germany and France (Plastics Europe, 2012). Or another comparison: a destitute waste collector may spend a day collecting a mass of 1 kg of plastic shopping bags when rummaging the open garbage tips of Asia's megacities, while the USA buries nearly 80 000 t of plastic in its landfills every day. While in the USA only about 8% of discarded plastics were recovered in 2010 (with the rate ranging from 23% for PET (polyethylene terephthalate) bottles to less than 1% for PP (polypropylene) waste), the EU's goal for 2020 is full diversion of plastic waste from landfills (EPRO, 2011). This would require a 50% increase of the 2010 recovery rate of 66%, roughly split between recycling and incineration for energy recovery. And, of course, waste recovery is not synonymous with recycling as significant shares of collected materials are not reused but landfilled (after volume reduction by shredding or compression).

Recycling should aim at maximum practicable rates and by far the most important universal step in that direction would not require exceptional arrangements or ruinous investment. Products should be designed with disassembly and recycling in mind, a task that has been made much easier by modern CAD (computer-assisted design) but one that is still rarely seen as important. Such rational, recycling-friendly design would be especially helpful in managing the rising mass of e-waste. An interesting innovation is the introduction of robot kiosks by ecoATM (more than 100 in the USA by the end of 2012) that assess the value of old mobile phones and offer instant cash for the devices that are then sold to bidders: the firm acts as a broker for companies that can refurbish and resell the phones or recycle them for components or precious metals (ecoATM, 2013). This model could be extended to all other e-devices.

On the other hand, I should note that, fortunately, some forecasts of future recycling requirements may turn out to be rather exaggerated. For example, Lu *et al.* (2010) forecast that by 2030 the annual generation of obsolete personal computers will reach as many as 700 million units in developing countries and up to 300 million in developed regions – but even in the short time since that paper was written (in 2009) it has become clear that smaller mobile devices will become increasingly more important as they fill many niches (ranging from e-mail to photo banks and journal reading) that previously required owning a PC. As a result, the composition of discarded e-waste will almost certainly look very different from the one portrayed by that rather mechanistic forecast.

As I showed in the second chapter, the history of material use is, to a large extent, the history of material substitutions. This process is now perhaps more important than ever and light-weighting is its most common demonstration. Light-weighting can be used with the greatest rewards in all forms of metal-based manufacturing – and it may not require abandoning traditional materials with higher specific weight. For example, weight reductions can be achieved by using advanced high-strength steels processed by cold forming or hot stamping, or by forming lighter yet stiffer sandwich materials, such as car roofs with less than 0.4 mm of polymers between two 0.2 and 0.3 mm steel sheets that weigh nearly 40% less (Hoffmann, 2012).

Substitutions of steel by aluminum have been progressing for decades, particularly in the car industry. In 1975, the average US vehicle contained 36 kg of aluminum, by 2012 the total was 155 kg, above all in engines, transmissions, drives, and wheels (Schultz, 2012). Still, considerable opportunities remain, and the IAI (2008) estimated that light-weighting in the transport sector has the potential to avoid 700 Mt of CO_2 emissions annually: every form of passenger and freight transportation (cars, trucks, trains, vessel, and aircraft) remains a rewarding candidate. Light-weighting using carbon-fiber-reinforced plastic (CFRP) is particularly appealing because it can reduce the mass of a car chassis by 50% compared to a steel structure and by 30% compared to Al alloys: BMW's i3, a plug-in hybrid, is the first moderately-priced car using CFRP (BMW, 2013).

But light-weighting opportunities are everywhere, although some of them face nontrivial perception/preference barriers. Putting wine into plastic pouches is a great example of such resistance – but in aggregate, and considering the volume of intercontinental wine trade with millions of bottles carried by container ships, this substitution would entail large energy and material savings. Wine bottles can be made lighter (by as much as 25%) and the quality of wine would not suffer if it were distributed in Tetra Paks, a paper container with a polymer lining that would (even when comparing 1 l pack with 0.75 l bottle) reduce mass by about 95% and whose life-cycle energy cost would be about 70% lower (Franklin Associates, 2006).

And just a reminder that light-weighting may not always lead to the best functional solution. Ranking of common beverage containers (from the lightest to the heaviest) moves from aluminum cans, plastic bottles, steel cans, to glass bottles; according to ascending energy intensity (MJ/kg of material) it is glass, steel, plastics, aluminum; and the energy/volume ranking ascends from steel (2.4 MJ/l), plastics (3.5–5.5 MJ/l), glass (about 8 MJ/l), to aluminum (9 MJ/l). Steel thus comes far ahead in the functional ranking: no other material costs less energy to contain a unit of fluid (ImpEE Project, 2008). Qualitative considerations are often most important as new materials can be endowed with properties that enable them to fill new niches: for example, DuPont's Hytrel TPC-ET thermoplastic elastomer has the flexibility of rubber but the strength of plastic (DuPont, 2013).

One of the most consequential substitutions has been electronics without lead (Li *et al.*, 2005). In Japan all new e-products have been lead free since 2005, in the EU since 2006: lead-free alloys (of tin, silver, and copper) and electrically conductive adhesives are used instead. An even more fundamental shift would be the end of silicon and a switch to semiconductor alloys designated as III–IVs because of their position on the periodic table of elements – gallium arsenide (GaAs), indium gallium arsenide (InGaAs), and indium antimonide (InSb) – whose charge-carrying capacity far surpasses that of Si, in the case of InSb by 2 orders of magnitude (Service, 2009).

6.3   New Materials and Dematerialization

Prospects for new materials, for better varieties of commonly used commodities and for more efficient (above all less energy-intensive) and less environmentally damaging means of their production and consumption are as encouraging as ever, and intensive R&D activities are addressing all of these goals. Detailed technical descriptions of some of the most important advances and promises for new materials for energy and environmental protection, including the promise of artificial photosynthesis, can be found in Ginley and Cahen (2012). A particular achievement would be the introduction of new materials for better (both in terms of density and discharge time) large-scale energy storage (Dunn *et al.*, 2011).

Here I will note just a few remarkable accomplishments and list a few desirable innovations that might find many important practical applications during the next 10–30 years: they range from the great promise of new nanomaterials to old materials endowed with new, remarkable properties, or deployed on unprecedented scales. Nanomaterials, engineered structures with at least one dimension of 100 nm or less, are already used as catalysts, in filters, semiconductors, and cosmetics and as drug carriers, and will find many future uses in several major industries and in a growing range of everyday products. Graphene is a two-dimensional (one-atom thick) carbon fabric that combines extreme mechanical strength with exceptionally high electronic and thermal conductivities (Novoselov *et al.*, 2012). Its tensile strength is 100 TPa, 5 orders of magnitude greater than that of steel, its stiffness (Young's modulus) is nearly equal to that of diamond, its electron mobility surpasses that of silicon by 2 orders of magnitude, and its thermal conductivity is an order of magnitude greater than that of silver (Savage, 2012).

Production of this material is very challenging, but its future applications will range from coatings and transparent conductive layers to high-frequency transistors and sensors. In 2013, the annual production capacity of carbon nanotubes (cylinders made from graphene) reached several thousand tons; they already go into such diverse products as rechargeable batteries, automotive parts, water filters, large-area coatings, supercapacitors, and electromagnetic shields (de Volder *et al.*, 2013). Their future uses should include new transistors operating as switches at molecular dimensions of a mere 9 nm (Franklin *et al.*, 2012).

There are also many opportunities to endow classical materials with new desirable properties. Self-cleaning glass surfaces have a thin coating of TiO_2 that helps to clean the surface by a combination of photocatalysis and hydrophilicity (Pilkingtons's Activ™ glass), or they are made from super-hydrophobic materials. TiO_2 can be also incorporated into exposed materials, giving them self-cleaning properties as the UV-activated dioxide molecules serve as a catalyst to produce free radicals: unprotected surfaces will be eventually invaded by bacteria, mosses, algae, and fungi but treated surfaces resist these degradations (Pilkington, 2013). Bendable concrete is another innovative twist on an ancient material.

Experiments have shown that fly-ash could replace as much as 65% of cement when combined with an adequate dose of superplasticizer. In addition to, or instead of, fly-ash it is also possible to substitute slag from metallurgical operations. Concrete's energy cost could also be greatly reduced by using a more durable variety of the material developed by Victor Li. This bendable material, more accurately known as ductile concrete or ECC (engineered cementitious composite), contains small polystyrene particles, and when a polymer fiber made from polyvinyl alcohol is added to the concrete tiny plastic fibers bridging microcrack planes provide load-bearing capacity (Li, 2012).

There is also a nascent field of self-healing materials (polymers and composites) that can repair cracks or tears and recover their original functionality (White *et al.*, 2011). Few future accomplishments would have such far-reaching consequences as making plastics not only completely biodegradable (oxo-degradables) but, even better, biocompostable, so they could be converted by microorganisms into CO_2 and H_2O in industrial composting plants in times similar to the decay of such common organic wastes (Shalaby and Burg, 2004). Of course, an even more radical step would be making biocompostable plastics, functionally identical to today's polymers, solely from plant feedstocks (for example, PE (polyethylene) and PP from ethanol derived from sugar cane). The next few decades should also see many mundane as well as unusual applications for new bio-inspired materials, structures designed by organisms but augmented by inserting synthetic materials (Meyers *et al.*, 2013) and the use of synthetic printed materials made to act like living cells (Reiffel *et al.*, 2013).

Lithium is perhaps the best example of an old material with an exceptionally promising future. This soft white metal has been used by the ceramic and glass industries, in lubricants, polymers, and pharmaceuticals, but its future use will be dominated by batteries, especially by rechargeable units in portable electronic gadgets (computers, tablets, cellphones) and tools and in transportation (Goonan, 2012). The metal's global production reached 28 100 t in 2010, with Chile and Australia producing about 70% of the total, and there is no shortage of exploitable resources (USGS, 2013). The R/P ratio in 2010 was in excess of 400 years, and Bolivia, at Salar de Uyuni, has more than half of the world's resources of the metal (Friedman-Rudovsky, 2011).

As always, new solutions will also bring new problems. There are concerns about the toxic potential of materials at the nanolevel because they can have undesirable interactions with living molecules or be outright toxic (Nel *et al.*, 2006). Extracting lithium in order to make hundreds of millions of new batteries for future electric cars, and then discarding the batteries, will introduce a new environmental concern (Tahil, 2006). Mass production of biopolymers will require considerable amounts of plant feedstock that will have to be grown in competition with food and feed crops and whose cultivation could have many unwelcome environmental impacts, as is well known from other intensive cropping practices.

Foresight, new regulations, and the self-interest of producers to offer more environmentally benign goods will determine the as yet uncertain extent of these new impacts. What is much more certain is that new materials will make an important contribution to the continuing trend of the relative dematerialization of products and services: Boeing 787s and BMW i3 are just the first notable applications of composite materials in transportation, and many other planes and vehicles, as well as trains and vessels, will follow. Light-weighting using new and traditional materials (whose many examples were described in the previous section) will combine with better design and smarter controls to offer better performance and superior functionality with reduced product mass, and although many descriptions of imminent infallible robotic diagnoses and personal assistance reflect wishful thinking more than they do tomorrow's quotidian realities, there is also no doubt that many services will entail mobilization of considerably smaller masses of materials.

But we can be also fairly confident in saying what these advances will not bring: their combined effect will not result in any absolute aggregate global dematerialization. I have not seen a single forecast that envisages any decline in global demand for any major material, and even low-case scenarios translate into considerable increases between 2010 and 2050. For example, low variants forecast by the IEA are 2.4 Gt for steel, 95 Mt for aluminum, 3.8 Gt for cement, and 800 Mt for paper (IEA, 2010), implying 2010–2050 gains of 70% for steel, 33% for aluminum, 15% cement, and 200% for paper. Similarly, even low- to moderate-growth scenarios foresee a 40–70% increase in the use of NPK fertilizers between 2010 and 2050 (Tilman *et al.*, 2011; Blanco, 2011), while a slower growth in demand for plastics (from about 3% in the recent past to 2%) would more than double the global output of polymers by 2050. Higher forecasts put the 2050 output for cement at 4.5 Gt, that of steel at 2.8 Gt, and of paper at 1.1 Gt (IEA, 2010).

National prospects are a different story. China's extraordinary post-1990 expansion of some key material flows cannot be sustained for two more generations, and expectations are that steel as well as cement production (hence also sand and stone extraction needed for concrete) will be substantially below the 2010 level by 2050, but plastics and paper and fertilizers may be up. China's aggregate material flow in 2050 is thus hard to predict, but there will be no absolute dematerialization in other modernizing economies. The fastest gain will be in countries that combine still fairly low wealth per capita with relatively high population growth rates: India, Pakistan, Nigeria, and Ethiopia are the best examples in this expansive category.

Affluent countries with declining and rapidly aging populations should be able to keep their aggregate material consumption from rising and they could see some substantial declines in specific categories: Japan and many nations within the EU are in this category and their demand for construction materials and fertilizers would be particularly affected. In contrast, in the USA, the necessity of accommodating continuing population growth (the USA is expected to have just over 400 million people by 2050 compared to 310 million in 2010) and eventually fixing its deteriorating infrastructure does not point to any decline in aggregate material demand before 2050.

The environment responds to absolute inputs and throughputs of pollutants resulting from these larger material flows: of course, without relative dematerialization the global and national atmospheric emissions, water pollution, land degradation, inappropriate waste disposal, and toxic burden would be even worse, but the combination of continuing relative dematerialization, moderated population growth, and improved environmental protection would have to be seen as a great success if it could just slow down the overall rate of degradation or if it could, in a few specific cases, stop any further growth and even bring some noticeable declines (akin to the already achieved turnaround in global emissions of SO_2 or elimination of lead from common consumer products).

During the past generation, dematerialization, defined in terms of declining consumption per GDP of energy or of goods, has generally persisted in both affluent and modernizing countries (Ausubel and Waggoner, 2008), and there are no insurmountable reasons why its pace should not be at least maintained, and even accelerated, in decades to come. But, as already stressed, this relative dematerialization will not result in any significant declines of aggregate global consumption, and just the opposite will be true as far many key fuels, raw materials, and manufactured products are concerned. And while further advances in electronics will be helpful they will not be able to turn the aggregate material tide.

I will illustrate these limits by a few interesting comparisons. In Chapter 5 (in the section 5.2 on relative dematerialization) I noted the list of devices whose ownership could be eliminated by owning an iPhone. Tupy (2012) drew this conclusion from his list:

> Dematerialization, in other words, should be welcome news for those who worry about the ostensible conflict between the growing world population on the one hand,

and availability of natural resources on the other hand. While opinions regarding scarcity of resources in the future differ, dematerialization will better enable our species to go on enjoying material comforts and be good stewards of our planet at the same time. That is particularly important with regard to the people in developing countries, who ought to have a chance to experience material plenty in an age of rising environmental concerns.

I cited the entire paragraph because it is a perfect encapsulation of the high hopes that people infatuated with e-gadgets have for the transformative powers of electronics in general and for its capacity as an agent of dematerialization in particular. How wrong that conclusion is can be best illustrated by looking at the material consequences of smartphones and other, concurrently diffusing, electronic devices. Nokia Communicator, released in 1996, was the first major smartphone and it took 16 years before the number of these devices surpassed 1 billion, which occurred during the third quarter of 2012 (mobiThinking, 2013). Even when assuming (as outlined in Chapter 5) that every one of these 1 billion devices has eliminated the need to buy a watch, an alarm clock, a small radio, a voice recorder, and a digital camera, the net aggregate savings attributable to the diffusion of mobile phones (about 400 g/phone: 500 g for the package of five devices compared to 110 g for a smartphone) would have reached 400 000 t (most of it aluminum, glass, and microprocessors) by the end of 2012.

But that is merely a theoretical construct, wishful thinking, as there have been no such eliminations, not even any substantial cuts in demand. During the first decade of the twenty-first century, worldwide sales of watches, and particularly of luxury models, continued to grow by 2–4% a year, and between 2000 and 2012 worldwide sales of digital cameras surpassed 1 billion units, and those of much more massive LCD TVs totaled just over 1.1 billion (Dash, 2012). Even when assuming that an average LCD TV weighs just 10 kg (for an 80-cm screen) these units incorporate more than 10 Mt of materials, mostly aluminum, glass, and microprocessors, marginalizing even the maximum theoretical material savings from purchases of smartphones.

And the chain of comparisons could be extended still further because soaring sales of watches, digital cameras, and LCD TVs have been accompanied by record sales of cars and luxury products many of which (private jetliners and yachts) represent some of the most concentrated embodiments of energy in select materials. Consequently, even if no alarm clocks, rolodexes, voice recorders, or digital cameras were ever bought because of the smartphone's multifunctionality, the material savings represented by such a loss of demand would be largely obliterated by the expanding claims for the same kind of resources ranging from aluminum and glass to wires and microprocessors.

There are other examples of astonishing claims of future dematerialization that can be contrasted with efforts to further expand material consumption. Qualcomm Tricorder X Prize will award $10 million for the design of a portable (maximum weight of 5 lbs) device "capable of capturing key health metrics and diagnosing a set of 15 diseases" (Qualcomm, 2013), a first step toward dematerializing routine doctor's office appointments. But at the same time, Matternet, the Internet of material products, is "creating the next paradigm for transportation using a network of unmanned aerial vehicles" (Matternet, 2013). Fleets of drones are envisaged to transport goods (starting with medicines and spare parts) in parts of the world that lack all-weather roads, a development that has an obvious huge potential for boosting material consumption. What would be (assuming they succeed commercially) the net material outcome of these two countervailing developments?

6.4 Chances of Fundamental Departures

Any serious consideration of future material needs must first acknowledge this: too many people still live in conditions of degrading and unacceptable material poverty that translates into poor nutrition, inadequate health care, frequent morbidity, fractured education, exhausting work, minimal leisure, absence of even small comforts, and shortened life expectancy. All of those people – and, depending on the specific category of that short list in the preceding sentence, they number between hundreds of millions and a few billion – need to consume more materials per capita in order to enjoy a decent life. But complications set in right at this point: what is a decent life? To what extent should we define a decent life by material consumption as opposed to measuring it by such universally valid indicators of physical quality of life as infant mortality, morbidity, and life expectancy? How do we include intangible indicators that measure quality of education or leisure? And what weight is to be given to subjective indicators of wellbeing, to satisfaction with life, or to the relative state of one's happiness?

And there is more. What is the actual link between material consumption and objective and subjective quality of life once the basic needs for food, clothes, shelter, and mobility are well satisfied? Going from material misery to modest material comfort will make many things in life better but, obviously, the link is not an endless escalator. But if so, where is the saturation point? Can such a level actually be quantified in a meaningful way? These questions must be asked even if there are no easy answers, mainly because of the situation that is the very opposite of the material poverty outlined at the beginning of this section: too many people live in the condition of material excess and this does not endow them with a higher physical quality of life than that enjoyed by moderate consumers and it does not make them exceptionally happy.

At the most fundamental level, the question is about the very nature of modern economies. All but a tiny minority of economists (those of ecological persuasion) see the constant expansion of output as the fundamental goal. And not just any expansion: economies should preferably grow at annual rates in excess of 2%, better yet 3%. This is the only model, the only paradigm, and the only precept, as the economists in command of modern societies cannot envisage a system that would deliberately grow at a minimum rate, even less so one that would experience zero growth, and the idea of a carefully managed decline appears to them to be outright unimaginable. The pursuit of endless growth is, obviously, an unsustainable strategy (Binswanger, 2009), and the post-2008 experience has shown how dysfunctional modern economies become as soon as the growth becomes negligible, ceases temporarily or when there is even a slight decline: rising unemployment, falling labor participation, growing income inequality, and soaring budget deficits.

But alternatives are imaginable and unorthodox economists and ecologists have long argued for the decoupling of energy use and material consumption from standard perceptions of progress, for a transition toward a low-growth, then a zero-growth society, and eventually even to a managed reduction of energy and material flows. The reason for this advocacy is obvious: the fundamental incompatibility of the growth imperative and of the second law of thermodynamics (Georgescu-Roegen, 1971, 1975). This perspective dictates that minimizing entropy should be the foremost goal for a rational society. Rephrased in plain terms using the titles of recent books on the subject, we should stop shoveling fuel for the runaway train of economic growth (Czech, 2000), confront consumption (Princen *et al.*, 2002), embrace the logic of sufficiency and break with the throwaway culture (Slade, 2006).

This calls for a new society where, once basic material needs are taken care of, the sense of wellbeing and satisfaction would be derived from experiences that are not at all, or only marginally, correlated with higher energy flows and expanding material possessions. Modern affluent societies would seem to be the perfect candidates for this shift, because studies in the USA, Europe, and Australia have found that materialism is negatively related to life satisfaction, that the pursuit of wealth and possessions is associated with diminished subjective wellbeing: Ryan and Dziurawiec (2001) list pre-2001 findings, Hudders and Pandaleare (2012) add more recent work.

This reality is easily confirmed by checking country rankings on now readily available global lists of happiness or satisfaction in life (Layard, 2005; White, 2006; Helliwell *et al.*, 2012). Germany, Europe's richest large economy, impresses with its material affluence and consumption of nonessentials: its private homes look substantial and well maintained, Germans drive their heavy Audis, BMWs, and Mercedes without speed limit on *Autobahnen* (a deliberately and extraordinarily wasteful activity), and they fly more frequently to distant locales than any other Europeans. But Germans rank 43rd on the global Happy Index (Helliwell and Wang, 2012), and they are actually less satisfied with their lives than people in Colombia, the Philippines, or even Rwanda, who enjoy a small fraction of German affluence.

Similarly, Asia's richest major economy has left behind any Buddhist simplicity in its post-World War II dash toward affluence: Japan's housing is a cramped affair, but those small houses and apartments are stuffed with an enormous variety of products, Japan's roads are clogged with vehicles, the Japanese enjoy the world's best functioning public transportation infrastructure and are indefatigable travelers to foreign lands – and yet, according to White's index, they are the least satisfied population in the affluent world (ranking 90th), behind dozens of poor countries including Papua New Guinea and Uzbekistan. In other comparisons they do better: they are 41st in average life satisfaction, but that still puts them behind such economic underperformers as Paraguay, Nicaragua, and Turkmenistan.

But where do these relatively unhappy nations rank on the consumption ladder? While continuing acquisition of material possessions does not increase happiness or satisfaction, does it eventually translate into consuming less? Perhaps only in relative terms as longer-term perspectives do not show any dramatic declines either in Germany or in France – Europe's second largest economy and (according to some rankings) an even less happy society. During the five pre-crisis years (2002–07) for which comparable EU statistics for direct material inputs are available (Eurostat, 2013), not-so-happy Germany (number 35, just behind Colombia) and unhappy France (number 62, behind Argentina and Mongolia) increased their overall demand by, respectively, 5 and 6%.

But comparisons between happy and unhappy countries hint at a fascinating difference, one that does not bode well for linking any restraints in material consumption with happiness. During 2002–07 Denmark, Finland, and Sweden – the three Nordic countries that stand at the top of virtually all satisfaction/happiness ladders and that enjoy equally high, or perhaps even higher, material quality of life than Germany or France or Japan – increased their direct material inputs by, respectively, 19, 16, and 20%.

None of this is surprising: as with all addictions, there is no easy way out of the habit of material acquisition, and the quest for possessions may have little to do with the ranking on the global satisfaction/happiness ladder. During his research into changing ownership and desires Easterlin (2003, pp. 1118-0–1118-1) found

> that as people acquire those goods for which aspirations were fairly high to start with (a home, a car, a TV), their aspirations increase for goods which were initially much less likely to be viewed as part of the good life ... This finding suggests that new aspirations arise as previous ones are satisfied, and, to judge from the mean number of goods desired, to about the same extent.

The economic downturn of 2008 and 2009 affected consumption in all affluent countries but we will have to wait many years before any new clear trends emerge – and before we could argue that any moderation (or decline) of material consumption is predicated on being chronically unhappy and dissatisfied in life, or that subjective happiness does not curtail continued material consumption. Still, while even clear signs of a slower growth of material demand would be one thing, a shift to substantially dematerialized economies quite another. Stahel (2011) writes about the service economy "where success is measured in wealth (stock) and its usage value" not as in today's arrangements, where success depends on throughput and its exchange value.

His objective is "wealth without resource consumption" – a goal that, as he admits, is of little interest to the prevailing "river" economy, and, as I would argue, of no real interest to all but a tiny fraction of citizens in any country, affluent or impoverished. Others make proposals for "economic de-growth" (Flipo and Schneider, 2008), for managing prosperous economies without growth (Victor, 2008; Jackson, 2009). And yet others argue that the time to begin these great transformations is now because the "denouement of exponentials," the end of exponential expansion of material use and of accelerating expansion of debt, is fast approaching (Morgan, 2010).

Some think not only that this conversion to a new paradigm is doable (Jackson, 2009) but that "the amount of wealth extracted from one unit of natural resources can quadruple. Thus we can live twice as well – yet use half as much" (von Weizsacker *et al.*, 1997, p. xv). The latter claim is a long-term possibility (using, of course GDP, that standard but in so many ways grossly misleading measure of wealth) but as a student of powerful inertial forces in history, I know we cannot come close to such a performance during the next 10–20 years. So what practical steps – short of miraculously shifting the global economic system from a river with rising energy and material flows to a cycle designed to manage stability – can be taken, first in order to limit consumption of materials, and then to begin to move to absolute dematerialization? And does the best path toward this goal lead through collective enforcement or individual enlightenment?

Of course, history offers many examples of societies whose rulers kept populations in material misery, and this was not even the worst they could expect as famines took an incomparably greater toll than absence of material comforts. And there is no need to go back into the distant past for illustrations: during the past three generations the most disturbing examples include Maoist China (and the Mao-made famine that killed more than 30 million people between 1959 and 1961) and the continuing madness of North Korea. But the evolution of modern societies has run in the opposite direction. Indeed, states now exist to a large extent in order to maintain and to promote economic, technical, and legal foundations and infrastructures of mass consumption. Even the former paragons of alternative options based on material restrictions or outright impoverishment – the Soviet Union and the Communist China – have embraced the consumption option with alacrity (the key difference being that in the Chinese case the Party has not lost its control).

Consequently, it is exceedingly unlikely that we will soon see a major economy whose leaders will promise its citizens less and less, or at least no more than there is. Of course, there is always the possibility that the enlightened self-interest of a large number of well-informed individuals will combine to subvert the old paradigm and that this will have a major long-term impact on material requirements. To this I would say: do not underestimate the appeal of possession, acquisition, ownership, and excess consumption. Without exaggeration, material acquisition in modern societies can be seen as a common form of the addictive behavior that is usually associated with alcohol, smoking, drugs, or gambling. A major difference is that this addictive behavior has become even more pervasive, as it has crossed all national and cultural barriers and evolved into a compulsive global phenomenon.

During the process it has become closely intertwined with perceptions of class ranking and social status, it has offered opportunities for assuming vicarious identities, it has promoted individual and inter-generational aspirations, it has become a tool of desire, passion, and gratification, and it has brought the rewards of personal indulgence. Because of this, it is no exaggeration to see a fundamentally mundane act of shopping as an expression of self-identity pursued to negate the rootlessness, role-lessness, and bureaucratic rationalities of a modern society that traps its subjects and delimits their freedoms (Latimer, 2001).

These realities have greatly blurred, if not erased, the boundaries between necessary and superfluous consumption, as yesterday's unattainables become tomorrow's indispensables. They have also converted human desires into a continuous process of collecting and discarding as mass production and consumption have succeeded in creating a new ethos of ephemerality (Cooper, 2001). Intentional obsolescence has become a key driver for many industries making consumer products, a development going back to GM's decision of the late 1920s to launch a new model every year (Slade, 2006). As Lefebvre (1971) noted decades ago, besides the obsolescence of products there is also the obsolescence of needs.

And, to make sure that the process does not stop, modern mass production is not only about pervasive quantity but also about mesmerizing abundance and variety, although that variety is mostly superficial, hiding essentially identical products in diversely packaged and promoted guises. Marsh (2012) estimated that the annual global output now comprises about 10 billion different products, and foresees that total (larger than the human population) to be greatly surpassed by the trend toward mass customization as the combination of CAD and 3-D printing will allow companies to produce items perfectly tailored to individual needs. If the prices of this option fall low enough, will this boost or lower the overall material demand as many people choose the cachet of unique items?

But there are many things we can do – while waiting for the combination of declining population growth, aging, and chronic unhappiness to put a sufficient downward pressure on material demand in affluent countries. There is no shortage of effective steps that could make a real difference, but the main problem with nearly all of them is the extent to which they will be broadly adopted and perpetuated by both producers and consumers in modern societies. Design for durability is perhaps the most obvious option. There is no reason why we could not design cars or refrigerators that last 20–25 years (their average life-span is now 10–12 years, while a well-maintained jetliner can operate safely for 30 years). Similarly, life-spans of electronic devices could be doubled.

The technical challenges of doing so are minimal, but the products would have to cost more and the new way of doing things would have to break with decades-long expectations of frequently changing models and designs and people. Is there a massive US or EU constituency for proudly driving 17-year old cars? And in the case of e-devices – where new designs are introduced in intervals measured in months rather than in years and where having the latest design is, to some people, worth queuing up for days before its release to be the first to own it – the idea of a durable cellphone or PC is an anathema. Not an impossible shift, but not one that is going to happen in a hurry.

An arrangement that could allow more frequent changes of models and that would assure complete recycling at the end of a product's life is not to sell any major manu-factured items but merely rent them on long-term service contracts and then return them to their makers for disassembly and reuse: this approach could be applied to products ranging from computers to car tires and from refrigerators to air conditioners. Arrange-ments aimed at higher intensity of use are another kindred path as leasing tries to address the pervasively high idling capacity of many products. Leasing, rather than owning, has come to dominate such expensive industrial operations as offshore drilling rigs, trans-portation of massive objects by special ships and cargo plans, and by 2012, more than a third of all large jetliners were also leased.

Leasing could be used to lower the pervasively high idling capacity of many con-sumer products ranging from cars and lawn mowers to many specialized tools. These concepts of new rentalism – peer-to-peer consumerism, fractional ownership, or social sharing – can be extended from machines and tools to housing, travel, and many ser-vices (McFedries, 2012). New arrangements for leasing and take-back of products to be recycled, and in some cases even up-cycled rather than down-cycled, include textiles, carpets, and mattresses (Braungart, 2013). At the same time, rental services for infre-quently used tools and machines have been around for decades, and so have car leases (guaranteed take-back after an agreed period) and time-shares in vacation housing – but these arrangements have captured only small shares of their respective markets. And for many people the very idea of sharing their precious cars or toolboxes is unthinkable.

Clearly, the dominant model of the endless-escalator economy has deep foundations in individual desire for material possessions, and given the global distribution of wealth this consumption escalator has a long way to run. As Easterlin (2003) confirmed, con-sumers in affluent countries keep coming up with new objects of desire. SUVs – machines embodying 2–3 t of metals, plastics, glass, and rubber – have been the most massive must-haves of the past quarter century. Among the recent *desiderata* are home gyms, large-screen TVs, and outdoor (patio or deck) furniture. The gyms are equipped with exercise machines weighing tens to hundreds of kilograms apiece (the shipping weight of Bowflex Ultimate Home Gym is 239 kg), a 150-cm (diagonal) screen HD TV has the mass of 26 kg, outdoor furniture has progressed from simple plastic or rattan chairs to large dining tables and full-size sofas, to say nothing of adjacent outdoor grills capable of preparing food for scores of people.

The greatest progressing wave of material consumption is now engulfing hundreds of millions of higher-income people in the cities of Asia and Latin America (and, to a much lesser extent, in Africa) whose often crassly ostentatious display of wealth surpasses many past European or North American excesses (Kamdar, 2011; People's Daily, 2013). The consumption pattern of China's *nouveaux riches* appears to be a particularly slavish copy of the worst American experience, as it consists of fake mansions within gated communities, obligatory large SUVs (Pierson, 2012), and obsessive amassment of luxury goods ranging from Swiss watches to custom-made yachts (KPMG China, 2013).

And right behind that wave, we can already see the crest of a much more massive flow forming, one that is beginning to elevate a few billion people in Asia and Latin America from bare subsistence in the countryside to incipient affluence in expanding cities: an overwhelming majority of those material claims will help to create higher quality of life, but the share of frivolous uses is bound to rise too. And, eventually, another wave will also sweep Africa's masses, and by the time its mass consumption begins to rise, the continent may have more than 2 billion people (the total was just over 1 billion in 2010). The numbers are clear: in 2013 perhaps about a billion people are (by any reasonable definition) fairly rich, another half a billion are approaching that level – but that still leaves more than 5 billion waiting to get on the material escalator.

The potential demand dwarfs everything that has been achieved so far, and comparisons of average rates of car ownership are one of its best indicators. In the rich countries, these rates have saturated at more than 500 vehicles per 1000 people (in 2010 the USA had nearly 800, Japan almost 600), but the rate was less than 300 in Brazil, less than 100 China, and about 20 in India. Even if the eventual ownership rates outside North America, Europe, Australia, and Japan saturated at just a third of the Japanese level (about 200/1000) the potential increase would amount to more than a billion new vehicles, more than doubling the total of about 1.1 billion vehicles that were on the roads in 2012. And similar calculations are easy to do for refrigerators, TVs, air conditioners, furniture, and so on.

And so mass consumption of materials will continue, made up of changing proportions of the inputs necessary to provide a decent quality of life and of materials turned into ephemeral junk; production of both will claim nonrenewable resources, will require considerable energy inputs (also for, now routinely, inter-continental transportation), and will burden the environment with pollutants. To moderate and eventually reduce this global wave of material consumption by appealing to reason – be it by stressing the untenability of existing economic arrangements, raising concerns about the environmental burdens of rising consumption, preaching inter-generational responsibility for preserving natural resources, or extolling monetary and security advantages restraint – remains a veritable Sisyphean task.

Deliberate individual choice of minimized material consumption – adoption of voluntary abnegation, limited wants, or utter simplicity – is much like voluntary adherence to nutritional minima. Very few people (actually less than 1%) are willing to stick with veganism, the strictest form of vegetarian diet that does not permit even consumption of dairy products or eggs, and even less restrictive forms of what I call hyphenated vegetarianism (lacto- or lacto-ovo- or even lacto-ovo-pisci-vegetarianism) are consistently practiced by only a few percent of Western populations (Smil, 2013). And so it will be with appeals to voluntary reductions of material demands.

As Kamdar (2011, p. 17) put it – writing about the pleasures of excess among India's newly rich – "voluntary simplicity can only appeal to those who have enough to choose to live with less." But that desire for simpler ways of life amidst material plenty is usually expressed in abandoning, largely in a symbolic way, just a few chosen claims and habits, and such actions will have only a marginal effect on the overall material demand. Perhaps we need powerful external drivers in order to significantly depress consumer demand and to change the course of the river economy: two of them are already with us, already causing concern but not yet perceived as sufficiently threatening to bring about relatively rapid and profound changes.

Global warming, after several decades of relatively steep temperature increases, paused its climb between 2000 and 2010 (Guemas *et al.*, 2013). Should this stagnation continue for another decade, it is highly unlikely that the major economies will take any drastic steps to lower the emissions of greenhouse gases (assorted Kyoto, Bali, and Copenhagen treaties and agreements have been ineffective and have done little to arrest their rise). But a return to a pronounced rise of global temperatures, and even more so their unprecedented decadal jump, would likely lead to measures whose effect would be to accelerate the relative dematerialization (and decarbonization) of the global economy and to reduce overall demand for materials.

The other decisive and unpredictable development would be an unprecedented economic crisis with consequences both much broader and much deeper than those following the down-turn that began in 2008 and that, for many countries, still shows no signs of ending. Perhaps the best way to imagine this is to see the global economy behaving for an extended period of time much like Japan's economy has been behaving since 1990: weak or nonexistent growth accompanied by deflation and endless budget deficits. And there are yet other crises and catastrophes that could end the rising consumption of materials, or at least set it back for generations (Smil, 2010).

Two of those ever-present risks – a global pandemic and Earth's collision with an asteroid – have been, finally, given closer attention in recent years because of the epidemics of SARS, H1N1, and because of a few relatively close approaches of several asteroids and the spectacular atmospheric disintegration of the Chelyabinsk meteor in February 2013. A new pandemic whose mortality would merely replicate the toll of the 1918–19 event would result in at least 150 million deaths worldwide, while the consequences of a collision with even a fairly small asteroid (200–400 m diameter) would depend on where it took place.

Atmospheric change (prolonged cooling engendered by lofting huge masses of dust into the atmosphere) would be much greater with an impact on land, while the economic consequences would obviously be more severe if an asteroid hit Western Europe or Eastern China than Eastern Africa or the Canadian Arctic. But modern societies, despite their vastly increased scientific understanding, think too little about unthinkable events and take too few steps toward improving their chances of coping with events that have very low probabilities but whose catastrophic impacts would surpass anything that humans have experienced during their recorded history.

In contrast, the last 50 years have seen the rise of concerns about the ultimate avail-ability of the Earth's resources and about the biosphere's ability to cope with the consequences of their mass extraction and use. The worries of the 1960s matured rapidly into the more probing inquiries of the 1970s, and these turned into systematic efforts during the 1980s, and since that time we have seen a constant flow of new studies and reports concerning global environmental change, resource depletion, destruction of biodi-versity, specific problems with water supply, toxic pollutants, soil erosion, deforestation, and acidification of the ocean. But only a very few of these concerns have resulted in determined and effective action.

A global ban on the use of chlorofluorocarbons, whose emissions were implicated in the destruction of ozone layer, has been perhaps the most important achievement; other notable improvements include the removal of lead from gasoline and absolute reduction of global SO_2 emissions, widespread adoption of at least primary water treatment in cities and, of course, all those substantial reductions of energy intensities of every major industrial product described in this book with the associated declines of both atmospheric and water pollutants per unit of output of every major material, be it steel or cement, ammonia or glass.

But, to stress the key point for the last time, these impressive achievements of relative dematerialization have not translated into any absolute declines of material use on the global scale – and, leaving the just outlined catastrophes aside, the global gap between the haves (approximately 1.5 billion people in 2013) and the have-nots (more than 5.5 billion in 2013) remains so large that even if the aspirations of the materially deprived four-fifths of the humanity were to reach only a third of the average living standard that now prevails in affluent countries, the world would be looking at the continuation of aggregate material growth for generations to come. Pretentious forecasts might spec-ify how many generations, but a responsible attitude is to be agnostic on this matter whose outcome will be determined by complex interactions of population dynamics, human desires and attitudes, economic growth and product costs, technical innovations, international relations, and global environmental change.

These will determine how long humanity can continue the obviously unsustainable practice of river economy. As yet, nothing has been irretrievably foreclosed, and at this point it is not difficult to imagine rational futures of moderated energy and material use aimed at maximizing global quality of life for a stationary, even slowly declining, population – nor that of a further indiscriminate quest for energy and materials that results, to a large extent, in wasted ephemeral consumption, perpetuates the great global gap in the average standard of living, and weakens the fundamental biospheric functions, the only irreplaceable foundations of any civilization.

We must hope that human ingenuity (so admirably deployed particularly during the past two centuries) and adaptability (displayed, unfortunately, not well ahead of antic-ipated crises but only when they are upon us) will, sooner rather than later, guide us along the first path – but even in that case the transformation of humanity's material uses will be a gradual and difficult process with an uncertain outcome.

Appendix A: Units and Unit Multiples

Units Used in the Text

a	ar (unit of area, $10 \times 10\,\text{m}$)
°C	degree celsius (unit of temperature)
g	gram (unit of mass)
m	meter (unit of length)
Pa	Pascal (unit of pressure)
t	ton (1000 kg)

Unit Multiples

Prefix	Abbreviation	Scientific notation
deka	da	10^1
hecto	h	10^2
kilo	k	10^3
mega	M	10^6
giga	G	10^9
tera	T	10^{12}
peta	P	10^{15}
exa	E	10^{18}
zeta	Z	10^{21}
yota	Y	10^{24}

Making the Modern World: Materials and Dematerialization, First Edition. Vaclav Smil.
© 2014 John Wiley & Sons, Ltd. Published 2014 by John Wiley & Sons, Ltd.

Submultiples

Prefix	Abbreviation	Scientific notation
deci	d	10^{-1}
centi	c	10^{-2}
milli	m	10^{-3}
micro	μ	10^{-6}
nano	n	10^{-9}

Appendix B: US Material Production, GDP and Population, 1900–2005

	Total	Construction materials	Wood	Metals	GDP	Population
1900	144	55	60.3	10.3	422.8	76.0
1910	271	125	78.8	23.1	533.8	92.0
1920	277	115	67.2	30.6	687.7	105.7
1930	425	255	49.5	31.2	892.8	122.8
1940	519	319	48.7	45.6	1166.9	131.7
1950	851	501	56.2	77.9	2006.0	150.7
1960	1540	1120	49.1	77.7	2830.9	179.3
1970	2110	1530	60.2	98.4	4269.9	203.2
1980	2140	1500	58.8	97.8	5009.0	226.6
1990	2540	1840	77.9	100.8	8033.9	248.7
2000	3400	2560	82.3	143.8	11226.0	281.4
2005	3720	2850	91.4	137.8	12623.0	296.5

All material flows are in megatons (Matos, 2009); annual GDPs are in billions of constant $(2005) (http://www.measuringworth.com/usgdp/); population numbers (except for 2005) are census totals in millions (USCB, 2012).

Making the Modern World: Materials and Dematerialization, First Edition. Vaclav Smil.
© 2014 John Wiley & Sons, Ltd. Published 2014 by John Wiley & Sons, Ltd.

Appendix C: Global Population, Economic Product, and Production of Food, Major Materials, and Fuels 1900–2010

	1900	1950	2000	2010
Population (billion)	1.6	2.5	6.1	6.9
Economic product ($1990 trillion)	1.9	5.3	36.7	72.1
Food and feed crops (Mt)	410	790	2850	3430
Biomaterials				
Roundwood (Gm^3)	1.3	1.7	3.4	3.4
Paper (Mt)	17.0	43.7	324.0	394.0
Cotton (Mt)	4.0	6.6	18.5	23.7
Metals				
Steel (Mt)	28.3	189.0	850.0	1440.0
Aluminum (Mt)	–	1.5	24.3	41.2
Copper (Mt)	0.5	2.4	13.2	16.1
Cement (Mt)	13	133	1660	3260
Plastics (Mt)	–	1.5	200.0	265.0
Ammonia (Mt N)	–	4.8	108.0	131.0
Fuels				
Coal (Mt)	825	1815	4693	7273
Crude oil (Mt)	21	537	3611	3913
Natural gas (Gm^3)	7	197	2179	2881

Population: NEAA (2013); economic product: Maddison (2007) and IMF (2013); food and feed crops: Smil (2013); biomaterials: FAO (2013); metals, cement, and ammonia: Kelly and Matos (2013); plastics: Europe Plastics (2011); fuels: Smil (2010), BP (2013), and NEAA (2013).

Making the Modern World: Materials and Dematerialization, First Edition. Vaclav Smil.
© 2014 John Wiley & Sons, Ltd. Published 2014 by John Wiley & Sons, Ltd.

Appendix D: Global Energy Cost of Major Materials in 2010

Material	Output[a] (Mt)	Energy intensity[b] (GJ/t)	Total energy cost (EJ)
Steel	1430	25	35.8
Aluminum	40.8	175	7.1
Chromium	23.7	50	1.2
Copper	15.9	90	1.4
Manganese	13.9	50	0.7
Zinc	12	50	0.6
Cement	3310	4	13.2
Glass	56	7	0.4
Paper	399	25	10.0
Plastics	265	80	21.2
Ammonia	59	40	6.3
Si wafers	9000 t	20 000	0.18

[a] USGS (2012), NSG (2011), Takiguchi and Morita (2011), and FAO (2013).
[b] IEA (2007), Worrell *et al.* (2007), and Smil (2008).

Making the Modern World: Materials and Dematerialization, First Edition. Vaclav Smil.
© 2014 John Wiley & Sons, Ltd. Published 2014 by John Wiley & Sons, Ltd.

Appendix E

Decarbonization and Desulfurization of Global Fossil Fuel Supply, 1900–2010

	Fuels(EJ)	Carbon (Gt C)	Carbon intensity (kg C/GJ)	Sulfur (Mt S)	Sulfur intensity (g S/GJ)
1900	21.5	0.53	24.8	10.0	465
1950	72.5	1.63	22.5	29.1	401
2000	303.3	6.75	22.3	53.4	176
2010	437.1	9.14	20.9	51.5	118

Decarbonization and Desulfurization of the World's Total Primary Energy Supply (TPES), 1900–2010

	TPES (EJ)	Carbon (Gt C)	Carbon intensity (kg C/GJ)	Sulfur (Mt S)	Sulfur intensity (g S/GJ)
1900	43.6	1.2	27.5	10.0	229
1950	100.7	2.5	24.8	29.1	289
2000	382.4	8.1	21.2	53.4	140
2010	544.0	10.5	19.3	51.5	95

Data sources: TPES: Smil (2010) and BP (2013); carbon: CDIAC (2013); sulfur: Smith *et al.* (2010) and Klimont *et al.* (2013).

Making the Modern World: Materials and Dematerialization, First Edition. Vaclav Smil.
© 2014 John Wiley & Sons, Ltd. Published 2014 by John Wiley & Sons, Ltd.

References

AA (The Aluminum Association) (2003) The Industry, AA, Washington, DC.

AA (2011) Auto Aluminum Usage Hits All-Time High, Expected to Aggressively Accelerate, New Survey of Automakers Says, http://www.aluminum.org/AM/Template.cfm?Section=Home&TEMPLATE=/CM/ContentDisplay.cfm&CONTENTID=31938 (accessed 22 May 2013).

AA (2012) U.S. Aluminum Beverage Can Recycling, http://www.aluminum.org/Content/NavigationMenu/NewsStatistics/StatisticsReports/UsedBeverageCanRecyclingRate/UBCRecyclingRate.pdf (accessed 22 May 2013).

AA (2013) The Infinitely Recyclable Aluminum Can, http://www.aluminum.org/Content/NavigationMenu/TheIndustry/PackagingConsumerProductMarket/Can/default.htm (accessed 22 May 2013).

Abbott, R.T. and Dance, S.P. (2000) *Compendium of Seashells: A Color Guide to More Than 4,200 of the World's Marine Shells*, El Cason, CA: Odyssey Publications.

Abe, S.S., Watanabe, Y., Onishi, T. *et al.* (2011) Nutrient storage in termite (Macrotermes bellicosus) mounds and the implications for nutrient dynamics in a tropical savanna Ultisol. *Soil Science and Plant Nutrition*, **57**: 786–795.

Abeles, P.W. (1949) *The Principles and Practice of Prestressed Concrete*, Charles Lockwood, London.

Abramson, A. (2003) *The History of Television, 1942 to 2000*, McFarland & Company, Jefferson, NC.

Adam, J.-P. (1994) *Roman Building: Materials and Techniques*, Routledge, London.

Adriaanse, A., Bringezu, S., Moriguchi, Y. *et al.* (1997) *Resource Flows: The Material Basis of Industrial Economies*, World Resources Institute, Washington, DC.

Adshead, S.A.M. (1997) *Material Culture in Europe and China, 1400–1800*, Macmillan, London.

Agrawal, G. (2010) *Fiber-Optic Communication Systems*, John Wiley & Sons, Inc., New York.

Air Products (2013) http://www.airproducts.com/ (accessed 22 May 2013).

Aker Solution (2013) Statfjord B, http://www.akersolutions.com/en/Global-menu/Projects/technology-segment/Engineering/Offshore-oil-and-gas-production-facilities1/Statfjord-B/ (accessed 22 May 2013).

Aktas, C.B. and Bilec, M.M. (2012) Impact of lifetime on US residential building results. *International Journal of Life Cycle Assessment*, **17**: 37–349.

Alcoa (2012) Comparative Life Cycle Assessment of Aluminum and Steel Truck Wheels, http://www.alcoawheels.com/alcoawheels/north_america/en/pdf/Alcoa_Comparative_LCA_of_Truck_Wheels_with_CR_statement.pdf (accessed 19 June 2013).

Allwood, J.M. and Cullen, J.M. (2012) *Sustainable Materials with Both Eyes Open*, UIT, Cambridge.

Almond, J.K. (1981) A century of basic steel: Cleveland's place in successful removal of phosphorus from liquid iron in 1879, and development of basic converting in ensuing 100 years. *Ironmaking and Steelmaking*, **8**: 1–10.

Almqvist, E. (2003) *History of Industrial Gases*, Springer, Berlin.

Amato, I. (2013) Concrete solutions. *Nature*, **494**: 300–301.

Ambrose, S.H. (1998) Late Pleistocene human population bottlenecks: volcanic winter, and the differentiation of modern humans. *Journal of Human Evolution*, **34**: 623–651.

Anderson, R. and Anderson, R.C. (1926) *A Short History of the Sailing Ship*, Robert M. McBride, New York.

Andrae, A.S.G. and Andersen, O. (2010) Life cycle assessments of consumer electronics – are they consistent? *International Journal of Life Cycle Assessment*, **15**: 827–836.

APA (2013) Concrete Forming: Design/Construction Guide, http://www.g-lumber.com/pdf/APA-Concrete-Formwork.pdf (accessed 19 June 2013).

Apple (2012) Apple Introduces iPhone 5, http://www.apple.com/pr/library/2012/09/12Apple-Introduces-iPhone-5.html (accessed 22 May 2013).

ASCE (American Society of Civil Engineers) (2013) 2013 Report Card for America's Infrastructure, http://www.infrastructurereportcard.org/a/#p/home (accessed 22 May 2013).

Atsushi, U. (1995) The riddle of Japan's quakeproof pagodas. *Japan Echo*, Spring: 70–77.

Atterbury, P. (ed.) (1982) *The History of Porcelain*, Morrow, New York.

Austin, P. (2010) Reducing Energy Consumption in Paper Making Using Advanced Process Control and Optimisation, http://www.lcmp.eng.cam.ac.uk/wp-content/uploads/Reducing-Energy-Consumption-in-Paper-Making-using-APC-and-Optimisation.pdf (accessed 22 May 2013).

Ausubel, J. (2003) *Decarbonization: The Next 100 Years*. Lecture at the 50th Anniversary Symposium of the Geology Foundation, April 25 2003, Jackson School of Geosciences, University of Texas, Austin, TX, http://phe.rockefeller.edu/AustinDecarbonization/AustinDecarbonization.pdf (accessed 22 May 2013).

Ausubel, J.H. and Waggoner, P.E. (2008) Dematerialization: variety, caution, and persistence. *Proceedings of the National Academy of Sciences of the United States of America*, **105**: 12774–12779.

AWWS (American Water Works Service) (2004) Deteriorating Buried Infrastructure Management Challenges and Strategies, USEPA, Washington, DC.

Ayres, R. (2008) The life-cycle of chlorine, Part I: chlorine production and the chlorine-mercury connection. *Journal of Industrial Ecology*, **1**: 81–94.

Ayres, R.U., Kneese, A.V. and Allen, V. (1969) Production, consumption and externalities. *American Economic Review*, **59**: 282–297.

Baekeland, L.H. (1909) Condensation product and method of making same. Specification of Letters Patent 942,809, Dec. 7, 1909, USPTO, Washington, DC.

Bajpai, P. (2010) *Environmentally Friendly Production of Pulp and Paper*, John Wiley & Sons, Inc., Hoboken, NJ.

Ball Packaging Europe (2012) Light as a Feather, Tough as Ever, http://www.ball-europe.com/businesscards/gb/687.html (accessed 22 May 2013).

Bandaranayake, S. (1974) *Sinhalese Monastic Architecture: The Viháras of Anurádhapura*, E.J. Brill, Leiden.

Barnes, D.K.A., Galgani, F., Thompson, R.C. and Barlaz, M. (2009) Accumulation and fragmentation of plastic debris in global environments. *Philosophical Transactions of the Royal Society B*, **364**: 1985–1998.

Barney, G.O. (ed.) (1980) *The Global 2000 Report to the President*, Council on Environmental Quality, Washington, DC.

Barrett, W. (1990) World bullion flows, 1450–1800, in *The Rise of Merchant Empires: Long Distance Trade in the Early Modern World, 1350–1750* (ed J.D. Tracy), Cambridge University Press, Cambridge, pp. 224–254.

Barreveld, W.H. (1989) Rural Use of Lignocellulosic Residues. FAO, Rome.

Barros, E., Pashani, B., Constantino, R. and Lavelle, P. (2002) Effects of land-use system on the soil macrofauna in western Brazilian Amazonia. *Biology and Fertility of Soils*, **35**: 338–347.

Beardsley, T.M. (2011) Peak phosphorus. *BioScience*, **61**: 91.

Beaumont, W.W. (1902 and 1906) *Motor Vehicles and Motors: Their Design Construction and Working by Steam Oil and Electricity*, Vol. **2**: Archibald Constable & Company, Westminster.

Beck, T.R. (2001) *Electrolytic Production of Aluminum*, Electrochemical Technology Corporation, Seattle, WA.

de Beer, J., Worrell, E. and Blok, K. (1998) Future technologies for energy-efficient iron and steel making. *Annual Review of Energy and the Environment*, **23**: 123–205.

Behrens, A., Giljum, S., Kovanda, J. and Niza, S. (2007) The material basis of the global economy. *Ecological Economics*, **64**: 444–453.

Bell, I.L. (1884) *Principles of the Manufacture of Iron and Steel*, George Routledge & Sons, London.

Bell, P. (2012) China: Ready for a New Era? International Cement Review (Mar 2012), http://www.cemnet.com/Articles/story/149175/china-ready-for-a-new-era-.html (accessed 22 May 2013).

Berard, A. (ed.) (1991) *History, Science, Theology, and the Shroud*, Man in the Shroud Committee of Amarillo, Amarillo, TX.

Berge, B. (2009) *The Ecology of Building Materials*, Elsevier, Amsterdam.

Bernardini, O. and Galli, R. (1993) Dematerialization: long-term trends in the intensity of use of materials and energy. *Futures*, **25**: 431–448.

Berry, R.S., Long, T.V. II, and Makino, H. (1975) An international comparison of polymers and their alternatives. *Energy Policy*, **3**: 144–155.

Berry, B., Ritt, A., and Greissel, M. (1999) A Retrospective of Twentieth-Century Steel. Iron Age New Steel (Dec 1999).

Bessemer, H. (1905) *Autobiography*, Offices of Engineering, London.

Billington, D.P. (1989) *Robert Maillart's Bridges: The Art of Engineering*, Princeton University Press, Princeton, NJ.

Binczewski, G.J. (1995) The point of a monument: a history of the aluminum cap of the Washington Monument. *JOM*, **47**(11): 20–25.

Binswanger, M. (2009) Is there a growth imperative in capitalist economies? A circular flow perspective. *Journal of Post Keynesian Economics*, **31**: 707–727.

BIR (Bureau of International Recycling) (2011) Annual Report 2011, BIR, Brussels, http://www.bir.org/assets/Documents/publications/brochures/2011-ARfinalE.pdf (accessed 22 May 2013).

BIR (2012) World Steel Recycling in Figures 2007–2011, BIR, Brussels, http://www.bir.org/assets/Documents/publications/brochures/WorldSteelinFiguresIIIFINLoRes.pdf (accessed 22 May 2013).

Birch, A. (1967) *The Economic History of the British Iron and Steel Industry 1784–1879*, Cass, London.

Birdsey, R.A. (1996) Carbon storage in United States forests, in *Forests and Global Change*, Vol. **II** (eds N. Sampson and D. Hair), American Forests, Washington, DC, pp. 1–25.

Blanc, A., Holmes, L.L. and Harbottle, G. (1998) *Lutetian Limestones in the Paris Region: Petrographic and Compositional Examination*, Brookhaven National Laboratory, Brookhaven, NY.

Blanco, M. (2011) Supply of and Access to Key Nutrients NPK for Fertilizers for Feeding the World in 2050. Agronomos ETSIAUPM, Madrid, http://eusoils.jrc.ec.europa.eu/projects/NPK/Documents/Madrid_NPK_supply_report_FINAL_Blanco.pdf (accessed 22 May 2013).

Boeckel, B. and Baumann, K.-H. (2008) Vertical and lateral variations in coccolithophore community structure across the subtropical frontal zone in the South Atlantic Ocean. *Marine Micropaleontology*, **67**: 255–273.

Boeing (2012) 747 Family, http://www.boeing.com/boeing/commercial/747family/747-8_facts.page (accessed 22 May 2013).

Boesch, C. and Tomasello, M. (1998) Chimpanzee and human cultures. *Current Anthropology*, **39**: 591–614.

Bolin, C.A. and Smith, S. (2011) Life cycle assessment of ACQ-treated lumber with comparison to wood plastic composite decking. *Journal of Cleaner Production*, **19**: 620–629.

Borchers, W. (1904) *Electric Smelting and Refining*, Charles Griffin and Company, London.

Börjesson, P. and Gustavsson, L. (2000) Greenhouse gas balances in building construction: wood versus concrete from lifecycle and forest land-use perspectives. *Energy Policy*, **28**: 575–588.

Bornehag, C.G. and Nanberg, E. (2010) Phthalate exposure and asthma in children. *International Journal of Andrology*, **33**: 333–345.

Boskey, A.L. (2003) Biomineralization: an overview. *Connective Tissue Research*, **44** (Suppl. 1): 5–9.

Boulding, K. (1970) *Economics as Science*, McGraw-Hill, New York.

BP (British Petroleum) (2013) Statistical Review of World Energy, http://www.bp.com/assets/bp_internet/globalbp/globalbp_uk_english/reports_and_publications/statistical _energy_review_2011/STAGING/local_assets/pdf/statistical_review_of_world_ energy_full_report_2012.pdf (accessed 22 May 2013).

Braungart, M. (2013) Upcycle to eliminate waste. *Nature*, **494**: 174–175.

Brice, G. (1752) *Description de La Ville de Paris, Et de Tout Ce Qu'elle Contient de Plus Remarquable*, Librairies Associés, Paris.

Bringezu, S. and Schütz, H. (2001) *Total Material Requirement of the European Union*, European Environment Agency, Copenhagen.

Broecker, W.S. (1970) A boundary condition on the evolution of atmospheric oxygen. *Journal of Geophysical Research*, **75**: 3553–3557.

Broudy, E. (2000) *The Book of Looms: A History of the Handloom from Ancient Times to the Present*, University Press of New England, Lebanon, NH.

Brown, K.S., Marean, C.W., Herries, A.I.R. *et al.* (2009) Fire as an engineering tool of early modern humans. *Science*, **325**: 859–862.

Browne, M.A., Crump, P., Niven, S.J. *et al.* (2011) Accumulation of microplastic on shorelines worldwide: sources and sinks. *Environmental Science and Technology*, **45**: 9175–9179.

Brunetta, L. and Craig, C.L. (2010) *Spider Silk: Evolution and 400 Million Years of Spinning, Waiting, Snagging, and Mating*, Yale University Press, New Haven, CT.

Brydson, J.A. (1975) *Plastic Materials*, Newnes-Butterworths, London.

BTS (Bureau of Transportation Statistics) (2013) Passenger, http://www.rita.dot.gov/bts /data_and_statistics/by_subject/passenger.html (accessed 22 May 2013).

Buckingham, D.A. (2006) *Steel Stocks in Use in Automobiles in the United States*, USGS, Denver, CO.

Buehler, E. and Teal G.K. (1956) Process for producing semiconductive crystals of uniform resistivity. US Patent 2,768,914, Oct. 30, 1956, USPTO, Washington, DC.

Burke, J. (2012) *Nest: The Art of Birds*, Allen & Unwin, Crows Nest, NSW.

Campbell, C.J. (1997) *The Coming Oil Crisis*, Multi-Science Publishing and Petroconsultants, Brentwood.

Canadian Wood Council (2004) *Wood-Frame Housing – A North American Marvel*. Ottawa: Canadian Wood Council, http://www.cwc.ca/documents/durability/BP5_ WoodFramesAndEarthquakes.pdf (accessed 22 May 2013).

Cathcart, R.B. (2011) Anthropic rock: a brief history. *History of Geo- and Space Sciences*, **2**: 57–74.

CARE (Carpet America Recovery Effort) (2012) CARE Annual Report 2011, http://www.carpetrecovery.org/pdf/annual_report/11_CARE-annual-rpt.pdf (accessed 22 May 2013).

Carmona, M. and Camiller, P. (2002) *Haussmann: His Life and Times and the Making of Modern Paris*, Ivan R. Dee, Lanham, MD.

Carothers, W.H. (1937) Linear condensation polymers. US Patent 2,071,250, Feb. 16, 1937, USPTO, Washington, DC.

Cartwright, J., Cheng, J, Hagan, J *et al.* (2011) Assessing the Environmental Impacts of Industrial Laundering: Life Cycle Assessment of Polyester/Cotton Shirts, Donald Bren School of Environmental Science and Management, Santa Barbara, CA, http://www .bren.ucsb.edu/research/documents/missionlinen_report.pdf (accessed 22 May 2013).

CCC (Canadian Chamber of Commerce) (2012) Canada's Rare Earth Deposits Can Offer a Substantial Competitive Advantage, http://www.chamber.ca/images/uploads/Reports/2012/201204RareEarthElements.pdf (accessed 22 May 2013).

CDIAC (Carbon Dioxide Information Center) (2013) Fossil-Fuel CO_2 Emissions, http://cdiac.ornl.gov/trends/emis/meth_reg.html (accessed 22 May 2013).

Cement Association of Canada (2006) A Life Cycle Perspective on Concrete and Asphalt Roadways: Embodied Primary Energy and Global Warming Potential, http://www.cement.ca/images/stories/athena%20report%20Feb.%202%202007.pdf (accessed 22 May 2013).

Chang, A., Pashikanti, K. and Liu, Y.A. (2012) *Refinery Engineering*, Wiley-VCH Verlag GmbH, Weinheim.

Chapman, P.F., Leach, G. and Slesser, M. (1974) The energy cost of fuels. *Energy Policy*, **2**: 231–243.

Charette, R.N. (2009) This Car Runs on Code. IEEE Spectrum, http://spectrum.ieee.org/green-tech/advanced-cars/this-car-runs-on-code/0 (accessed 22 May 2013).

Cherry, C.R. and Gottesfeld, P. (2009) Plans to distribute the next billion computers by 2015 create lead pollution risk. *Journal of Cleaner Production*, **17**: 1620–1628.

China Daily (2011) China to Check Overcapacity of Flat Glass Production. China Daily (Oct 5, 2011), http://www.chinadaily.com.cn/business/2011-05/10/content_12480549.htm (accessed 19 June 2013).

Clark, D. (2011) Peak Stuff: The Data. The Guardian (Nov 1, 2011), http://www.theguardian.com/environment/blog/2011/nov/01/peak-stuff-consumption-data.

CMI (Can Manufacturers Institute) (2012) Aluminum Beverage Cans, http://www.cancentral.com/pdf/AluminumBeverageCanFacts.pdf (accessed 22 May 2013).

CMI (2013) Steel Food Cans, http://www.cancentral.com/pdf/SteelCanFoodFacts.pdf (accessed 22 May 2013).

Cobb, H.M. (2010) *The History of Stainless Steel*, ASM International, Materials Park, OH.

Cole, M., Lindeque, P., Halsband, C. and Galloway, T. (2011) Microplastics as contaminants in the marine environment: A review. *Marine Pollution Bulletin*, **62**: 2588–2597.

Composites World (2006) Advanced Materials for Aircraft Interiors, http://www.compositesworld.com/articles/advanced-materials-for-aircraft-interiors (accessed 22 May 2013).

Cooper, E. (2000) *Ten Thousand Years of Pottery*, British Museum, London.

Cooper, R. (2001) Interpreting mass: collection/dispersion, in *The Consumption of Mass* (eds N. Lee and R. Munro), Blackwell Publishers, Oxford, pp. 16–43.

Courland, R. (2011) *Concrete Planet*, Prometheus Books, New York.

Cuenot, F. (2009) CO_2 emissions from new cars and vehicle weight in Europe; how the EU regulation could have been avoided and how to reach it? *Energy Policy*, **37**: 3832–3842.

Cwalina, B. (2008) Biodeterioration of concrete. *Architecture, Civil Engineering, Environment*, **4**: 133–140.

Czech, B. (2000) *Shoveling Fuel for a Runaway Train: Errant Economists, Shameful Spenders, and a Plan to Stop Them All*, University of California Press, Berkeley, CA.

Czochralski, J. (1918) Ein neues Verfahren zur Messung des Kristallisationsgeschwindigkeit der Metalle. *Zeitschrift der Physikalischen Chemie*, **92**: 219–221.

Daigo, I., Igarashi, Y., Matsuno, Y. and Adachi, Y. (2009) Material stocks and flows accounting of copper and copper-based alloys in Japan. *Resources Conservation and Recycling*, **53**: 208–217.

Daigo, I., Matsuno, Y. and Adachi, Y. (2010) Substance flow analysis of chromium and nickel in material flow of stainless steel in Japan. *Resources Conservation and Recycling*, **54**: 851–863.

Daimler (2008) New Generation of 4-Cylinder Inline Engines, OM 651, http://vip .5053.com.cn/CarResource/B-%E5%A5%94%E9%A9%B0/%E5%85%A8%E9%83% A8%E8%BD%A6%E5%9E%8B/mediadb/mb_Acrobat/einf/204/einf_br204_cdi _OM651__en.pdf (accessed 19 June 2013).

Daniel, G. (1980) Megalithic monuments. *Scientific American*, **263**(1), 78–90.

Darwin, C.R. (1881) *The Formation of Vegetable Mould, Through the Action of Worms, With Observations on Their Habits*, John Murray, London.

Das, S. (2011) Life cycle assessment of carbon fiber-reinforced polymer composites. *International Journal of Life Cycle Assessment*, **16**: 268–282.

Dash, S. (2012) Large Size LCD Market Tracker, http://www.isuppli.com/Abstract /P28680_20130226081617.pdf (accessed 22 May 2013).

Davidovits, J. (2002) *Ils ont bâti les Pyramides*, Jean-Cyrille Godefroy, Paris.

Davidson, I. and McGrew, W.C. (2005) Stone tools and the uniqueness of human culture. *Journal of the Royal Anthropological Institute*, **11**: 793–817.

Davis, S.C., Diegle, S.W. and Boundy, R.G. (2011) *Transportation Energy Data Book: Edition 30*, Oak Ridge National Laboratory, Oak Ridge, TN.

Deffeyes, K.S. (2001) *Hubbert's Peak: The Impending World Oil Shortage*, Princeton University Press, Princeton, NJ.

Des Cars, J. (1988) *Haussmann, La Gloire du Second Empire*, Perrin, Paris.

De Carvalho, C.S.M., Sales-Campos, C. and De Andrade, M.C.N. (2010) Mushrooms of the Pleurotus genus: a review of cultivation techniques. *Interciencia*, **35**: 177–182.

Destination Oise (2013) The "Stone-Heritage Centre" in Saint-Maximin, http://www .oisetourism.co.uk/Must-sees/Savoir-faire/The-stone-heritage-centre-in-Saint-Maximin/The-stone-heritage-centre-in-Saint-Maximin (accessed 22 May 2013).

Di Blasi, C., Branca, C. and A. Galgano. 2010. Biomass screening for the production of furfural via thermal decomposition. *Industrial and Engineering Chemistry Research* **49**: 2658-2671.

Diderot, D. and D'Alembert, J.L.R. (1751–1777) *L'Encyclopedie ou Dictionnaire Raisonne des Sciences des Arts et des Metiers*, Ave Approbation and Privilege du Roy, Paris.

Diesel, R. (1913) *Die Entstehung des Dieselmotors*, Verlag von Julius Springer, Berlin.

Doggett, M. (2010) Sustainability of the Global Copper Industry: Learning from the Past Three Decades, http://www.amebc.ca/documents/roundup/Tech%20Session/Presentations/1505%20Doggett%20Mon.pdf (accessed 22 May 2013).

dotMobi (2013) Global Mobile Statistics 2013 Part A: Mobile Subscribers; Handset Market Share; Mobile Operators, http://mobithinking.com/mobile-marketing-tools/ latest-mobile-stats/a#subscribers (accessed 23 May 2013).

Duffar, T. (ed.) (2010) *Crystal Growth Processes Based on Capillarity: Czochralski, Floating Zone, Shaping and Crucible Techniques*, Wiley-Blackwell, Oxford.

Duffy, C. (1985) *The Fortress in the Age of Vauban and Frederick the Great, 1660–1789*, Routledge & Kegan Paul, London.

Dunn, B., Kamath, H. and J-M. Tarascon. 2011. Electrical energy storage for the grid: a battery of choices. *Science* **334**: 928–935.

DuPont (2013) DuPont™ Hytrel® TPC-ET Thermoplastic Polyester Elastomer, http://www2.dupont.com/Plastics/en_US/Products/Hytrel/Hytrel.html (accessed 23 May 2013).

Durrer, R. (1948) Sauerstoff-frischen in gerlafingen. *Von Roll Werkzeitung*, **19**: 73–74.

EAA (European Aluminium Association) (2012) Two Out of Three Aluminum Beverage Cans Recycled in Europe, http://www.alueurope.eu/wp-content/uploads/2011/08/Press-Release-Alu-bevcans-recycling-2010final_16July2012.pdf (accessed 23 May 2013).

EAA (2013) Aluminium Facts and Figures, http://www.alueurope.eu/about-aluminium/facts-and-figures/ (accessed 23 May 2013).

Easterlin, R.A. (2003) Explaining happiness. *Proceedings of the National Academy of Sciences of the United States of America*, **100**: 1176–1183.

EC (European Commission) (2009) LCA Tools, Services, Data and Studies, http://lca.jrc.ec.europa.eu/lcainfohub/download.vm (accessed 23 May 2013).

ecoATM (2013) Instant Cash for Phones and Other Electronics, http://www.ecoatm.com/ (accessed 23 May 2013).

Ecoinvent (2013) Swiss Center for Life Cycle inventories, http://www.ecoinvent.ch/ (accessed 23 May 2013).

Edinburgh Centre for Carbon Management (2006) Forestry Commission Scotland Greenhouse Gas Emissions Comparison Carbon benefits of Timber in Construction, http://www.forestry.gov.uk/pdf/Carbonbenefitsoftimberinconstruction.pdf/$file/Carbonbenefitsoftimberinconstruction.pdf (accessed 23 May 2013).

EEA (European Environmental Agency) (2013) Heating Degree Days, http://www.eea.europa.eu/data-and-maps/indicators/heating-degree-days-1 (accessed 23 May 2013).

Ehrlich, P.R. and Holdren, J. (eds) (1988) *The Cassandra Conference: Resources and the Human Predicament*, Texas A&M University Press, College Station, TX.

Elvidge, C.D., Tuttle, B.T., Sutton, P.C. *et al.* (2007) Global distribution and density of constructed impervious surfaces. *Sensors*, **7**: 1962–1979.

EMCORE (2013) EMCORE Solar Panels Power the Orbital-Built LDCM Satellite, http://investor.emcore.com/releasedetail.cfm?ReleaseID=741286 (accessed 23 May 2013).

Endler, J.A., Endler, L.C. and Doerr, N.R. 2010. Great bowerbirds create theaters with forced perspective when seen by their audience. *Current Biology* **20**: 1679–1684.

Energetics (2000) Energy and Environmental Profile of the U.S. Chemical Industry, Energetics, Columbia, MD, http://www1.eere.energy.gov/manufacturing/resources/chemicals/pdfs/profile_chap1.pdf (accessed 23 May 2013).

Engel, E. (1857) Die productions- und consumtionsverhaltnisse des königreichs sachsen. *Zeitschrift des Statistischen Bureaus des Königlich-Sächsischen, Ministerium des Innern*, **8/9**: 1–54.

EPN (Environmental Paper Network) (2007) Understanding Recycled Fiber, http://www
.greenpressinitiative.org/documents/recycledfiberfactsheet-EPN.pdf (accessed 23 May
2013).

EPRO (European Association of Plastics Recycling and Recovery Organisations) (2011)
Plastic Packaging Statistics 2010, http://www.plasticseurope.org/documents/document
/20111107101127-final_pe_factsfigures_uk2011_lr_041111.pdf (accessed 23 May
2013).

Eriksson, P.-E. (2004) Comparative LCAs for Wood and Other Construction Methods,
http://www.ewpa.com/Archive/2004/jun/Paper_032.pdf (accessed 23 May 2013).

European Commission (2001) Economy-Wide Material Flow Accounts and Derived Indicators a Methodological Guide, http://epp.eurostat.ec.europa.eu/cache/ITY_OFFPUB
/KS-34-00-536/EN/KS-34-00-536-EN.PDF (accessed 23 May 2013).

Eurostat (2009) Economy Wide Material Flow Accounts: Compilation Guidelines for
Reporting to the 2009 Eurostat Questionnaire Version 01 – June 2009, http://epp
.eurostat.ec.europa.eu/portal/page/portal/environmental_accounts/documents/Eurostat
%20MFA%20compilation%20guide%20for%202009%20reporting.pdf (accessed 23
May 2013).

Eurostat (2012) In the EU27, Almost Half of Renewable Energy Comes from Wood
and Wood Waste, http://epp.eurostat.ec.europa.eu/cache/ITY_PUBLIC/5-29112012-
AP/EN/5-29112012-AP-EN.PDF (accessed 23 May 2013).

Eurostat (2013) Material Flow Accounts, http://epp.eurostat.ec.europa.eu/statistics_expl-
ained/index.php/Material_flow_accounts#Per_capita_levels (accessed 23 May 2013).

FAO (Food and Agriculture Organization). (1990) Energy Conservation in the Mechanical Forest Industries, FAO, Rome, http://www.fao.org/docrep/T0269E/T0269E00.htm
(accessed 23 May 2013).

FAO (2011) World Statistical Compendium for Raw Hides and Skins, Leather and
Leather Footwear 1992–2011, FAO, Rome.

FAO (2013) FAOSTAT, http://faostat.fao.org/site/291/default.aspx (accessed 23 May
2013).

Felton, N. (2008) Consumption Spreads Faster Today. *New York Times* (Feb 10, 2008),
http://www.nytimes.com/imagepages/2008/02/10/opinion/10op.graphic.ready.html
(accessed 19 June 2013).

Fernandes, S.D., Trautmann, N.M., Streets, D.G. *et al.* (2007) Global biofuel
use, 1850–2000. *Global Biogeochemical Cycles*, **21**: GB2019. doi: 10.1029
/2006GB002836

Fernández, E., Boyd, P., Holligan, P.M. and Harbour, D.S. (1993) Production of organic
and inorganic carbon within a large-scale coccolithophore bloom in the northeast
Atlantic Ocean. *Marine Ecology-Progress Series*, **97**: 271–285.

Fernández-González, F. (2006) *Ship Structures under Sail and under Gunfire*. Madrid:
Universidad Politécnica de Madrid, http://oa.upm.es/1520/1/PONEN_FRANCISCO_
FERNANDEZ_GONZALEZ_01.pdf (accessed 23 May 2013).

Fischer-Kowalski, M., Krausmann, F., Giljum, S. *et al.* (2011) Methodology and indicators of economy-wide material flow accounting. *Journal of Industrial Ecology*, **15**:
855–876.

Fitchen, J. (1961) *The Construction of Gothic Cathedrals: A Study of Medieval Vault
Erection*, The University of Chicago Press, Chicago.

Fleming, P.A. and Loveridge, J.P. (2003) Miombo woodland termite mounds: resource islands for small vertebrates? *Journal of Zoology, London*, **259**: 161–168.

Flipo, F. and Schneider, F. (eds) (2008) Proceedings of the 1st International Conference on Economic De-Growth for Ecological Sustainability and Social Equity, http://events.it-sudparis.eu/degrowthconference/appel/Degrowth%20Conference%20-%20Proceedings.pdf (accessed 23 May 2013).

FLSmidth (2011) Rotary Kilns for Cement Plants, http://www.flsmidth.com/~/media/Brochures/Brochures%20for%20kilns%20and%20firing/RotaryKilnsforcementplants.ashx (accessed 23 May 2013).

Flynn, D.O. and Giráldez, A. (1995) Born with a "Silver Spoon": the origin of world trade in 1571. *Journal of World History*, **6**: 202–221.

Föll, H. (2000) *Electronic Materials*, University of Kiel, Kiel.

Forest Products Laboratory (2010) *Wood Handbook*. Madison, WI: USDA, http://www.fpl.fs.fed.us/documnts/fplgtr/fpl_gtr190.pdf (accessed 23 May 2013).

Forth Bridges (2013) Forth Rail Bridge Facts and Figures, http://archive.is/C09r (accessed 3 July 2013).

Frank, A.G. (1998) *ReOrient: Global Economy in the Asian Age*, University of California Press, Berkeley, CA.

Franklin, A.D., Luisier, M., Han, S.-J. *et al.* (2012) Sub-10 nm carbon nanotube transistor. *Nano Letters*, **12**: 758–762.

Franklin Associates (2006) Life Cycle Inventory of Container Systems for Wine, http://www.tetrapak.com/se/Documents/WineContainers_report%5B1%5D.pdf (accessed 23 May 2013).

Franz, J.E. (1974) N-phosphonomethyl-glycine Phytotoxicant Compositions. US Patent 3,799,758, Mar. 26, 1974, USPTO, Washington, DC.

Friedman-Rudovsky, J. (2011) Dreams of lithium empire. *Science*, **334**: 896–897.

Frison, G.C. (1987) Prehistoric, plains-mountain, large-mammal, communal hunting strategies, in *The Evolution of Human Hunting* (eds M.H. Nitecki and D.V. Nitecki), Plenum Press, New York, pp. 177–223.

Fruehan, R. (ed.) (1998) *The Making, Shaping and Treating of Steel Steelmaking and Refining Volume*, The AISE Steel Foundation, Pittsburgh, PA.

Gao, F., Nie, Z.R. and Wang, Z.H. (2009) Greenhouse gas emissions and reduction potential of primary aluminum production in China. *Sciences in Chin Series E: Technological Sciences*, **52**: 2161–2166.

Georgescu-Roegen, N. (1971) *The Entropy Law and the Economic Process*, Harvard University Press, Cambridge, MA.

Georgescu-Roegen, N. (1975) Energy and economic myths. *Ecologist*, **5**: 164–174, 242–252.

Geschichte-Club VÖEST (1991) *Geschichte der VÖEST: Rückblick auf die Wechselhaften Jahre des Grössten Österreichischen Industrieunternehmens*, Geschichte-Club VÖEST, Linz.

GHK (2006) A Study to Examine the Benefits of the End of Life Vehicles Directive and the Costs and Benefits of a Revision of the 2015 Targets for Recycling, Re-Use and Recovery Under the ELV Directive, http://ec.europa.eu/environment/waste/pdf/study/final_report.pdf (accessed 23 May 2013).

Gibson, I., Rosen, D.W. and Stucker, B. (2010) *Additive Manufacturing Technologies: Rapid Prototyping to Direct Digital Manufacturing*, Springer, New York.

Giljum, S., C. Lutz, A. Jungnitz *et al.* 2008. *Global Dimensions of European Natural Resource Use*. Vienna: Sustainable Europe Research Institute, http://seri.at/wp-content/uploads/2009/08/SERI-Working-Paper-7.pdf (accessed 23 May 2013).

Ginley, D.S. and Cahen, D. (eds) (2012) *Fundamentals of Materials for Enrgy and Environmental Sustainability*, Cambridge University Press, Cambridge.

Ginsburg, M. (ed.) (1991) *The Illustrated History of Textiles*, Studio Editions, London.

Global Environmental Facility (2012) Energy Efficiency Improvements in the Indian Brick Industry, http://www.resourceefficientbricks.org/background.php (accessed 23 May 2013).

Glulam (2013) Why Glulam? http://www.lilleheden.dk/uk/glulam.asp (accessed 23 May 2013).

Goeller, H.E. and Weinberg, A.M. (1976) The age of substitutability. *Science*, **191**: 683–689.

Gold, B., Peirce, W.S., Rosegger, G. and Perlman, M. (1984) *Technological Progress and Industrial Leadership: The Growth of the U.S. Steel Industry, 1900–1970*, Lexington Books, Lexington, MA.

Gong, X., Nie, Z., Wang, Z. *et al.* (2012) Life cycle energy consumption and carbon dioxide emission of residential building designs in Beijing. *Journal of Industrial Ecology*, **16**: 576–587.

Goodman, S.H. (ed.) (1998) *Handbook of Thermoset Plastics*, Noyes Publications, Westwood, NJ.

Goonan, T.G. (2012) Lithium Use in Batteries, http://pubs.usgs.gov/circ/1371/pdf/circ1371_508.pdf (accessed 23 May 2013).

Gordon, R.B., Bertram, M. and Graedel, T.E. (2006) Metals stocks and sustainability. *Proceedings of the National Academy of Sciences of the United States of America*, **103**: 1209–1214.

Gould, J.L., Gould, C.G. and Grant, C. (2007) *Animal Architects: Building and the Evolution of Intelligence*, Basic Books, New York.

Graedel, T.E. (2011) On the future availability of the energy metals. *Annual Review of Material Research*, **41**: 323–35.

Graedel, T.E., Allwood, J.M., Birat, J.-P. *et al.* (2011) Recycling Rates of Metals, http://www.unep.org/resourcepanel/Portals/24102/PDFs/Metals_Recycling_Rates_110412-1.pdf (accessed 23 May 2013).

Grantham, J. (2012) Be persuasive. Be brave. Be arrested (if necessary). *Nature*, **491**: 303.

Green Delta (2013) OpenLCA Case Study of a Beer Bottle: Aluminium Can vs PET Bottle, Green Delta, Berlin, http://www.openlca.org/c/document_library/get_file?uuid=2f1207d9-2481-416b-8ba1-a59ef821e0d7&groupId=15473 (accessed 23 May 2013).

Greenpeace (2013) Go PVC-Free, http://www.greenpeace.org/usa/en/campaigns/toxics/go-pvc-free/ (accessed 23 May 2013).

Grimes, S., Donaldson, J., and Gomez, G.C. (2008) Report on Environmental Benefits of Recycling, BIR, Brussels, http://www.bir.org/assets/Documents/publications/brochures/BIR_CO2_report.pdf (accessed 23 May 2013).

Grotte, J. and Marrey, B. (2000) *Freyssinet: la Précontrainte et l'Europe, 1930–1945*, Éditions du Linteau, Paris.

GSM Arena (2013) Mobile Phone Evolution: Story of Shapes and Sizes, http://www .gsmarena.com/mobile_phone-evolution-review-493.php (accessed 23 May 2013).

Guemas, V., Doblas-Reyes, F.J., Andreu-Burillo, I. and Asif, M. (2013) Retrospective prediction of the global warming slowdown in the past decade. *Nature Climate Change*. doi: 10.1038/nclimate1863

Gunston, B. (1986) *World Encyclopedia of Aero Engines*, Patrick Stephens, Wellingborough.

Gypsum Association (2013) Gypsum and Sustainability, http://www.gypsumsustainability .org/recycled.html (accessed 23 May 2013).

Haberl, H., Weisz, H., Amann, C. *et al.* (2006) The energetic metabolism of the European Union and the United States. *Journal of Industrial Ecology*, **10**: 151–171.

Habu, J. (2004) *Ancient Jomon of Japan*, Cambridge University Press, Cambridge.

Hahn, P.O. (2001) The 300 mm silicon wafer – a cost an technology challenge. *Microelectronic Engineering*, **56**: 3–13.

Hall, E.C. (1996) *Journey to the Moon: The History of the Apollo Guidance Computer*, American Institute of Aeronautics and Astronautics, Washington, DC.

Hansell, M.H. (2007) *Built by Animals: The Natural History of Animal Architecture*, Oxford University Press, Oxford.

Harpending, H.C., Batzer, M.A., Gurven, M. *et al.* (1998) Genetic traces of ancient demography. *Proceedings of the National Academy of Sciences of the United States of America*, **95**: 1961–1967.

Harper, D. (1992) *Eutrophication of Freshwaters: Principles, Problems and Restoration*, Springer-Verlag, New York.

Hasanbeigi, A., Price, L.; Aden, N., *et al.* 2012. *A Comparison of Iron and Steel Production Energy Intensity in China and the U.S*. Berkeley, CA: Lawrence Berkeley National Laboratory http://china.lbl.gov/sites/china.lbl.gov/files/US_China_Steel_ Comparison_Report_EN_19July2011.pdf (accessed 23 May 2013).

Hasanbeigi, A., Price, L. and Lin, E. (2012) Emerging energy-efficiency and CO_2 emission-reduction technologies for cement and concrete production: a technical review. *Renewable & Sustainable Energy*, **16**: 6220–6238.

Hatayama, H., Daigo, I., Matsuno, Y. and Adachi, Y. (2009) Assessment of the recycling of aluminum in Japan, the United States, Europe and China. *Materials Transactions*, **50**: 650–656.

Hatayama, H., Daigo, I., Matsuno, Y. and Adachi, Y. (2010) Outlook of the world steel cycle based on the stock and flow dynamics. *Environmental Science & Technology*, **44**: 6457–6463.

Heal, D.W. (1975) Modern perspectives on the history of fuel economy in the iron and steel industry. *Ironmaking and Steelmaking*, **4**: 222–227.

Hebbert, F.J. (1990) *Soldier of France: Sébastien le Prestre de Vauban, 1633–1707*, P. Lang, New York.

Heinberg, R. (2010) *Peak Everything: Waking Up to the Century of Declines*, New Society Publishers, Gabriola Island, BC.

Helliwell, R., Layard, R., and Sachs, J. (2012) World Happiness Report, http://www
.earth.columbia.edu/sitefiles/file/Sachs%20Writing/2012/World%20Happiness%20
Report.pdf (accessed 23 May 2013).

Helliwell, R. and Wang, S. (2012) The state of world happiness, in *World Happiness
Report* (eds R. Helliwell, R. Layard and J. Sachs), University of British Columbia,
Vancouver, BC, pp. 10–57.

Herman, A. (2012) *Freedom's Forge: How American Business Produced Victory in World
War II*, Random House, New York.

Herman, R., Ardekani, S.A. and Ausubel, J.H. (1990) Dematerialization. *Technological
Forecasting and Social Change*, **37**: 333–347.

Hermes, M.E. (1996) *Enough for One Lifetime: Wallace Carothers, Inventor of Nylon*,
American Chemical Society and the Chemical Heritage Foundation, Washington, DC.

Herold, A. (2003) Comparison of CO_2 Emission Factors for Fuels Used in Greenhouse
Gas Inventories and Consequences for Monitoring and Reporting Under the EC Emissions Trading Scheme, http://acm.eionet.europa.eu/docs/ETCACC_TechnPaper_2003_
10_CO2_EF_fuels.pdf (accessed 23 May 2013).

Hills, R.L. (1993) *Power from Steam: A History of the Stationary Steam Engine*,
Cambridge University Press, Cambridge.

Hodgson, E. (1990) *Safer Insecticides Development and Use*, CRC Press, Boca Raton,
FL.

Hoffmann, O. (2012) Steel Lightweight Materials and Design for Environmental Friendly
Mobility, ThyssenKrupp Steel Europe, http://www.industrialtechnologies2012.eu/sites
/default/files/presentations_session/03_Oliver_Hoffmann.pdf (accessed 23 May 2013).

Hogan, W.T. (1971) *Economic History of the Iron and Steel Industry in the United States*,
Lexington Books, Lexington, MA.

Hölldobler, B. and Wilson, E.O. (1990) *The Ants*, Harvard University Press, Cambridge, MA.

Holliday, R., Harper, T. and J. Heber. 2012. *Simply No Substitute?* Birmingham:
Cientifica, http://www.cientifica.com/wp-content/uploads/downloads/2012/09/Simply-
No-Substitute.pdf (accessed 23 May 2013).

Holt, J.A. and Easy, J.F. (1993) Numbers and biomass of mound-building termites
(Isoptera) in a semiarid tropical woodland near Charters Towers, North Queensland,
Australia. *Sociobiology*, **21**: 281–286.

Hong, S., Candelone, J.P., Patterson, C.C. and Boutron, C.F. (1994) Greenland ice
evidence of hemisphere lead pollution two millennia ago by Greek and Roman civilizations. *Science*, **265**: 1841–1843.

Horn, R.E., Grant, T. and Verghese, K. (2009) *Life Cycle Assessment: Principles, Practice
and Prospects*, CSIRO Publishing, Canberra.

Hosford, W.F. and Duncan, J.L. (1994) The aluminum beverage can. *Scientific American*,
271: 48–53.

Hudders, L. and Pandelaere, M. (2012) The silver lining of materialism: the impact
of luxury consumption on subjective well-being. *Journal of Happiness Studies*, **13**:
411–437.

IAI (International Aluminium Institute) (2007) Life Cycle Assessment of Aluminium:
Inventory Data for the Primary Aluminium Industry.

IAI (2008) Transport and Aluminium, http://www.world-aluminium.org/media/filer_public/2013/01/15/none_21 (accessed 23 May 2013).

IAI (2012) 75% of All Aluminium Ever Produced is Currently Still in Productive Use, http://recycling.world-aluminium.org/index.php?id=126 (accessed 23 May 2013).

Ibeh, C. (2011) *Thermoplastic Materials: Properties, Manufacturing Methods, and Applications*, CRC Press, Boca Raton, FL.

IEA (2007) Tracking Industrial Energy Efficiency and CO_2 Emissions. IEA, Paris, http://www.iea.org/publications/freepublications/publication/tracking_emissions.pdf (accessed 23 May 2013).

IEA (2010) Energy Technology Perspectives, http://www.iea.org/publications/freepublications/publication/etp2010.pdf (accessed 23 May 2013).

IEA (2013) Statistics and Balances, IEA,Paris, http://www.iea.org/stats/index.asp (accessed 19 June 2013).

IFIA (International Fertilizer Industry Association) (2013) Phosphorus and "Peak Phosphate", http://www.fertilizer.org/ifa/HomePage/SUSTAINABILITY/Phosphorus-peak-phosphate (accessed 23 May 2013).

IFIAS (International Federation of Institutes for Advanced Study) (1974) Energy Analysis Workshop on Methodology and Conventions, Stockholm, IFIAS.

ILZSG (International Lead and Zinc Study Group) (2013) http://www.ilzsg.org/static/home.aspx (accessed 23 May 2013).

IMF (2013) Data and Statistics, http://www.imf.org/external/data.htm (accessed 23 May 2013).

ImpEE Project. 2013. *Plastic Waste in the UK*. University of Cambridge: The Cambridge-MIT Institute, http://www-g.eng.cam.ac.uk/impee/topics/RecyclePlastics/files/Recycling%20Plastic%20v3%20PDF.pdf (accessed 23 May 2013).

Inoue, T., Takematsu, Y., Hyodo, F. *et al.* (2001) The abundance and biomass of subterranean termites (Isoptera) in a dry evergreen forest of northeast Thailand. *Sociobiology*, **37**: 41–52.

Intel (2011) From Sand to Silicon, http://download.intel.com/newsroom/kits/chipmaking/pdfs/Sand-to-Silicon_32nm-Version.pdf (accessed 23 May 2013).

Intel (2012) The Intel Xeon Phi, http://www.intel.com/content/dam/www/public/us/en/documents/solution-briefs/high-performance-xeon-phi-coprocessor-brief-2.pdf (accessed 23 May 2013).

Intel (2013) Intel Timeline: A History of Innovation, http://www.intel.com/content/www/us/en/history/historic-timeline.html (accessed 23 May 2013).

IPCC (Intergovernmental Panel on Climate Change) (2007) *Fourth Assessment Report of the Intergovernmental Panel on Climate Change*, IPCC, Geneva.

IRA (International Rubber Association) (2013) Natural Rubber Statistics, http://www.intrubberassoc.org/v2/ (accessed 23 May 2013).

IRGC (International Risk Governance Council) (2013) The Rebound Effect: Implications of Consumer Behaviour for Robust Energy Policies, IRGC, Lausanne, http://www.irgc.org/wp-content/uploads/2013/04/IRGC_ReboundEffect-FINAL.pdf (accessed 23 May 2013).

ISO (International Organization for Standardization) ISO 14040:2006. (2006a) *Environmental Management – Life Cycle Assessment – Principles and Framework*, ISO,

Geneva, http://www.iso.org/iso/catalogue_detail?csnumber=37456 (accessed 23 May 2013).

ISO ISO 14044:2006. (2006b) *ISO 14044:2006 Environmental Management – Life Cycle Assessment – Requirements and Guidelines*, ISO, Geneva, http://www.iso.org/iso/catalogue_detail?csnumber=38498 (accessed 23 May 2013).

Jackson, T. (2009) *Prosperity without Growth: Economics for Finite Planet*, Earthscan, London.

Jacobs and IPST (Institute of Paper Science and Technology) (2006) *Pulp and Paper Industry: Energy Bandwidth Study*, Georgia Institute of Technology, Atlanta, GA.

James, H. (2001) Use and production of solid sawn timber in the United States. *Forest Products Journal*, **51**: 23–28.

Japan Paper Association (2013) Paper and Paperboard Production, http://www.jpa.gr.jp/en/industry/data02/ (accessed 23 May 2013).

Jenkins, D. (ed.) (2003) *The Cambridge History of Western Textiles*, Cambridge University Press, Cambridge.

Jevons, W.S. (1865) *The Coal Question: An Inquiry Concerning the Progress of the Nation, and the Probable Exhaustion of our Coal Mines*, Macmillan, London.

Jolliet, O., Saade, M., Shaked, S. and Jolliet, A. (2013) *Environmental Life Cycle Assessment*, Waterstones, London.

Jönsson, Å., Tillman, A.M. and Svensson, T. (1997) Life cycle assessment of flooring materials: case study. *Building and Environment*, **32**: 245–255.

Kahhat, R. and Williams, E. (2012) Materials flow analysis of e-waste: domestic flows and exports of used computers from the United States. *Resources Conservation and Recycling*, **67**: 67–74.

Kalliala, E.M. and Nousiainen, P. (1999) Environmental profile of cotton and polyester-cotton fabrics. *AUTEX Research Journal*, **1**: 8–20.

Kamdar, M. (2011) The pleasures of excess. *World Policy Journal*, **28**: 15–19.

Karg, S. (2011) New research on the cultural history of the useful plant Linum usitatissimum L. (flax), a resource for food and textiles for 8,000 years. *Vegetation History and Archaeobotany*, **20**: 507–508.

Kato, K., Murata, A. and Sakuta, K. (1997) Energy payback time and life-cycle CO_2 emission of residential PV power system with silicon PV module. *Progress in Photovoltaics:Research and Applications*, **6**: 105–115.

Kauppi, P.E., Rautiainen, K.T., Korhonen, A. *et al.* (2010) Changing stock of biomass carbon in a boreal forest over 93 years. *Forest Ecology and Management*, **259**: 1239–1244.

Kawaoka, K., Tsuda, A., Matsuoka, Y. *et al.* (2006) Latest Blast Furnace Relining Technology at Nippon Steel. Nippon Steel Technical Report 94, pp. 127–132.

Kawashima, C. (1986) *Minka: Traditional Houses of Rural Japan*, Kodansha, Tokyo.

Kaye, G.D. (2002) Using GIS to estimate the total volume of Mauna Loa Volcano, Hawai'i, 98th Annual Meeting, May 13-15, 2002, https://gsa.confex.com/gsa/2002CD/finalprogram/abstract_34712.htm (accessed 23 May 2013).

Kelly T.D. and G. Matos. 2011. *Historical Statistics for Mineral and Material Commodities in the United States*, Data Series, Vol. **140**: Washington, DC: U.S. Geological Survey, http://minerals.usgs.gov/ds/2005/140/ (accessed 23 May 2013).

Kelly, T.D. and Matos, G.R. (2013) Historical Statistics for Mineral and Material Commodities in the United States, http://minerals.usgs.gov/ds/2005/140/ (accessed 23 May 2013).

Kenny, C. (2012) Paving Paradise. *Foreign Policy* (Jan/Feb 2012), pp. 31–32.

Kenny, J.F., Barber, N.L., Hutson, S.S. *et al.* (2009) *Estimated Use of Water in the United States in 2005*, USGS, Washington, DChttp://www.epa.gov/otaq/fetrends.htm#report (accessed 23 May 2013).

Kermeli, K., Worrell, E. and E. Masanet. 2011. *Energy Efficiency Improvement and Cost Saving Opportunities for the Concrete Industry*. Berkeley, CA: Ernest Orlando Lawrence Berkeley National Laboratory.

Kerr, R.A. (2012) Is the world tottering on the precipice of peak gold? *Science*, **335**: 1038–1039.

King, C.D. (1948) *Seventy-Five Years of Progress in Iron and Steel*, American Institute of Mining and Metallurgical Engineers, New York.

Klare, M.T. (2012) The end of easy everything. *Current History*, **111**: 24–28.

Klein Goldewijk, K., Beusen, A. and P. Janssen. 2010. Long-term dynamic modelling of global population and built-up area in a spatially explicit way: HYDE 3.1. *The Holocene* **20**: 565–573.

Klimont, Z., Smith, S.J. and J. Cofala, 2013. The last decade of global anthropogenic sulfur dioxide: 2000–2011 emissions. *Environmental Research Letters* **8**:1–6.

Kole, C. and Michler, C. (2010) *Transgenic Crop Plants*, Springer, Berlin.

Kovanda, J. and Hak, T. (2011) Historical perspectives of material use in Czechoslovakia in 1855–2007. *Ecological Indicators*, **11**: 1375–1384.

KPMG China (2013) Global Reach of China Luxury, KPMG China, Hong Kong, http://www.kpmg.com/CN/en/IssuesAndInsights/ArticlesPublications/Documents/Global-Reach-China-Luxury-201301.pdf (accessed 23 May 2013).

Krausmann, F. (2008) The global socio-metabolic transition: past and present metabolic profiles and their future trajectories. *Journal of Industrial Ecology*, **12**: 637–656.

Krausmann, F., Gingrich, S., Eisenmenger, N. *et al.* (2009) Growth in global materials use, GDP and population during the 20th century. *Ecological Economics*, **68**: 2696–2705.

Krausmann, F., Gingrich, S. and Nourbakhch-Sabet, R. (2011) The metabolic transition in Japan. *Journal of Industrial Ecology*, **15**: 877–892.

Kuriki, S., Daigo, I., Matsuno, Y. *et al.* (2010) Recycling potential of platinum group metals in Japan. *Journal of Japan Institute of Metals*, **74**: 801–805.

Kvavadze, E., Bar-Yosef, O., Belfer-Cohen, A. *et al.* (2009) 30,000-year-old wild flax fibers. *Science*, **325**: 1359.

Lampert, L., Quéguiner, B., Labasque, L. *et al.* (2002) Spatial variability of phytoplankton composition and biomass on the eastern continental shelf of the Bay of Biscay (north-east Atlantic Ocean). Evidence for a bloom of Emiliania huxleyi (Prymnesiophyceae) in spring 1998. *Continental Shelf Research*, **22**: 1225–1247.

Landsberg, H.H. (1964) *Natural Resources for U.S. Growth: A Look Ahead to the Year 2000*, Johns Hopkins Press, Baltimore, MD.

Landsberg, H.H., Fischman, L.L. and J.L. Fisher. 1963. *Resources in America's Future: Patterns of Requirements and Availabilities, 1960-2000*. Baltimore, MD: Johns Hopkins Press.

Latimer, J. 2001. All-consuming passions: materials and subjectivity in the age of enhancement. In: N. Lee and R. Munro, eds., *The Consumption of Mass*, Oxford, Blackwell Publishers, pp. 158–173.

Lauk, C., Haberl, H., Erb, K.-H. *et al.* (2012) Global socioeconomic carbon stocks in long-lived products 1900–2008. *Environmental Research Letters*, **7**: 034023 http://iopscience.iop.org/1748-9326/7/3/034023 (accessed 23 May 2013).

Layard, R. (2005) *Happiness*, The Penguin Press, New York.

Leach, E.R. (1959) Hydraulic society in Ceylon. *Past and Present*, **15**: 2–26.

Lefebvre, H. (1971) *Everyday Life in the Modern World*, Allen Lane, London.

Lenzen, M. and Dey, C.J. (2000) Truncation error in embodied energy analysis of basic iron and steel products. *Energy*, **25**: 577–585.

Lenzen, M. and Treloar, G. (2003) Differential convergence of life-cycle inventories toward upstream production layers, implications for life-cycle assessment. *Journal of Industrial Ecology*, **6**: 3–4.

Lei, Q. (2011) The Development of China's Cement Industry, http://www.tcma.org.tr /images/file/Cin%20Cimento%20Birligi%20Baskani%20Lei%20QIANZHI%20.pdf (accessed 23 May 2013).

Li, J. (2012a) World Bank Experience on Clean Brick Production in South Asia Region, http://www.ine.gob.mx/descargas/dgcenica/2012_ladrilleras_pon_s6_jli_eng.pdf (accessed 23 May 2013).

Li, V. (2012b) Can concrete be bendable? *American Scientist*, **100**: 484–493.

Li, Y., Richardson, J.B., Walker, A.K. *et al.* (2006) PTCLP heavy metal leaching of personal computer components. *Journal of Environmental Engineering*, **2006**: 497–504.

Li, C., Zhang, Q., Krotkov, N.A. *et al.* (2010) Recent large reduction in sulfur dioxide emissions from Chinese power plants observed by the Ozone Monitoring Instrument. *Geophysical Research Letters*, **37**: L08807. doi: 10.1029/2010GL042594

Lifset, R.J., Gordon, R., Graedel, T. *et al.* (2002) Where has all the copper gone: the stocks and flows project, Part 1. *Journal of Metallurgy*, **2002**: 21–26.

Linde, C.P. (1916) *Aus Meinem Leben und von Meiner Arbeit*, R. Oldenbourg, München.

Linebaugh, P. (1993) *The London Hanged: Crime and Civil Society in the Eighteenth Century*, Cambridge University Press, Cambridge.

Longhurst, A.H. (1979) *The Story of the Stūpa*, Asian Educational Services, New Delhi.

Lowenstam, H.A. (1981) Minerals formed by organisms. *Science*, **211**: 1126–1130.

Lu, Z., Streets, D.G., Zhang, Q. *et al.* (2010) Sulfur dioxide emission in China and sulfur trends in East Asia since 2000. *Atmospheric Chemistry and Physics*, **10**: 6311–6331.

Luo, Z. and Soria, A. (2008) Prospective Study of the World Aluminium Industry, European Commission, Seville, ftp://ftp.jrc.es/pub/EURdoc/JRC40221.pdf (accessed 23 May 2013).

MacBean, C. (2013) *The Pesticide Manual: A World Compendium*, CABI, Wallingford.

Macfarlane, A. and Martin, G. (2002) *Glass: A World History*, University of Chicago Press, Chicago.

Maddison, A. (2007) *Contours of the World Economy, 1-2030 AD*, Oxford University Press, Oxford.

Madlool, N.A., Saidur, R., Hossian, M.S. and Rahim, N.A. (2011) A critical review on energy use and savings in the cement industries. *Renewable and Sustainable Energy Reviews*, **15**: 2042–2060.

Magnaghi, G. (2011) Recovered Paper Market in 2010, http://www.bir.org/assets/Documents/industry/MagnaghiReport2010.pdf (accessed 23 May 2013).

Malpass, D.B. (2010) *Introduction to Industrial Polyethylene: Properties, Catalysts, and Processes*, Wiley-Scrivener, New York.

Marchetti, C. (1985) Nuclear plants and nuclear niches. *Nuclear Science and Engineering*, **90**: 521–526.

Markets and Markets (2011) Hydrogen Generation Market, http://www.marketsandmarkets.com/Market-Reports/hydrogen-generation-market-494.html (accessed 23 May 2013).

Marsh, P. (2012) *The New Industrial Revolution: Consumers, Globalisation and the End of Mass Production*, Yale University Press, New Haven, CT.

Matos, G.R. (2009) *Use of Minerals and Materials in the United States from 1900 through 2006*, USGS, Washington, DChttp://pubs.usgs.gov/fs/2009/3008/ (accessed 23 May 2013).

Matos, G. and Wagner, L. (1998) Consumption of materials in the United States, 1900–1995. *Annual Review of Energy*, **23**: 107–122http://pubs.usgs.gov/annrev/ar-23-107/aerdocnew.pdf (accessed 23 May 2013).

Matternet (2013) http://matternet.us/ (accessed 23 May 2013).

Matthews, E., Bringezu, S., Fischer-Kowalski, M. *et al.* (2000) *The Weight of Nations: Material Outflows from Industrial Economies*, World Resources Institute, Washington, DC.

McFedries, P. (2012) Consumption 2.0. *IEEE Spectrum*, **2012**: 26.

McLaren, D.J. and Skinner, B.J. (eds) (1987) *Resources and World Development*, John Wiley & Sons, Ltd, Chichester.

Meadows, D.H., Meadows, D.L., Randers, J. *et al.* (1972) *The Limits to Growth*, Universe Books, New York.

Meyers, M.A., McKittrick, J. and P. Chen. 2013. Structural biological materials: critical mechanics-materials connections. *Science* **339**: 773–779.

Miettinen, A., Sarmaja-Korjonen, K., Sonninen, E. *et al.* (2008) *The Palaeoenvironment of the Antrea Net Find*, Finnish Antiquarian Society, Helsinki.

McKelvey, V.E. (1973) Mineral resource estimates and public policy, in *United States Mineral Resources* (eds D.A. Brobst and W.P. Pratt), USGS, Washington, DC, pp. 9–19.

McWhan, D. (2012) *Sand and Silicon: Science that Changed the World*, Oxford University Press, New York.

Meil, J., Bushi, L., Garrahan, P. *et al.* (2009) *Status of Energy Use in the Canadian Wood Products Sector*, Forintek, Vancouver, BChttp://www.forintek.ca/public/pdf/Public_Information/technical_rpt/Status%20on%20Energy%20use%20in%20Canadian%20Wood%20Products%20sector%20-%202010.pdf (accessed 23 May 2013).

Mendoza, J., Domenico, A., Fatemeh, J. *et al.* (2012) Life cycle assessment of granite application in sidewalks. *International Journal of Life Cycle Assessment*, **17**: 580–592.

Menzel, P. (1995) *Material World: A Global Family Portrait*, Sierra Club, San Francisco, CA.

Mersiowsky, I. (2002) Long-term fate of PVC products and their additives in landfills. *Progress in Polymer Science*, **27**: 2227–2277.

Metz, B., O. Davidson, H. de Coninck, *et al.* (eds) 2005. *Carbon Dioxide Capture and Storage*. Cambridge: Cambridge University Press.

Mitscherlich, G. (1963) *Zustand, Wachstum und Nutzung des Waldes im Wandel der Zeit*, Freiburger Universitätsreden, Freiburg.

Mitsubishi (2005) LCA Results of Aluminum Can, Mitsubishi Materials, Tokyo, http://www.jia-page.or.jp/environment/epd/directory/img/LCAresult_e.pdf (accessed 23 May 2013).

mobiThinking (2013) Global Mobile Statistics 2013 Part A: Mobile Subscribers; Handset Market Share; Mobile Operators, http://mobithinking.com/mobile-marketing-tools/latest-mobile-stats/a#phone-shipments (accessed 23 May 2013).

Moll, S., Popescu, C., and Nickel, R. (2012) EU's Resource Productivity on the Increase, http://epp.eurostat.ec.europa.eu/cache/ITY_OFFPUB/KS-SF-12-022/EN/KS-SF-12-022-EN.PDF (accessed 23 May 2013).

Montgomery Ward & Company (1895) *Catalogue and Buyers' Guide 1895*, Skyhorse Publishing, New York 2008.

Moore, G. (1965) Cramming more components onto integrated circuits. *Electronics*, **38**: 114–117.

Moore, G.E. (1975) Progress in digital integrated electronics. Technical Digest, IEEE International Electron Devices Meeting, pp. 11–13.

Moore, C.J. (2008) Synthetic polymers in the marine environment: a rapidly increasing long-term threat. *Environmental Research*, **108**: 131–139.

Moore, C.G. and Phillips, C. (2011) *Plastic Ocean*, Penguin, London.

Morgan, T. 2010. *End-game: The Denouement of Exponentials*. Tullett Prebon, London, http://www.tullettprebon.com/Documents/strategyinsights/tp0610d_tpsi_006.pdf (accessed 23 May 2013).

Morita, Z. and Emi, T. (eds) (2003) *An Introduction to Iron and Steel Processing*, Kawasaki Steel 21st Century Foundation, Tokyo.

Mukerji, C. (1983) *From Graven Images: Patterns of Modern Materialism*, Columbia University Press, New York.

Muradov, N.Z. and Nejat Veziroğlu, T. (eds) (2012) *Carbon-Neutral Fuels and Energy Carriers*, CRC Press, Boca Raton, FL.

NAHB (National Association of Home Builders) (2010) Smaller homes to remain popular even after recession's end, http://www.nahb.org/news_details.aspx?newsID=11485&fromGSA=1 (accessed 19 June 2013).

NAS (National Academy of Sciences) (1969) *Resources and Man*, W.H. Freeman, San Francisco, CA.

National Archives (2012) British Empire: Living in the British Empire – India, http://www.nationalarchives.gov.uk/education/empire/pdf/g2cs4s2.pdf (accessed 23 May 2013).

NBSC (National Bureau of Statistics of China) (2013) Statistical Data, http://www.stats.gov.cn/english/ (accessed 23 May 2013).

Ndoro, W. (1997) The Great Zimbabwe. *Scientific American*, **277**: 94–99.

NEAA (Netherlands Environmental Assessment Agency) (2013) HYDE site: History Database of the Global Environment, http://themasites.pbl.nl/tridion/en/themasites/hyde/ (accessed 19 June 2013).

Nel, A., Tian, X., Lutz, M. *et al.* (2006) Toxic potential of materials at the nanolevel. *Science*, **311**: 622–627.

Nemoto, T., Tanaka, Y., Tsujioka, S. *et al.* (2011) Resource recycling for sustainable industrial development. *Hitachi Review*, **60**: 335–340.

Newby, F. (ed.) (2001) *Early Reinforced Concrete*, Ashgate Publishing Ltd, Burlington, VY.

Nielsen, S. (2003) Boeing turns to Russian programming talent in massive database project. *Serverworld*, **1**: 1–4.

Norgate, T. and Jahanshahi, S. (2011) Reducing the greenhouse gas footprint of primary metal production: where should the focus be? *Minerals Engineering*, **24**: 1563–1570.

Novoselov, K.S., Falko, V.I., Colombo, L. *et al.* (2012) A roadmap for graphene. *Nature*, **490**: 192–200.

NREL (National Renewable Energy Laboratory) (2009) U.S. Life Cycle Inventory Batabase Roadmap, http://www.nrel.gov/lci/pdfs/45153.pdf (accessed 23 May 2013).

NREL (2013a) Best Research-Cell Efficiencies, http://www.nrel.gov/ncpv/images/efficiency_chart.jpg (accessed 23 May 2013).

NREL (2013b) U.S. Life Cycle Inventory Database, http://www.nrel.gov/lci/ (accessed 23 May 2013).

NRC (2009) Energy Benchmarking: Canadian Potash Production Facilities, NRC, Ottawa, http://oee.nrcan.gc.ca/publications/infosource/pub/cipec/Potash-production/7132 (accessed 23 May 2013).

NSG (2011) NSG Group and the Flat Glass Industry 2011, http://www.nsg.co.jp/~/media/NSG/Site%20Content/Temporary%20Downloads/Japanese/NSGFGI_2011%20EN2.ashx (accessed 23 May 2013).

Nuttall, G.H. (1967) *Theory and Operation of the Fourdrinier Paper Machine*, Phillips, London.

Oakes, J. (2005) *Inuit annuraangit. Our Clothes: A Travelling Exhibition of Inuit Clothing*, Aboriginal Issues Press, Winnipeg.

O'Brien, C.J., Peloquin, J.A., Vogt, M. *et al.* (2012) Global marine phytoplankton functional type biomass distributions: coccolithophores. *Earth Systems Scientific Data Discussions*, **5**: 491–520.

OEA (Organization of European Aluminium Refiners and Remelters) (2010) Aluminium Recycling in Europe, http://www.alueurope.eu/wp-content/uploads/2011/08/Aluminium-recycling-in-Europe-2007.pdf (accessed 23 May 2013).

OECD (Organization for Economic Cooperation and Development) (1995) The Life Cycle Approach: An Overview of Product/Process Analysis, OECD, Paris.

OECD (2013) Energy Intensity, http://www.oecd-ilibrary.org/docserver/download/3011041ec047.pdf?expires=1366555903&id=id&accname=guest&checksum=BDB5DB8A6AD551F0E60F1C824BF3C719 (accessed 23 May 2013).

ONS (Office for National Statistics) (2012) UK Environmental Accounts, 2012, http://www.ons.gov.uk/ons/dcp171778_267211.pdf (accessed 23 May 2013).

Owens, B. (2013) Mining: extreme prospects. *Nature*, **495**: S4–2S6.

Pant, S. (1976) *The Origin and Development of Stūpa Architecture in India*, Bharata Manisha, New Delhi.

Paper Recycling Promotion Center (2009) Statistics of Recovered Paper in Japan, http://www.prpc.or.jp/menu05/pdf/statics-of-recovered-paper2008.pdf (accessed 19 June 2013).

Paper Recycling Promotion Center (2012) 2011 Statistics of Recovered Paper in Japan, http://www.prpc.or.jp/linkfile/english-paperrecycling.pdf (accessed 23 May 2013).

Patterson, C.C. (1972) Silver stocks and losses in ancient and medieval times. *The Economic History Review*, **25**: 205–233.

PCA (Portland Cement Association) (2011) Concrete Design and Production, http://www.cement.org/tech/cct_concrete_prod.asp (accessed 23 May 2013).

PE Americas (2010) Life Cycle Impact Assessment of Aluminum Beverage Cans, PE Americas, Boston, MA, http://www.container-recycling.org/assets/pdfs/aluminum /LCA-2010-AluminumAssoc.pdf (accessed 23 May 2013).

Penner, M., Power, K., Muhairwe, C. *et al.* (1997) *Canada's Forest Biomass Resources: Deriving Estimates from Canada's Forestry Inventory*, Pacific Forestry Centre, Victoria, BChttp://cfs.nrcan.gc.ca/publications/?id=4775 (accessed 19 June 1023).

People's Daily (2013) China's Luxury Consumption Hits $46 Billion. People's Daily Online (Jan 15), http://english.peopledaily.com.cn/90778/8093172.html (accessed 23 May 2013).

Peray, K.E. (1986) *The Rotary Cement Kiln*, Chemical Publishing, New York.

Petersen, A.K. and Solberg, B. (2002) Greenhouse gas emissions, life-cycle inventory and cost-efficiency of using laminated wood instead of steel construction. Case: beams at Gardermoen airport. *Environmental Science and Policy*, **5**: 169–182.

Petersen, A.K. and Solberg, B. (2004) Greenhouse gas emissions and costs over the life cycle of wood and alternative flooring materials. *Climatic Change*, **64**: 143–167.

Pierson, D. (2012) SUVs are Big in China. Los Angeles Times (Apr 24), http://articles.latimes.com/2012/apr/24/business/la-fi-china-suv-20120425 (accessed 23 May 2013).

Pilkington, L.A.B. (1969) Review lecture. The ploat glass process. *Proceedings of the Royal Society of London, Series A: Mathematical and Physical Sciences*, **314**: 1–25.

Pilkington, L.A.B. (2010) Pilkington and the Flat Glass Industry 2010 http://www .pilkington.com/resources/pfgi2010.pdf (accessed 23 May 2013).

Pilkington, L.A.B. (2013) Self-Cleaning, http://www.pilkington.com/products/bp/by-benefit/selfcleaning/home.htm.

Pilz, H., Brandt, B. and Fehringer, R. (2010) *The Impact of Plastics on Life Cycle Energy Consumption and Greenhouse Gas Emissions in Europe*, Denkstatt, Vienna http://www.plasticseurope.org/documents/document/20111107113205-e_ghg _packaging_denkstatt_vers_1_1.pdf (accessed 23 May 2013).

Plastics Europe (2011) Plastics – The Facts, 2011, Plastics Europe, Brussels, http://www.plasticseurope.org/documents/document/20111107101127-final_pe _factsfigures_uk2011_lr_041111.pdf (accessed 23 May 2013).

Plastics Europe (2012) Plastics – The Facts 2012, http://www.plasticseurope.org/cust /documentrequest.aspx?DocID=54693 (accessed 23 May 2013).

Porada, E. (1965) *The Art of Ancient Iran: Pre-Islamic Cultures*, Crown Publishers, New York.

Potočnik, J. (2012) Any Future for the Plastic Industry in Europe? http://europa.eu/rapid /press-release_SPEECH-12-632_en.htm (accessed 23 May 2013).

Prak, M. (2011) Mega-structures of the middle ages: the construction of religious build-ings in Europe and Asia, c. 1000-1500. *Journal of Global History*, **6**: 381–406.

President's Materials Policy Commission (1952) *Resources for Freedom, A Report to the President by the President's Materials Policy Commission*, US Government Printing Office, Washington, DC.

Princen, T., Maniates, M. and Conca, K. (eds) (2002) *Confronting Consumption*, MIT Press, Cambridge, MA.

Qualcomm (2013) Qualcomm Tricorder X Prize, http://www.qualcommtricorderxprize .org/competition-details/overview (accessed 23 May 2013).

Rafiqul, I., Weber, C., Lehmann, B. and Voss, A. (2005) Energy efficiency improvements in ammonia production – perspectives and uncertainties. *Energy*, **30**: 2487–2504.

Reeves, G.M., Sims, I. and J. C. Cripps. 2006. *Clay Materials used in Construction*. London: Geological Society.

Reiffel, A.J., Kafka, C., Hernandez, K.A. *et al.* (2013) High-fidelity tissue engineering of patient-specific auricles for reconstruction of pediatric microtia and other auricular deformities. *PLoS ONE*, **8**: e56506. doi: 10.1371/journal.pone.0056506 (Epub 2013 Feb 20).

Research and Markets (2013) Global and Chinese Industrial Gases Industry Report, 2012–2013, Research and Markets, Dublin.

Richardson, B.A. (1978) *Wood Preservation*, Construction Press, New York.

Robinson, B.H. (2009) E-waste: an assessment of global production and /environmental impacts. *Science of the Total Environment*, **408**: 183–191.

Roche, D. (2000) *A History of Everyday Things: The Birth of Consumption in France, 1600-1800*, Cambridge University Press, Cambridge.

Ruthven, D.M., Farooq, S. and Knaebel, K.S. (1993) *Pressure Swing Adsorption*, John Wiley & Sons, Inc., New York.

Ryan, L. and Dziurawiec, S. (2001) Materialism and its relationship to life satisfaction. *Social Indicators Research*, **55**: 185–197.

Salmon, E.T. (1999) *Roman Coins and Public Life under the Empire*, The University of Michigan Press, Ann Arbor, MI.

Savage, N. (2012) Graphene. *Nature*, **483**: 530–531.

Savage, N. (2013) Mine, all mine! *Nature*, **495**: 52–53.

Schandl, H. and Eisenmenger, N. (2006) Regional patterns in global resource extraction. *Journal of Industrial Ecology*, **10**: 133–147.

Schandl, H., Poldy, F., Turner, G.M. *et al.* (2008) Australia's resource use categories. *Journal of Industrial Ecology*, **12**: 669–685.

Scheel, H.J. and Fukuda, T. (eds) (2004) *Crystal Growth Technology*, John Wiley & Sons, Inc., New York.

Schmidt, M., Hottenroth, H., Schottler, M. *et al.* (2012) Life cycle assessment of silicon wafer processing for microelectronic chips and solar cells. *International Journal of Life Cycle Assessment*, **17**: 126–144.

Schoer, K., Kovanda, J., and Lauwigi, C. (2012) Economy-Wide Material Flow Accounts (EW-MFA), http://epp.eurostat.ec.europa.eu/portal/page/portal/environmental _accounts/documents/Economy-wide%20material%20flow%20accounts%20compilat-ion%20guide%20%20-.pdf (accessed 23 May 2013).

Schoer, K., Weinzettel, J., Kovanda, J. *et al.* (2012) Raw material consumption of the European Union –concept, calculation method, and results. *Environmental Science and Technology*, **46**: 8903–8909.

Schultz, D. (2012) The Past, Present and Future of Aluminum in North American Light Vehicles, http://www.drivealuminum.org/research-resources/PDF/Speeches%20and%20Presentations/2012/The-Past-Present-and-Future-of-Aluminum.pdf (accessed 19 June 2013).

Schutz, B. (2002) *Great Cathedrals*, Abrams, New York.

Semon, W.L. (1933) Synthetic rubber-like composition and method of making same. US Patent 1,929,453, USPTO, Washington, DC.

SERI (Sustainable Europe Research Institute) (2013) The Online Portal for Material Flow Data, http://www.materialflows.net/ (accessed 23 May 2013).

Service, R.F. (2009) Is silicon's reign nearing its end? *Science*, **323**: 1000–1002.

Sethuraj, M.R. and Mathew, N.M. (eds) (1992) *Natural Rubber: Biology, Cultivation, and Technology*, Elsevier, New York.

Sevenster, A. (2008) Eco-profiles and environmental declarations for PVC. *Plastic, Rubber and Composites*, **37**: 403–405.

SFA (State Forestry Administration) (2009) *China Forestry Statistical Yearbook 2009*, China Forestry Publishing House, Beijing.

Shadman, F. and McManus, T.J. (2004) Comment on "The 1.7 kilogram microchip: energy and material use in the production of semiconductor devices". *Environmental Science and Technology*, **38**: 1915.

Shaeffer, R.E. (1992) *Reinforced Concrete: Preliminary Design for Architects and Builders*, McGraw-Hill, New York.

Shafizadeh, F. (1981) Basic principles of direct combustion, in *Biomass Conversion Process for Energy and Fuels* (eds S.S. Sofer and O.R. Zabrosky), Plenum Press, New York, pp. 103–112.

Shalaby, W.S. and Burg, K.J.L. (eds) (2004) *Absorbable and Biodegradable Polymers*, CRC Press, Boca Raton, FL.

Shephard, S. (2000) *Pickled, Potted, and Canned: How the Art and Science of Food Preserving Changed the World*, Simon & Schuster, New York.

Shumaker, R.W., Walkup, K.R. and B.B. Beck. 2011. *Animal Tool Behavior: The Use and Manufacture of Tools by Animals*. Baltimore, MD: The Johns Hopkins University Press.

SIA (Semiconductor Industry Association) (2013) Historical Billings Reports, http://www.sia-online.org/clientuploads/directory/IndustryStatistics/GSR1976-December%202012.xls (accessed 23 May 2013).

Sibley, S.F. (2011) *Overview of Flow Studies for Recycling Metal Commodities in the United States*, USGS, Washington, DChttp://pubs.usgs.gov/circ/circ1196-AA/pdf/circ1196-AA.pdf (accessed 23 May 2013).

Sika. 2005. *Concrete Handbook*. Zurich: Sika. http://gbr.sika.com/content/dam/Corporate/01_General/publications/sika_concrete_handbook.pdf (accessed 23 May 2013).

Simon, J. (1981) *The Ultimate Resource*, Princeton University Press, Princeton, NJ.

Simon, J. (ed.) (1995) *The State of Humanity*, Blackwell-Wiley, New York.

Sims, G. and Rusmana, Y. (2011) Tin Gets Scarcer Amid Growing Global Demand. Bloomberg, http://tininvestingnews.com/933-tin-gets-scarcer-amid-growing-global-demand.html (accessed 23 May 2013).

Singh, R. (2006) Lac Culture, Department of Zoology, Udai Pratap Autonomous College, Varanasi.

Singh, S.J., Krausmann, F., Gingrich, S. *et al.* (2012) India's biophysical economy, 1961–2008. Sustainability in a national and global context. *Ecological Economics*, **76**: 60–69.

Sivak, M. and Tsimhoni, O. (2009) Fuel efficiency of vehicles on US roads: 1936–2006. *Energy Policy*, **37**: 3168–3170.

Slade, G. (2006) *Made to Break: Technology and Obsolescence in America*, Harvard University Press, Cambridge, MA.

Smil, V., Nachman, P. and T.C. Long, II., 1983. *Energy Analysis in Agriculture*. Boulder, CO: Westview.

Smil, V. (1994) *Energy in World History*, Westview, Boulder, CO.

Smil, V. (1999) Crop residues: agriculture's largest harvest. *BioScience*, **49**: 299–308.

Smil, V. (2000) *Feeding the World*, The MIT Press, Cambridge, MA.

Smil, V. (2001) *Enriching the Earth*, The MIT Press, Cambridge, MA.

Smil, V. (2003) *Energy at the Crossroads*, The MIT Press, Cambridge, MA.

Smil, V. (2005) *Creating the Twentieth Century*, Oxford University Press, New York.

Smil, V. (2006a) *Transforming the Twentieth Century*, Oxford University Press, New York.

Smil, V. (2006b) Peak oil: a catastrophist cult and complex realities. *World Watch*, **19**: 22–24.

Smil, V. (2007) Light behind the fall: Japan's electricity consumption, the environment, and economic growth. *Japan Focus* http://www.vaclavsmil.com/wp-content/uploads /docs/smil-article-light-behind-fall-20070000-apj.pdf (accessed 23 May 2013).

Smil, V. (2008) *Energy in Nature and Society*, The MIT Press, Cambridge, MA.

Smil, V. (2010) *Prime Movers of Globalization*, The MIT Press, Cambridge, MA.

Smil, V. (2012) Jeremy Grantham, Starving for Facts. The American (Dec 5), http://www .vaclavsmil.com/wp-content/uploads/Jeremy-Grantham-Starving-for-Facts – The-American-Magazine.pdf (accessed 23 May 2013).

Smil, V. (2013) *Harvesting the Biosphere*, The MIT Press, Cambridge, MA.

Smith, D.C. (1970) *History of Papermaking in the United States*, Lockwood Publishing, New York.

Smith, S.J., van Aardenne, J., Klimont, Z. *et al.* (2010) Anthropogenic sulfur dioxide emissions 1850–2005. *Atmospheric Chemistry and Physics Discussions*, **10**: 16111–16151.

Solarplaza (2013) Top 10 World's Most Efficient Solar PV Mono-Crystalline Cells, http: //www.solarplaza.com/top10-monocrystalline-cell-efficiency/ (accessed 23 May 2013).

Sommath, S.S., Davalis, J.F., Zipfel, M.G. *et al.* (1995) A review of wood railroad ties performance. *Forest Products Journal*, **45**: 55.

Stahel, W.R. (1997) The service economy: 'wealth without resource consumption'? *Philosophical Transactions of the Royal Society London A*, **355**: 1309–1319.

Stanley, S.M., Ries, J.B. and Hardie, L.A. (2005) Seawater chemistry, coccolithophore population growth, and the origin of Cretaceous chalk. *Geology*, **33**: 593–596.

STAP (Scientific and Technial Advisory Panel) (2011) Marine Debris as a Global Environmental Problem, http://www.thegef.org/gef/sites/thegef.org/files/publication /STAP%20MarineDebris%20-%20website.pdf (accessed 23 May 2013).

Steinberger, J.K., Krausmann, F. and Eisenmenger, N. (2010) Global patterns of materials use: a socioeconomic and geophysical analysis. *Ecological Economics*, **69**: 1148–1158.

Stenkjaer, N. (2009) Straw Ovens, http://www.folkecenter.net/gb/rd/biogas/technologies /straw_ovens/ (accessed 23 May 2013).

Stephan, D. 2012. The world's largest low-density polyethylene plant inaugurated in Qatar. *Process Worldwide*, http://www.process-worldwide.com/engineering_construc- tion/materials/polymers/articles/386413 (accessed 23 May 2013).

Stern, B.R. and Lagos, G. (2008) Are there health risks from the migration of chemical substances from plastic pipes to drinking water? A review. *Human and Ecological Assessment*, **14**: 753–779.

Strawbale (2013) http://www.strawbale.com/ (accessed 23 May 2013).

Strom, E.T. and Rasmussen, S.C. (eds) (2011) *100+ Years of Plastics: Leo Baekeland and Beyond*, Oxford University Press, New York.

Stuart, M. (2012) Concrete Deterioration, PDH Center, Fairfx, http://www.pdhonline.org /courses/s155/s155content.pdf (accessed 23 May 2013).

Sui, T. (2010) China's Cement Industry Towards Sustainability, http://www.italcementig- roup.com/NR/rdonlyres/4FFAD881-52B9-439B-8D51-965D7CC4588F/0/Tongbo_ UK.pd (accessed 23 May 2013).

Sullivan, D.E., Sznopek, J.L. and L.A. Wagner. 2000. *20th Century U.S. Min- eral Prices Decline in Constant Dollars*. Washington, DC: USGS. http://pubs .er.usgs.gov/publication/ofr00389 (accessed 23 May 2013).

Sun, W., Cai, J. and Ye, Z. (2013) Advances in energy conservation of china steel industry. *Scientific World Journal*, **2013**: 247035. doi: 10.1155/2013/247035

Suwanmanee, U., Varabuntoonvit, V., Chaiwutthinan, P. *et al.* (2013) Life cycle assess- ment of single use thermoform boxes made form polystyrene (PS), polylactic acid (PLA), and PLA starch: cradle to consumer gate. *International Journal of Life Cycle Assessment*, **18**: 401–417.

Tahil, W. (2006) The Trouble with Lithium: Implications of Future PHEV Production for Lithium Demand, http://tyler.blogware.com/lithium_shortage.pdf (accessed 23 May 2013).

Takiguchi, H. and Morita, K. (2011) Global flow analysis of crystalline silicon, in *Crys- talline Silicon –Properties and Uses* (ed S. Basu), InTech, pp. 329–355.

Tanner, A.H. (1998) *Continuous Casting: A Revolution in Steel*, Write Stuff Enterprises, Fort Lauderdale.

Taylor, P., Funk, C. and A. Clark. 2006. *Luxury or Necessity? Things We Can't Live Without: The List Has Grown in the Past Decade*. Washington, DC: Pew Research Center, http://www.pewsocialtrends.org/files/2010/10/Luxury.pdf (accessed 23 May 2013).

Terakado, R., Takahashi, K., Daigo, I. *et al.* (2009) In-use stock of copper in Japan esti- mated by bottom-up approach. *Journal of the Japan Institute of Metals*, **73**: 713–719.

The Economist (2011) Commodities: Crowded Out. The Economist (Sep 24, 2011), http://www.economist.com/node/21528986 (accessed 19 June 2013).

The Linde Group (2004) 125 Years of Linde, http://www.the-linde-group.com/internet. global.thelindegroup.global/en/images/chronicle_e%5B1%5D14_9855.pdf (accessed 23 May 2013).

The Linde Group (2012) Cryogenic Air Separation: History and Technological Progress, The Linde Group, Wiesbaden.

Thieme, H. (1997) Lower Paleolithic hunting spears from Germany. *Nature*, **385**: 807–810.

Thomas, J. (ed.) (1979) *Energy Analysis*, Westview Press, Boulder, CO.

Thomsen, C.J. (1836) *Ledetraad Til Nordisk Oldkyndighed, udg. af det Kongelige Nordiske Oldskrift-selskab*, S. L. Møllers, Copenhagen.

Thorpe, R.S. and Williams-Thorpe, O. (1991) The myth of long-distance megalith transport. *Antiquity*, **65**: 64–73.

ThyssenKrupp (2012) Crew Celebrates Milestone: Europe's Biggest Blast Furnace Produces 70 Millionth Ton of Steel, http://www.thyssenkrupp.com/en/presse/art_detail. html&eid=TKBase_1328095396431_284885812 (accessed 23 May 2013).

Tickner, J.A., Schettler, T., Guidotti, T. *et al.* (2001) Health risks posed by use of di-2-ethylhexyl phthalate (DEHP) in PVC medical devices: a critical review. *American Journal of Inustrial Medicine*, **39**: 100–111.

Tilahun, A., Kebede, F., Yamoah, C. *et al* (2012) Quantifying the masses of Macrotermes subhyalinus mounds and evaluating their use as a soil amendment. *Agriculture, Ecosystems and Environment*, **157**: 54–59.

Tilman D., Balzer, C., Hill J., Befort, B.L. 2011. Global food demand and the sustainable intensification of agriculture. *Proceedings of the National Academy of Sciences* **108**: 20260–20264.

Tréguer, P., Nelson, D.M., van Bennekom, A.J. *et al.* (1995) The silica balance in the world ocean: a reestimate. *Science*, **268**: 375–379.

Trinkhaus, E. (2005) Early modern humans. *Annual Review of Anthropology*, **34**: 207–30.

Tupy, M.L. (2012) Why iPhone 5 and Siri are Good for Capitalism, http://www.cato.org /blog/miracle-iphone-or-how-capitalism-can-be-good-environment (accessed 23 May 2013).

Turkenburg, W.C., Arent, D.J., Bertani, R. *et al.* (2000) Renewable energy technologies, in *World Energy Assessment: Energy and the Challenge of Sustainability* (eds J. Goldemberg *et al.*), UNDP, New York, pp. 219–227.

Turner, J.S. (2000) *The Extended Organism*, Harvard University Press, Cambridge, MA.

Twede, D. (2002) The packaging technology and science of ancient transport amphoras. *Packaging Technology and Science*, **15**: 181–195.

Twede, D. (2005) The cask age: the technology and history of wooden barrels. *Packaging Technology and Science*, **18**: 253–264.

Ulm, F.-J. (2012) Innovationspotenzial Beton:Von Atomen zur Grünen Infrastruktur. *Beton- und Stahlbetonbau*, **107**: 504–509.

UN (United Nations) (2013) Energy Statistics, http://unstats.un.org/unsd/energy/ (accessed 23 May 2013).

UNDESA (United Nation Department of Economic and Social Affairs) (2013) Detailed Indicators, http://esa.un.org/wpp/unpp/panel_indicators.htm (accessed 19 June 2013).

Underwood, J.R. (2001) Anthropic rocks as a fourth basic class. *Environmental and Engineering Geoscience*, **7**: 104–110.

UNDP (United Nations Development Programme) (2013) International Human Development Indicators, http://hdr.undp.org/en/statistics/ (accessed 23 May 2013).

UNECE (United Nations Economic Commission for Europe) (2010) Forest Product Conversion Factors for the UNECE Region, UNECE, Geneva, http://www.unece.org/fileadmin/DAM/timber/publications/DP-49.pdf (accessed 23 May 2013).

UNEP (2009) Distribution of Marine Litter, http://www.unep.org/regionalseas/marinelitter/about/distribution/default.asp (accessed 23 May 2013).

Unger, P.W. (ed.) (1994) *Managing Agricultural Residues*, Lewis Publishers, Boca Raton, FL.

University of Stuttgart (2007) *The Sustainability of Packaging Systems for Fruit and Vegetable Transport in Europe Based on Life-Cycle Analysis*, University of Stuttgart, Stuttgart.

USCB (United States Census Bureau) (2012) The 2012 Statistical Abstract, http://www.census.gov/compendia/statab/ (accessed 23 May 2013).

USDA (United States Department of Agriculture) (2001) *U.S. Forest Facts and Historical Trends*, Forest Service, Washington, DC.

USDA (2003) *U.S. Timber Production, Trade, Consumption, and Price Statistics 1965–2002*, Forest Service, Washington, DC.

USDA (2012) 2012 Agricultural Statistics Annual, http://www.nass.usda.gov/Publications/Ag_Statistics/2012/index.asp (accessed 23 May 2013).

USDOE (US Department of Energy) (2005) Light Vehicle Market Share by Size Class, 1975–2005, http://www1.eere.energy.gov/vehiclesandfuels/facts/2005/fcvt_fotw387.html (accessed 23 May 2013).

USDOE (2007) U.S. Energy Requirements for Aluminum Production Historical Perspective, Theoretical Limits and Current Practices, http://www1.eere.energy.gov/manufacturing/resources/aluminum/pdfs/al_theoretical.pdf (accessed 23 May 2013).

USDOT (US Department of Transport) (2012) History of the Interstate Highway System, http://www.fhwa.dot.gov/interstate/history.htm (accessed 23 May 2013).

USEIA (United States Energy Information Agency) (2013) Sources and Uses, http://www.eia.gov/ (accessed 23 May 2013).

USEPA (US Environmental Protection Agency) (2003) *Background Document for Life-Cycle Greenhouse Gas Emission Factors for Clay Brick Reuse and Concrete Recycling*, USEPA, Washington, DC.

USEPA. 2011a. *Municipal Solid Waste Generation, Recycling, and Disposal in the United States Tables and Figures for 2010*. Washington, DC: US EPA. http://www.epa.gov/wastes/nonhaz/municipal/pubs/msw_2010_data_tables.pdf (accessed 23 May 2013).

USEPA (2011b) Paper Making and Recycling, http://www.epa.gov/osw/conserve/materials/paper/basics/papermaking.htm (accessed 23 May 2013).

USEPA (2012a) Statistics on the Management of Used and End-of-Life Electronics, http://www.epa.gov/osw/conserve/materials/ecycling/manage.htm (accessed 23 May 2013).

USEPA (2012b) *Light-Duty Automotive Technology, Carbon Dioxide Emissions, and Fuel Economy Trends: 1975 Through 2011*, US EPA, Washington, DChttp://www.epa.gov/otaq/fetrends.htm#report (accessed 23 May 2013).

USEPA (2013) National Emissions Inventory (NEI) Air Pollutant Emissions Trends Data, http://www.epa.gov/ttnchie1/trends/ (accessed 23 May 2013).

USGS (2001) Obsolete Computers, "Gold Mine," or High-Tech Trash? http://pubs.usgs .gov/fs/fs060-01/ (accessed 23 May 2013).

USGS (2005) Metals Stocks in Use in the United States, http://pubs.usgs.gov/fs/2005 /3090/2005-3090.pdf (accessed 23 May 2013).

USGS (2006a) Materials in use in U.S. Interstate Highways, http://pubs.usgs.gov/fs/2006 /3127/2006-3127.pdf (accessed 23 May 2013).

USGS (2006b) Recycled Cell Phones – A Treasure Trove of Valuable Materials, http://pubs.usgs.gov/fs/2006/3097/ (accessed 23 May 2013).

USGS. 2008. Demand for minerals in the United States. *The Encyclopedia of Earth*. Washington, D.C.: Environmental Information Coalition, National Council for Science and the Environment, http://www.eoearth.org/article/Demand_for_minerals_in_the_ United_States?topic=49532 (accessed 23 May 2013)

USGS (2009) *Copper – A Metal for the Ages*, USGS, Reston, VAhttp://pubs.usgs.gov/fs /2009/3031/ (accessed 23 May 2013).

USGS (2012) Mineral Commodity Summaries 2012, http://minerals.usgs.gov/minerals /pubs/mcs/2012/mcs2012.pdf (accessed 23 May 2013).

USGS (2013) Materials Flow, http://minerals.usgs.gov/minerals/mflow/ (accessed 23 May 2013).

Vale (2013) The Vale Fleet, http://www.vale.com/EN/business/logistics/shipping/Pages /default.aspx (accessed 23 May 2013).

van der Voet, E., van Oers, L. and Nikolic, I. (2005) Dematerialization: not just a matter of weight. *Journal of Industrial Ecology*, **8**: 121–137.

Vandiver, P., Soffer, O., Klima, B. and Svoboda, J. (1989) The origins of ceramic technology at Dolni-Vestonice, Czechoslovakia. *Science*, **246**: 1002–1008.

Van Kauwenbergh, S.J. (2010) World phosphate rock reserves and resources. http://pdf.usaid.gov/pdf_docs/PNADW835.pdf (accessed 23 May 2013).

Vasconcellos, A. (2010) Biomass and abundance of termites in three remnant areas of Atlantic forest in northeastern Brazil. *Revista Brasileira de Entomologia*, **54**: 455–461.

de Volder, M.F.L., Tawfick, S.H., Baughman, R.H. and Hart, A.J. (2013) Carbon nanotubes: present and future commercial applications. *Science*, **339**: 535–538.

Verbraeck, A. (ed.) (1976) *The Energy Accounting of Materials, Products, Processes and Services*, TNO (Netherlands Institute for Applied Scientific Research), Rotterdam.

Victor, P. (2008) *Managing Without Growth: Slower By Design, Not Disaster*, Edward Elgar, Cheltenham.

Vinylfacts (2013) Plastic Myths Unwrapped, http://www.vinylfacts.com/ (accessed 23 May 2013).

Vlachopoulos, J. 2009. *An Assessment of Energy Savings from Mechanical Recylcing of Polyethylene Versus New Feedstock*. Washington, DC: World Bank, https: //cdm.unfccc.int/filestorage/N/P/L/NPLRYFK7SEWG4O01ZJ3MX58VHC62AU /Assessment%20of%20energy%20savings%20in%20recycling%20versus%20new%

20feedstock.pdf?t=Z2p8bWxtMW84fDCrwXFgQdsVIyozX6A6hUbG (accessed 23 May 2013).

Wärtsilä (2013) Wärtsilä RT-flex96C/RTA96C, http://www.wartsila.com/en/engines/low-speed-engines/RT-flex96C (accessed 23 May 2013).

Watts, P. (1905) *The Ships of the Royal Navy as They Existed at the Time of Trafalgar*, Institution of Naval Architects, London.

von Weizsacker, E., Lovins, A.B. and Lovins, L.H. (1997) *Factor Four Doubling Wealth, Halving Resource Use by*, Earthscan, London.

WBCSD (World Business Council for Sustainable Development) (2004) *Mobility 2030: Meeting the Challenges of Sustainability*, WBCSD, Geneva.

Weisberg, D.E. (2008) Engineering Design Revolution, http://www.cadhistory.net/ (accessed 23 May 2013).

Weisz, H., Krausmann, F., Amann, C. *et al.* (2006) The physical economy of the European Union: cross-country comparison and determinants of material consumption. *Ecological Economics*, **58**: 676–698.

Wendel, J.F., Small, R.L., Cronn, R.C. and Brubaker, C.L. (1999) Genes, jeans, and genomes: reconstructing the history of cotton, in *Plant Evolution in Man-Made Habitats* (eds L.W.D. van Raamsdonk and J.C.M. den Nijs), University of Amsterdam, Amsterdam, pp. 133–161.

Wesseler, J.H.H. (ed.) (2005) *Environmental Costs and Benefits of Transgenic Crops*, Springer, Berlin.

White, A. (2006) Happiness is Being Healthy, Wealthy and Wise, http://www.le.ac.uk /ebulletin-archive/ebulletin/news/press-releases/2000-2009/2006/07/nparticle.2006-07-28.html (accessed 23 May 2013).

White, S.R., Blaiszik, B.J., Kramer, S.L.B. *et al.* (2011) Self-healing polymers and composites. *American Scientists*, **99**: 392–399.

Whiten, A., Goodall, J. and McGrew, W.C. (1999) Cultures in chimpanzees. *Nature*, **399**: 682–685.

Wilburn, D.R. and Goonan, T. (2013) *Aggregates from Natural and Recycled Sources*, USGS, Washington, DChttp://pubs.usgs.gov/circ/1998/c1176/c1176.html (accessed 23 May 2013).

Wilkins, J., Schoville, B.J., Brown, K.S. and Chazan, M. (2012) Evidence for early hafted hunting technology. *Science*, **338**: 942–946.

Wilkinson, P.H. (1936) *Diesel Aircraft Engines*, P.H. Wilkinson, Brooklyn, NY.

Wilkinson, C.F. and Lamb, J.C. VI, (1999) The potential health effects of phthalate esters in children's toys: a review and risk assessment. *Regulatory Toxicology and Pharmacology*, **30**: 140–155.

Williams, E. (2003) Forecasting material and economic flows in the global production chain for silicon. *Technological Forecasting and Social Change*, **70**: 341–357.

Williams, E. (2004) Energy intensity of computer manufacturing: hybrid assessment combining process and economic input–output methods. *Environmental Science and Technology*, **38**: 6166–6174.

Williams, E. (2011) Environmental effects of information and communication technologies. *Nature*, **479**: 354–358.

Williams, E., Ayres, R.U. and Heller, M. (2002) The 1.7 kilogram microchip:? Energy and material use in the production of semiconductor devices. *Environmental Science and Technology*, **36**: 5504–5510.

Wilson, A. and Boehland, J. (2005) Small is beautiful: U.S. house size, resource use, and the environment. *Journal of Industrial Ecology*, **9**: 277–287.

Williams, M. (2006) *Deforesting the Earth: From Prehistory to Global Crisis*, The University of Chicago Press, Chicago.

van Winkle, T.L., Edeleanu, J., Prosser, E.A. and Walker, C.A. (1978) Cottom versus polyester. *American Scientist*, **66**: 280–290.

Winter, N.B. (2012) Understanding Cement, http://www.understanding-cement.com /understanding-cement-book.html (accessed 23 May 2013).

Wirsenius, S. (2000) *Human Use of Land and Organic Materials*, Chalmers University of Technology, Göteborg.

Wood, R., Lenzen, M. and B. Foran. 2009. A material history of Australia: evolution of material intensity and drivers of change. *Journal of Industrial Ecology* **13**: 847–862.

World Bank (2007) Export Marketing of Gum Arabic from Sudan, http://siteresources .worldbank.org/INTAFRMDTF/Resources/Gum_Arabic_Policy_Note.pdf (accessed 23 May 2013).

World Bank (2013a) Countries and Economies, http://data.worldbank.org/country (accessed 23 May 2013).

World Bank (2013b) Motor Vehicles (Per 1,000 People), http://data.worldbank.org /indicator/IS.VEH.NVEH.P3 (accessed 23 May 2013).

Worrell, E. and Galitsky, C. (2008) *Energy Efficiency Improvement and Cost Saving Opportunities for Cement Making*, Ernst Orlando Lawrence Berkeley National Laboratory, Berkeley, CAhttp://www.energystar.gov/ia/business/industry/LBNL-54036.pdf (accessed 23 May 2013).

Worrell, E., Neelis, M., Price, L. *et al.* (2008) *World Best Practice Energy Intensity Values for Selected Industrial Sectors*, Ernst Orlando Lawrence Berkeley National Laboratory, Berkeley, CAhttp://ies.lbl.gov/iespubs/62806.pdf (accessed 23 May 2013).

Wrangham, R.W., McGrew, W.C., de Waal, F. B. M., and Heltne, P. G., eds. 1996. *Chimpanzee Cultures*. Cambridge, MA: Harvard University Press.

WSA (World Steel Association) (2011a) Steel and Raw Materials, WSA, Brussels, http://www.worldsteel.org/dms/internetDocumentList/fact-sheets/Fact-sheet_Raw-materials2011/document/Fact%20sheet_Raw%20materials2011.pdf (accessed 23 May 2013).

WSA (2011b) Methodology Report: Life Cycle Inventory Study for Steel Products, http://www.worldsteel.org/dms/internetDocumentList/bookshop/LCA-Methodology-Report/document/LCA%20Methodology%20Report.pdf (accessed 23 May 2013).

WSA (2013) Statistics Archive, http://www.worldsteel.org/statistics/statistics-archive .html (accessed 23 May 2013).

Wu, X., Zhang, C., Goldberg, P. *et al.* (2012) Early pottery at 20,000 years ago in Xianrendong Cave, China. *Science*, **336**: 1696–1700.

Yellishetty, M., Ranjith, P.G. and Tharumarajah, A. (2010) Iron ore and steel production trends and material flows in the world: Is this really sustainable? *Resources Conservation and Recycling*, **54**: 1084–1094.

Yogananda, P. 1946. *Autobiography of a Yogi*. New York: The Philosophical Library, http://www.crystalclarity.com/yogananda/chap30.php (accessed 23 May 2013).

Zeist, W. and Bakker-Heeres, J.A.H. (1975) Evidence for linseed cultivation before 6000 BC. *Journal of Archaeological Science*, **2**: 215–219.

Ziegler, K. (1963) Consequences and Development of an Invention, Nobel Lecture, December 12, 1963, The Nobel Prize in Chemistry 1963, Nobel e-Museum, Stockholm, http://www.nobelprize.org/nobel_prizes/chemistry/laureates/1963/ziegler-lecture.html (accessed 19 June 2013).

Zulehner, W. (2003) Historical overview of silicon crystal pulling development. *Materials Science and Engineering*, **73**: 1–12.

Index

Making the Modern World: Materials and Dematerialization, First Edition. Vaclav Smil.
© 2014 John Wiley & Sons, Ltd. Published 2014 by John Wiley & Sons, Ltd.